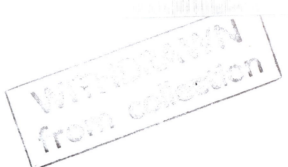

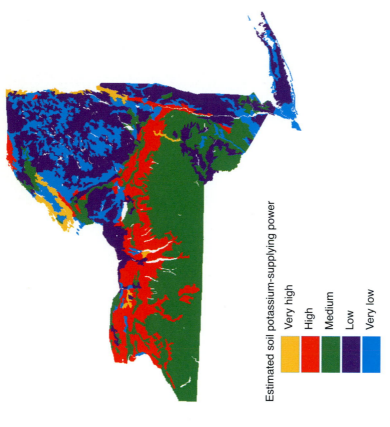

Plate 1. Estimated potassium-supplying power of New York State soils, based on fine-grained mica composition (provided by W.S. Reid and C.J. Post, Cornell University).

GRASS FOR DAIRY CATTLE

Grass for Dairy Cattle

Edited by

J.H. Cherney
Department of Soil,
Crop and Atmospheric Sciences
Cornell University, New York, USA

and

D.J.R. Cherney
Department of Animal Science
Cornell University, New York, USA

CABI *Publishing*

CABI *Publishing* – a division of CAB INTERNATIONAL

CABI *Publishing*
CAB INTERNATIONAL
Wallingford
Oxon OX10 8DE
UK

CABI *Publishing*
10 E 40th Street
Suite 3203
New York, NY 10016
USA

Tel: +44 (0)1491 832111
Fax: +44 (0)1491 833508
Email: cabi@cabi.org

Tel: +1 212 481 7018
Fax: +1 212 686 7993
Email: cabi-nao@cabi.org

A catalogue record for this book is available from the British Library, London, UK.

Library of Congress Cataloging-in-Publication Data
Grass for dairy cattle / edited by J.H. Cherney and D.J.R. Cherney.
    p.   cm.
    Includes index.
    ISBN 0-85199-288-9 (alk. paper)
    1. Dairy cattle -- Feeding and feeds   2. Grasses. 3. Pastures -
    - Management.   4. Grazing - - Management.   I. Cherney, Jerome Henry.
II. Cherney, D.J.R.
SF203.G745   1998
636.2'08'6- -dc21   98-22773
                                                        CIP

ISBN 0 85199 288 9

Typeset in Garamond by AMA Graphics Ltd, UK
Printed and bound in the UK at University Press, Cambridge

Contents

Contributors

T.W. Bruulsema, *Potash and Phosphate Institute, Guelph, Ontario, Canada.*

M.D. Casler, *Department of Agronomy, University of Wisconsin-Madison, 1574 Linden Drive, Madison, Wisconsin 53706-1597, USA.*

D.F. Chapman, *Department of Agriculture and Resource Management, University of Melbourne, Victoria, Australia.*

D.J.R. Cherney, *Department of Animal Science, 327 Morrison Hall, Cornell University, Ithaca, New York 14853-4801, USA.*

J.H. Cherney, *Department of Soil, Crop and Atmospheric Sciences, 153 Emerson Hall, Cornell University, Ithaca, New York 14853-1901, USA.*

D.A. Clark, *Dairying Research Corporation Ltd, Private Bag 3123, Hamilton, New Zealand.*

E.A. Clark, *Department of Crop Science, University of Guelph, Guelph, Ontario N1G 2W1, Canada.*

R.T. Cowan, *Australian Tropical Dairy Institute, University of Queensland, Gatton, Queensland 4345, Australia.*

G.J. Cuomo, *West Central Experiment Station, State Highway B26, Morris, Minnesota 56267, USA.*

S.L. Fales, *Department of Agronomy, 16 Agricultural Science Industries Building, The Pennsylvania State University, University Park, Pennsylvania 16802, USA.*

G.W. Fick, *Department of Soil, Crop and Atmospheric Sciences, 505 Bradfield Hall, Cornell University, Ithaca, New York 14853-1901, USA.*

J.B. Hacker, *ATFGRC, CSIRO Tropical Agriculture, 306 Carmody Road, St Lucia, Queensland 4069, Australia.*

L. Jank, *CNGGC/EMBRAPA, CxP 154, 79002-970 Campo Grande, Mato Grosso do Sul, Brazil.*

S.C. Jarvis, *Institute of Grassland and Environmental Research, North Wyke Research Station, Okehampton, Devon EX20 2SB, UK.*

V.R. Kanneganti, *USDA-Agricultural Research Service, US Dairy Forage Research Center, Department of Agronomy, University of Wisconsin-Madison, 1925 Linden Drive West, Madison, Wisconsin 53706, USA.*

K.F. Lowe, *Australian Tropical Dairy Institute, Department of Primary Industries, M. S. 825, Peak Crossing, Queensland 4306, Australia.*

B.W. Mathews, *College of Agriculture, University of Hawaii at Hilo, 200 W. Kawili St, Hilo, Hawaii 96720-4091, USA.*

D.R. Mertens, *USDA-Agricultural Research Service, US Dairy Forage Research Center, 1925 Linden Drive West, Madison, Wisconsin 53706, USA.*

S.C. Miyasaka, *College of Tropical Agriculture and Human Resources, University of Hawaii at Manoa, Beaumont Branch Station, 461 W. Lanikaula St, Hilo, Hawaii 96720-4037, USA.*

K.C. Moore, *Department of Agricultural Economics, 226B Mumford Hall, University of Missouri, Columbia, Missouri 65211, USA.*

R.E. Muck, *USDA-Agricultural Research Service, US Dairy Forage Research Center, 1925 Linden Drive West, Madison, Wisconsin 53706, USA.*

L.D. Muller, *Department of Dairy and Animal Sciences, 324 Henning Building, The Pennsylvania State University, University Park, Pennsylvania 16802, USA.*

P. O'Kiely, *Teagasc, Grange Research Center, Dunsany, Co. Meath, Ireland.*

C. Ohlsson, *Department of Forage Crops and Potatoes, Danish Institute of Agricultural Science, Research Center Foulum, PO Box 21, DK-8830, Tjele, Denmark.*

A.J. Parsons, *AgResearch, Grasslands Research Centre, Tennent Drive, Private Bag 11008, Palmerston North, New Zealand.*

P. Seguin, *Department of Agronomy and Plant Genetics, University of Minnesota, 1991 Buford Circle, 411 Borlaug Hall, St Paul, Minnesota 55108, USA.*

C.C. Sheaffer, *Department of Agronomy and Plant Genetics, University of Minnesota, 1991 Buford Circle, 411 Borlaug Hall, St Paul, Minnesota 55108, USA.*

J.P. Tritschler II, *Applied Epidemiology Inc., PO Box 2424, Amherst, Massachusetts 01004, USA.*

Preface

The pressures of a growing human population and the opportunities of new technologies are forcing complex and far-reaching changes on most aspects of human society. Agriculture, and especially animal-based agriculture, is certainly affected by these changes. Those segments of agriculture that do not find and refine more effective, efficient, and sustainable methods will not survive.

In response to the current interest in environmental and economic sustainability of dairy farms, grass forage crops have emerged as potential solutions to some of the nutrient management challenges now encountered on intensively managed dairy farms. On the other hand, intensive grass management can generate additional nutrient management concerns, such as excess potassium in dry cow diets. Although the term 'grass' often refers to more than just the grass family *Poaceae* (or *Gramineae*), this book specifically focuses on these grasses, and secondarily on grass–legume mixtures. Some notable improvements have been made in the breeding and management of grass for dairy cattle, making grass a more desirable forage option. The reintegration of grass-based systems into North American dairying will require a major paradigm shift, and many of the same economic, social and ecological factors involved in this change apply around the world.

International experts were selected to provide an up-to-date assessment of the breeding, management, storage, feeding and economics of grass for both lactating and non-lactating dairy cows. Chapters were not meant to provide comprehensive literature reviews, but instead focus on highlighting the state of the art in specific topical areas of grass for dairy cattle.

Jerome H. Cherney
Department of Soil, Crop and Atmospheric Sciences
Cornell University
Ithaca, New York, USA

Debbie J.R. Cherney
Department of Animal Science
Cornell University
Ithaca, New York, USA

The Future of Grass for Dairy Cattle

G.W. Fick[1] and E.A. Clark[2]

[1]Department of Soil, Crop and Atmospheric Sciences,
505 Bradfield Hall, Cornell University, Ithaca, New York
14853-1901, USA
[2]Department of Crop Science, University of Guelph, Guelph,
Ontario N1G 2W1, Canada

Introduction

The objective of this opening chapter is to establish the context and rationale for taking a fresh look at the role of grass for dairy cattle, including a
whole range of interacting factors that will influence decision making by
dairy farmers in the coming millennium. In contrast to the rest of the world,
the notion that grass could or should play a major role in dairy nutrition is
new in North America. The low regard for grass, and especially for grass
pastures, is shown in this quotation from Bromfeld (1945):

> In the past and even today the average farmer chooses for his permanent
> pasture the land which he finds least valuable to his farm. Mistakenly, he does
> not regard pasture as a crop but usually as wasteland where he can turn his
> cattle in summer and let them feed themselves.

While researchers in Western Europe, New Zealand, and elsewhere have
spent the last several decades refining understanding of grass physiology,
sward dynamics, and grazing management to enhance *per hectare* production of milk solids on pasture (Hodgson and Illius, 1996; Pearson and Ison,
1987), North American researchers concentrated on increasing production
per cow from lucerne (*Medicago sativa* L.), maize (*Zea mays* L.), and other
conserved feedstuffs (Fick and Cox, 1995; Clark and Poincelot, 1996). Thus,
reintegration of grass-based systems into the mainstream of North American
dairying will require a major paradigm shift, with implications for every-
thing from tillage intensity to dairy genetics. The economic, social and

ecological factors that underlie change in North America are similar around the world, and the role of grass in dairying warrants a fresh and comprehensive re-evaluation.

Historical Treatment of the Development of Grass-based Dairying

Because the past and present establish the context (and sometimes the limits) of the future, an historical summary of grass-based dairy cattle farming is in order. The term 'grass' is often used in a generic sense, meaning the herbaceous covering of grasslands in the world. The grass family (*Poaceae* or *Gramineae*) dominates, with legumes (*Fabaceae* or *Leguminosae*) and composites (*Asteraceae* or *Compositae*) being prominently represented among grassland flora (Weaver, 1968). The co-evolution of herb and herbivore in grassland ecosystems is nowhere better reflected than in the morphological and physiological adaptations of the grasses. Many grass species are well adapted to grazing, with a reservoir of growing points protected near the soil surface to quickly replace defoliated leaf surface (Fig. 1.1).

The wild ancestors of the modern dairy cow were equally well adapted to consuming grass tissues, with grinding teeth and a ruminant digestive system based on symbiotic microorganisms to extract nutrition from the cellulose- and hemicellulose-rich cell walls of the grasses (Van Soest, 1994). The grass-digesting symbiosis of ruminant livestock was first domesticated about 10,000 years ago, initially in the Near East (Minc and Vandermeer, 1990). Traces of cereal crops, grain legumes and livestock in the archeological record of that region indicate a mixed agricultural system with animals providing mainly meat for the human diet. This kind of agriculture spread in Asia, Europe, and Africa, but it remained surprisingly unchanged for almost 5000 years. Because it was run primarily on human power, it exploited a limited soil base that was probably showing serious long-term soil deterioration by the end of that first agricultural era (Sherratt, 1981).

By 3000 BC, a new system of agriculture characterized by the plough and pastoralism had emerged in the Near East. In this second phase of agriculture, humans learned to exploit livestock for energy (in transportation and draught power), for fibre (from wool), and for milk (by dairying), adding these secondary products to the already well-established use of livestock for meat. Sherratt (1981) called this transition the 'Secondary Products Revolution'. The recognition that livestock could be employed for energy greatly expanded the land accessible to agriculture. Land resources that were difficult to till with human power alone could now be ploughed with animal traction. Furthermore, the integration of livestock into cropping systems transformed grazing lands from 'wild land' used for hunting and gathering into 'agricultural land' (Sherratt, 1981).

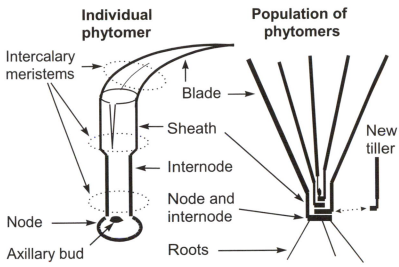

Fig. 1.1. Stylized representations of the structure of a vegetative grass plant. The grass shoot is made up of a series of phytomer units, each with a blade, sheath, internode, node and axillary bud. A developing branch from an axillary bud at the base of the plant is called a tiller. Specific adaptations of grasses to grazing include the stacking of phytomers (as shown on the right), often with very short internodes so that a large number of leaves can be initiated below the grazing height near the soil surface. In addition, intercalary meristems at the base of the blade, the sheath and the internode allow continued growth in length and height of blades after removal of the ends of the leaves. Axillary buds may also develop as new, erect tillers or as horizontal stems that extend the plant laterally. The horizontal stem are called rhizomes when below ground or stolons when above ground. All rhizomes and some stolons are also out of reach of grazing.

The use of livestock to produce milk for human consumption is a particularly intriguing innovation because lactose intolerance (which is the common human condition) prevents milk from being a satisfactory human food beyond the age of weaning (Sherratt, 1981; Cheeke, 1993). Dairy products such as yogurt and cheese would have reduced the problem, but the development of dairying must have been accompanied by a genetic change in the now lactose-tolerant human populations of Europe and central Asia (Sherratt, 1981).

The adoption of dairy-based agriculture had two major ecological consequences. One would have been the increase in energy conversion efficiency in food production systems because of the longer lifespan of food-producing animals. The second would have been an increase in agricultural sustainability brought by enhanced nutrient cycling through grass-consuming livestock (Sherratt, 1981).

This second point is worth closer examination. Most long-term agricultural systems capitalize on the natural accumulation of nutrients in wetlands. The flood-based systems of Egypt and the Near East, the 'floating gardens' of Mexico, and the rice (*Oryza sativa* L.) systems of eastern Asia are prominent examples. However, the mixed cropping systems that spread from the Near East to Europe and central Asia after the Secondary Products Revolution allowed agriculture to become less directly dependent upon wetlands. For example, in the forested zones of Europe, livestock were grazed on grassy meadows that were seasonally flooded wetlands. Penning the stock at night served to move soil nutrients from the grasslands and concentrate them in the form of livestock manure, which could then be applied to croplands (Vasey, 1992). Winter feeding based on hays made from the same wet grasslands further amplified nutrient transfer and accumulation. Indeed, Vasey (1992) described pre-Medieval agricultural systems of north-western Europe as 'sedentary pastoralism with cultivation'.

By 500–100 BC, dairying had apparently become the dominant type of agriculture in what is now in modern Denmark, the Netherlands, and possibly parts of Switzerland. Pirtle (1926a) reported that the Friesians had a pastoral–dairying economy by 300 BC. In agricultural communities near the North Sea, a typical structure was the 'aisled building', which was a combination barn and family dwelling. Archeological studies reveal floor plans (Fig. 1.2) with stalls and feed troughs arrayed along a central aisle, like modern stanchion barns, with feed storage areas and family living space at one end (Todd, 1972). As recently as the 1870s, Fussell (1965) reported that 'hay, family and servants, peat, and all the cows and sheep lived under one roof' in northern Europe, and examples of humans and livestock sharing housing can still be found there today.

In the Roman period, historians noted the importance of cattle to their northern Germanic neighbours. Julius Caesar recorded that the Germans lived on 'milk, cheese, and fish'. Tacitus noted the small size of German cows (Pirtle, 1926b). The Germanic tribes that later invaded the Roman Empire were also reported to have brought their cattle with them, both for milk and meat (Pirtle, 1926b).

The importance of dairy cattle and grass is also recorded in Europe through the Middle Ages and into the 20th century. Specialized dairy regions developed in Friesland and the lower Rhine Valley, Bavaria, Tyrol, Karinthia, Lombardy, and the Piedmont and higher elevations of the Alps (Fussell, 1965). Cows were milked from May to October, with herd size a function of the ability of the farmers to provide winter fodder (Fussell, 1966; Grigg, 1974). Riparian grasslands or 'wet meadows' (Vasey, 1992) were prized components of the agricultural system. They were closed to grazing from as early as February to as late as early May and hand-harvested with a scythe for hay in May to June. Natural seasonal flooding of the wet meadows replaced the nutrients removed in the hay, thus maintaining the productivity of the land resource (Fussell, 1966).

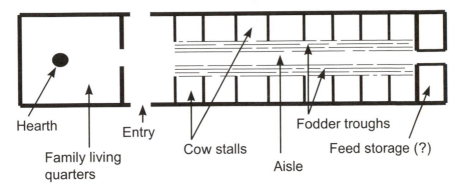

Hearth

Entry

Family living
quarters

Cow stalls

Aisle

Fodder troughs

Feed storage (?)

Fig. 1.2. The floorplan of an aisled building uncovered in archeological studies at Wesermünde, Germany (redrawn from Todd (1972)). The structure dates about the beginning of the Christian era and indicates the importance of livestock at that time in Europe.

From the Middle Ages through the 1600s in Western Europe, Fussell (1965; 1966) estimated that 0.4–1.2 ha (1–3 acres) were required for summer pasture for a dairy cow producing perhaps 750 kg of milk (200 US gallons) in a year. Another hectare was needed to produce the estimated 1.8 Mg (2 tons) of hay needed for winter feeding. In many areas of Europe, fallowed grain fields from the 'three-field' cropping systems and the 'wild grass' from wastelands would have supplied much of the summer grazing for dairy cows (Fussell, 1965; 1966). Herd size was small, even in the dairy-dominated areas (Fussell, 1965; Grigg, 1974), but by the 800s, the Danes and Swedes were exporting butter, and by the 1100s, the Dutch were exporting cheese (Grigg, 1974).

The development of improved dairying methods in England has been summarized in the book by Fussell (1966). By 1500, pulses such as peas (*Pisum sativum* L.) were sometimes mentioned as a supplemental feed, with beans (*Phaseolus* spp.) added by the 1600s. The 1600s also provide the first English records of ryegrass (*Lolium perenne* L.) and white clover (*Trifolium repens* L.) on the 'marl land' and importation of clover seed from Holland for sowing ley pastures in the English cereal–forage rotations. Turnip (*Brassica napus* L. and *B. rapa* L.) also became an important winter feed in the 17th century in England (Fussell, 1966). In 1739, Samuel Trowell published the first English instructions for rotational stocking of pastures, recommending rest periods of 21–30 days (Fussell, 1966). Also in the 1700s, cabbage (*B. oleracea* L.) was recommended as a winter feed for cattle, and the first English reference to forage quality was made, with pasture given a 'value' four times that of hay.

Fussell (1965) reported that the crop rotation known as *Koppelwirtschaft* spread from Schleswig Holstein in the 1700s. It had 5 years of grass out of 11 years in the rotation cycle, and it produced better quality hay

> **Box 1.1.** The characteristic features of the sustainable dairy farming system of the past derived from the analysis of Grigg (1974).
>
> 1 Farms and the herds were relatively small, mostly fewer than 12 cows.
> 2 Economic returns were good and more stable than many other agricultural systems.
> 3 Tenancy was uncommon because few tenants were prepared to give the cows the attention they needed. Dairying was mostly a family-based activity.
> 4 Dairying was often combined with other enterprises, for example pigs to consume skim milk and cereals to be fertilized with livestock manure.
> 5 Cheese and butter were the main products because they could be stored.
> 6 Cows were milked in summer while pasture was available and dried up for winter.
> 7 'The ideal system [was] based on grass', used in summer for grazing and in winter for hay.

and root crops for winter feeding. In the 1800s there are records of the practice of 'bone manure' soil fertilization in England (Fussell, 1966), and in general there was rapid adoption of agriculture based on scientific knowledge throughout Europe.

The 2500 years of dairying in Europe outlined above convey a sense of sustainability and durability of old-style dairy farming, although sustainability could be challenged on the basis of low yields. Grigg (1974) has tabulated some of the characteristic features of those old systems (Box 1.1), making an interesting reference and counterpoint as we consider more recent changes and the impact of those changes on system sustainability. In the present context, it is noteworthy that Grigg (1974) identified grass (meaning grazed and harvested perennial forages) as the 'ideal' feed for dairy systems. However, the evolution of dairy farming in the last 50 years has seen increasing emphasis on other kinds of feeds in many areas of the world.

The Present and Future: Productivity, Profitability, Sustainability

Many textbooks, including Barnes *et al.* (1995) and Webster (1993), describe the 20th-century development of science-based technology in forage and dairy production. Clark and Poincelot (1996) cover the more recent history of dairy farming in North America, particularly in the Dairy Belt around the Great Lakes. There, increased productivity has been achieved by confinement feeding with rations based on forages from silage (especially maize and lucerne) and concentrates from grain crops. Feed grains are imported to the farm when cropland is limiting.

In general, increased productivity has been associated with declining profitability because of both rising input costs and static demand for milk. In Canada, the aggregate farm price index increased by 16.5% between 1981 and 1989, while the aggregate farm product price index increased by only 0.2% (Ferguson, 1991). Producer response has been to increase herd size, or in the case of New Zealand farmers, to reduce costs to maintain net returns. The most recent summary for representative dairy farms in New York (Knoblauch *et al.*, 1996) showed that milk production per cow has increased from 7372 kg in 1986 to 9202 kg in 1995 (from 16,237 to 20,269 pounds). For the same period, farm size increased from 117 ha to 162 ha (288 to 399 acres) and herd size from 95 to 160 milking cows. Profit per cow (based on return to labour and management per operator) has fluctuated widely over the 10-year period, but it has fallen from an average of US$115 for the first 5 years to US$63 for the second 5 years.

Increasing purchased inputs are often justified on the basis of the economy of size and marginal efficiencies. However, the notion that high input agriculture is the most efficient way to produce food has been challenged by Kaffka (1989), Boyle (1993), and others. The New Zealand approach of low purchased feed and capital costs allows milk production at half the cost of Ireland, one-third the cost of the USA, and a quarter the cost of Canada (Boyle, 1993, cited in Nation, 1995). Increased international trade in dairy products will force changes in the structure of the industry.

While productivity and profitability have become dominant factors in the evolution of dairy farming, and will remain important in the future, sustainability is emerging as a third key factor. Opinion is divided on the prognosis for the resource-intensive approaches that have sustained North American dairying over the past several decades. For many, the future means 'megafarming' where production units of several thousand cows are managed under factory-like confinement systems. Cheeke (1993) listed the following advantages of large-scale, industrial-style livestock systems:

- improved animal welfare based on a vested interest in using optimal animal housing and on-site veterinary services;
- improved nutrition by total control of feed quality under the management of staff nutritionists;
- better worker benefits like those in any other type of modern employment;
- effective waste disposal to control pollution, which is affordable only to large corporations;
- consistent, high quality products because of total control of all aspects of production.

Others, however, have focused on the corollary – larger farms mean fewer farm families and declining rural communities (Berry, 1986; Jackson, 1994) and greater apparent risk to the environment (Perelman and Shea, 1972; Soule and Piper, 1992). Some argue that there are fewer herd health problems (Goldberg *et al.*, 1992), adequate nutrition at less cost (Murphy *et al.*, 1986), less pollution of soil, water, and air, and better working conditions with systems based on grazing (Jackson-Smith *et al.*, 1996; Mueller, 1996).

In the future, grass–legume mixtures could contribute to profitable and sustainable dairying in two divergent but complementary ways. In geographic regions favouring New Zealand-style grazing systems, dairy operators would rely on grass as the major feedstuff year-around for both milking and replacement herds. In Wisconsin, for example, Jackson-Smith *et al.* (1996) noted that nearly half of Wisconsin dairy producers employ pasture to some extent, with 14% or 3850 dairy farms employing management intensive rotational grazing (MIRG)[1] in 1994 – up from just 7% in 1992. Thus rate of adoption of MIRG technology is growing rapidly among Wisconsin dairy operators. At the same time, in regions favouring confinement-based dairying, we could see a stronger emphasis on perennial grasses to supply at least part of the energy and protein now derived from maize and lucerne, respectively. Grass-based swards may also absorb and recycle manure nutrients more effectively, reducing the risk of groundwater contamination, and extend stand life more reliably, reducing the need for frequent, costly reseeding of winter-killed stands. The global trend toward increased variability in the weather and increased incidence of severe weather events (Peters and Lovejoy, 1992; Subcommittee on Global Change Research, 1997) may also reduce the expected stand life of species such as lucerne. Under such circumstances, the high costs of frequent reseeding may outweigh concerns about the somewhat lesser yield and quality of better adapted and more persistent perennial grasses. In addition, the yield and quality of perennial grasses can probably be increased with research and improved management.

Impacts of Grass-based Dairying on the Environment and Society

Increasing utilization of perennial grasses for dairying will have major impacts on the environment. This will require fundamental changes to both crop and livestock agriculture and to agricultural research itself. A strong rationale is needed to justify the significant effort that will have to be made

---

[1] Sum of 'fully intensive' (moved cows at least once a week and relied on pasture as the primary source of total feed for milking cows during the grazing months) and 'semi-intensive' graziers (moved cows at least once a week, but did not rely on pasture as primary source of total feed during the grazing months).

Table 1.1. Impacts of grass-based dairy farming on the environment.

Natural resource	Positive impacts	Negative impacts
Air and climate	More soil organic matter (less CO_2 in atmosphere) Fisher *et al.*, 1994; Scharpenseel and Becker-Heidmann, 1997	Possibly more CH_4 produced Johnson and Johnson, 1995; McAllister *et al.*, 1996; Wilkerson *et al.*, 1995
Water	Cleaner, purer water with grass Coyne *et al.*, 1995; Flaherty and Drelich, Jr, 1997; Liaghat and Prasher, 1996; Sharp *et al.*, 1995	Potential contamination of surface water Goss *et al.*, 1993; Larsen *et al.*, 1994; Nelson *et al.*, 1996; Reaume *et al.*, 1994
Soil	Less soil erosion with grass Reid, 1985 Sharp *et al.*, 1995 Better soil structure with grass Carter *et al.*, 1994; Carter and Kunelius, 1997	
Biodiversity	Can be increased with proper management Clubine, 1995; Meister and Koch, 1997; Tilman *et al.*, 1996	

by both producers and researchers to undertake such a large-scale change in system design and management. The fundamental changes must anticipate the reality that agriculturalists of the future will have to address increasingly substantive societal and environmental challenges while attempting to maintain or enhance productivity and profitability. Research designed to address the multidimensional and divergent needs of various stakeholders is a much more complex process than simply increasing herd average, number of cows, or kind of feed. Some of the known linkages between grass-based dairying and the environment are profiled in Table 1.1. The implications of grass-based dairying for crop production systems, livestock nutrition and socioeconomic issues are reviewed in Table 1.2.

Air and climate

Increasing the hectarage sown to perennial grasses for dairy nutrition could influence the concentration of atmospheric constituents, including the greenhouse gases CO_2 and methane. Increases in CO_2 have had the greatest impact on global warming, accounting for 61% of the total change in radiative forcing since 1765 (Duxbury *et al.*, 1993). Withholding land from cultivation, particularly under a perennial grass sod, allows the accumulation of organic matter in the soil, effectively sequestering CO_2 from the atmosphere and reducing risk of global warming. Grass roots appear to be

Table 1.2. Effects of grass-based dairy farming on agricultural and human/socioeconomic issues.

	Positive impacts	Negative impacts

Crop production	Less purchased nitrogen fertilizer
	  Follett and Wilkinson, 1995;
	  Knoblauch *et al.*, 1980
	Tighter nutrient cycling
	  Bacon *et al.*, 1990;
	  Lanyon, 1995; Russelle *et al.*,
	  1997
	Less fossil fuel and biocide use
	  Knoblauch *et al.*, 1980; Olson
	  and Olson, 1985
	Possible biomass energy
	by-products
	  Cherney *et al.*, 1991; Landstrom
	  et al., 1996
		Possibly more nitrogen losses
		  Misselbrook *et al.*, 1996;
		  Ryden *et al.*, 1984; Stout
		  et al., 1997

Livestock nutrition	Less soluble protein
	  Hansen *et al.*, 1992;
	  Kohn and Allen, 1995
	Better levels of K
	  Cherney and Cherney, 1997;
	  Mayland and Cheeke, 1995
		More fibre, slower digesting
		  Buxton and Mertens, 1995;
		  Van Soest, 1973
		Lower levels of P, Ca and Mg
		  Follett and Wilkinson,
		  1995; Mayland and
		  Cheeke, 1995
		Possibly more antiquality
		factors
		  Hoveland, 1993;
		  Mayland and Cheeke, 1995

Human and socioeconomic issues	More vitamin E with pastures
	  Atwal *et al.*, 1991;
	  Flachowsky *et al.*, 1993
	Possible increased profits
	  Moore and Gerrish, 1995; Olson
	  and Olson, 1985; Parker *et al.*,
	  1997; Rust *et al.*, 1995
	Possibly stronger rural communities
	  Hassanein and Kloppenburg Jr,
	  1995; McSweeny, 1987
		Possible competition for food
		grains
		  Cheeke, 1993

particularly effective in stabilizing and increasing the soil organic matter fractions most resistant to degradation.

Conversely, increasing the proportion of the dairy ration contributed by grasses could increase production of methane. At present, domestic livestock account for roughly 24% of anthropogenic methane production, or 15% of total global methane generation (Duxbury *et al.*, 1993). The amount of methane produced in ruminant digestion depends upon the fermentation time in the rumen, which is in turn influenced by the quality of the grass. Increased utilization of low quality grasses could increase methane production. However, the high quality grasses needed for lactating dairy cows would give a smaller increase in methane production. Because each unit of methane has a global warming potential roughly 60 times greater than a unit of CO_2 (Duxbury *et al.*, 1993), small changes to methane production can materially influence risk of global warming.

Water and soil

One of the greatest benefits from increased reliance upon perennial grasses for dairy nutrition could be improvements in soil and water quality resulting from reductions in the need for annual tillage. The fibrous root system of grasses is widely recognized as a powerful force in soil and water conservation (Sharp *et al.*, 1995). The build-up of soil organic matter under a perennial sod improves soil structure and tilth, with beneficial effects on rooting, macropore continuity, aeration and microbial activity. With proper management, grasses can improve water infiltration, enhancing the quality and uniformity of both surface water and groundwater supplies. Greater infiltration means less runoff and soil erosion, and hence, less surface movement of nutrients, pathogens and agricultural chemicals that adhere to sediment particles (Table 1.1).

Nutrients, pathogens and sediment can move from pastured land or following surface application of manure to crop land. Cattle accessing watercourses can deposit faeces and urine in or near the water. However, Goss *et al.* (1993) also reported movement of liquid manure into field drains under tilled cropland with subsequent impairment of water quality in 75% of the manure spreading events investigated. Reaume *et al.* (1994) documented significant movement of bacteria into field drains within 2 h of spreading liquid manure. Management of intensive grazing, including timing and magnitude of grazing pressure, reinforced watercourse crossings, alternative watering options to decrease reliance on stream water, and placement of minerals can limit sediment delivery and resuspension from livestock accessing watercourses (various, cited in Clark, 1998). Similarly, grass buffer strips can retard movement of sediment and labile nutrients into surface watercourses.

Biodiversity

Above- and below-ground biodiversity will likely benefit by increasing the hectarage sown to perennial grasses within the agricultural landscape of intensively cropped land (Table 1.1). Enhancement of biodiversity is most likely if pasture and hay crops account for a larger share of the agricultural mosaic in areas like the North American Corn Belt, which are now dominated by monocrops and simple crop rotations. Less tillage and increased soil organic matter will promote increased biodiversity in the soil, just as the greater range of species present in perennial grass swards will support more complex food webs than monoculture grain crops (Doran and Werner, 1990).

Crop production

Increasing reliance on perennial grasses will significantly change many aspects of crop production for dairy nutrition. Of perhaps greatest environmental consequence will be improvements in nutrient cycling. A major problem in western Europe and parts of North America is the over-application of manure nutrients generated from confinement feeding of imported feedstuffs. In Holland, for example, Kauppi (1990) documented the annual excess of import (as feed and fertilizer) over export (as crop and livestock products) per hectare of agricultural land as 231 kg N and 39 kg P. The practice of feed import for confinement feeding means that soil nutrients from Nebraska are now becoming excessive in the soils of Pennsylvania, contributing to eutrophication of both surface and groundwater. Chapters 7 and 8 discuss these topics in greater depth.

In many places, lucerne, white and red clover (*Trifolium pratense* L.), and other legumes sown with grasses fix enough nitrogen to meet animal and crop demands, but import of soybean (*Glycine max* (L.) Merr.) and other supplements for rumen bypass protein unbalances the on-farm N budget. Production of more grass and less legume could enhance N utilization efficiency and cycling on individual dairy farms, with reduced risks of leaching excess N that comes from degraded lucerne protein and passes to urine (Table 1.2).

Perennial grass sods reduce risk of leaching from manure application, and in so doing, provide the double benefit of both tighter nutrient cycling and lesser need for purchased fertilizer (Table 1.2). Cooler soil temperatures under grass sods help to synchronize nutrient supply from mineralization with demand from growing plants. The widening of the C : N ratio under grass also serves as an effective sink for labile nutrients. Risk of leaching from high nutrient urine patches has been identified by some authors and discounted by others (Table 1.2). Nutrient cycling on perennial swards, under both conservation and grazing regimes, is an area requiring additional research attention.

Perennial forage grasses, such as switchgrass (*Panicum virgatum* L.), have been widely studied as a source of biomass energy (Sanderson *et al.*, 1996). With modest changes to production strategies, perennial grasses could be viewed as a renewable source of on-farm energy (Table 1.2). For example, excess forage could be grown to reduce feed-supply risk, with the option of harvesting the excess for on-farm energy generation. Such a practice would cycle CO_2 already in the atmosphere rather than releasing fossil carbon stored in wood or petroleum products.

Replacing annual grain crops with perennial grasses can also reduce dependence upon both fossil fuel energy and biocides. Establishment energy – approximately 3 million kcal ha^{-1} for lucerne hay (Heichel and Martin, 1980) – is amortized over the lifespan of a perennial sward, rather than annually in the case of grain crops. These energy savings are partially offset by the increased energy costs of mechanical harvesting for conserved feed systems, but are more fully realized in systems where herbage is harvested by grazing stock. Enterprise budgets for New York State showed that the power and equipment costs in dollars for lucerne hay were 80–108% of those for maize, while those for pasture (unspecified grass–legume mixture) were only 2% those of maize (Knoblauch *et al.*, 1980; Olson and Olson, 1985). Fertilizer costs for lucerne hay were 55–63% of those of maize, while fertilizer costs for pasture were 45–69% of those for maize, depending on use of N fertilizer. Biocide costs for lucerne hay were 24–45% of those for maize, compared with 'nil' for pastures. As has been well demonstrated with lucerne grown for hay, pastures composed of monoculture grass or simple mixtures of one grass and one legume can be vulnerable to encroachment by weeds. Weed encroachment into lucerne hay fields reduces stand life in North America, while perennial ryegrass is affected by Argentine stem weevil (*Listronotus bonariensis* (Kuschel)) and nematodes in New Zealand. More complex pasture swards, however, appear to have greater resilience to pest build-up, and as a result, are dependent upon few biocide applications beyond herbicides at seeding.

Livestock nutrition

Replacing lucerne with perennial grasses in the dairy ration will affect a wide range of nutritional factors. Increased emphasis on grasses would reduce soluble protein and facilitate management of N in the rumen (see also Chapters 4 and 13), as well as affecting levels of K in the diet (Chapter 6). However, more grasses would also change the amount and chemistry of dietary fibre, potentially reducing digestibility. Phosphorus, calcium and magnesium levels are also typically lower in grasses than in legumes grown on the same soil, potentially increasing the risk of hypomagnesaemia and other nutrient imbalances (Table 1.2). Species-specific issues which will require research and management attention include the fungal endophytes which can be hosted by species such as tall fescue (*Festuca arundinacea* Schreb.) and perennial ryegrass (Ball *et al.*, 1993), the alkaloids which are characteristic of species such as reed canarygrass (*Phalaris arundinacea* L.), and the cyanogenic glycosides which are produced by some *Sorghum* and related species (Vogel *et al.*, 1987).

Human and socioeconomic issues

The current practice of feeding grain to livestock is seen by some as detracting from the food available to support humankind. This perception is based upon a literal reading of the ecological principle which states that only a small fraction of digestible energy can pass from one trophic level to another (Reid, 1970). Hence, converting grain into milk or other commodities via livestock necessarily reduces energy flow to humankind. However, livestock, and especially ruminant livestock, consume and convert to human food vast quantities of nutrients otherwise unavailable to humans. Oltjen and Becket (1996) compared dairy rations that were high and low in grain content. The ratio of humanly-digestible outputs to humanly-digestible inputs ranged from 0.57 to 1.28 for energy and 0.96 to 2.76 for protein. The low-grain rations gave values greater than 1.0, indicating increased food supply for humans. The study also showed that grains can be used strategically to enhance energy and protein capture from fibre-rich forages. Feeding trials with very high quality grass are showing that more grain supplementation may be needed to maintain high milk production levels per cow than would be needed with lucerne (Jonker *et al.*, 1997).

Food derived from animals is recognized as an excellent source of protein and certain minerals and vitamins (Cheeke, 1993). In fact, there are no known sources of vitamin B_{12} from plant foods. However, food from animals can also be high in fat and cholesterol, both of which are regarded as undesirable in rich societies. Research by Shaish *et al.* (1995) and Tebib *et al.* (1997) has profiled the protective effect of antioxidants such as vitamin E in diets high in fat and cholesterol. Because fresh herbage is a good source of vitamin E, and vitamin E is concentrated in milk fat, the negative effects of cholesterol and fat in milk may be reduced or eliminated when dairy cattle graze fresh forage (Table 1.2).

One of the primary driving forces behind the move toward grass-based dairying has been producer concerns about profitability and sustainability, for both their own farms and their rural communities (Clark and Poincelot, 1996). In a 2-year study involving 500 operators in 1993 and 1200 in 1995, Jackson-Smith *et al.* (1996) compared MIRG operations with full confinement (FC) operations in Wisconsin (Table 1.3). While MIRG herds were

Table 1.3. Comparison of MIRG (management intensive rotational grazing) and full confinement dairy operations in Wisconsin (adapted from Jackson-Smith *et al.*, 1996).

Management	Herd size (number)	Rolling herd average		Total farm income (US$)	Total returns to assets, excluding labour/man-agement (US$)
		(kg cow^{-1})	(lbs cow^{-1})		
MIRG	41	7310	16,100	84,400	14,400
Full confinement	67	8400	18,500	146,100	10,680

smaller (61% of FC), achieved lower levels of production per cow (87% of FC), and lower levels of total farm income (58% of FC), total returns to assets were 35% higher on MIRG than on FC dairies. Early adopters of MIRG in Wisconsin have already shown that grass-based dairying may be less capital intensive and less productive but can still be economically competitive.

Other studies reported in Table 1.2 support the premise that productivity may decline in the switch from full confinement to pasture-based dairying, but both farm income and quality of life can improve. Value-added, farm-based processing has been examined in other studies, seeking to strengthen both the farm and general rural economy in the northeastern USA (Kaffka, 1989). However, enhancing sustainability will not necessarily result in higher rural populations and stronger rural communities. Gillespie (1995) has argued that ecological sustainability might be accomplished with practices that continue the trend to fewer, larger farms. The social component, especially as affected by scale of operation, warrants further research (Fick, 1996).

Summary

The foregoing discussion illustrates the complexity and comprehensiveness of the changes associated with an alteration in the way we provide forages for dairying. The future is thus very difficult to predict, but we believe some important conclusions emerge from our discussion. The first is that the history of dairy farming shows a long period of sustainability with grass-based forage systems the hallmark linking soils, crops, livestock and humans in a relatively efficient nutrient cycle that relied on natural processes for maintenance. As we design systems for the future, we must look again at the soil, crop, livestock and human elements to see that they retain that functional equilibrium of sustainability.

Our abbreviated analysis of the effect of grass-based dairy farming on basic environmental and agricultural resources showed some of the many possible impacts of feeding more grass. We believe that two of these possibilities are very likely to occur and constitute strong arguments for increased development of grass-based dairying. The first is the effect of such a change on soil and water conservation. The use of more perennial grasses will protect and enhance our soil and water resources. The other probable outcome is an enhancement of whole-farm nutrient management. This will also have important positive effects on soil and water, but in addition, it will make it easier to have healthy, productive cows and farms with reduced nutrient waste.

The economic and social implications are less clear, but they point to options that should be developed and explored. This is particularly true for dairy farming in North America where confinement feeding of maize, lucerne and soybean, often with concentrates imported to the farm, has placed producers in an increasingly unprofitable, unsustainable and untenable position. The capitalization and equipment required to grow, manage, harvest, store and feed both grains and forages, as well as housing cattle and managing manure, has become burdensome, encouraging ever larger operations that necessarily exclude an increasing number of dairy families. In addition, society is increasingly aware of the adverse environmental implications of annual cropping, particularly with crops that depend intrinsically on biocides for pest control, and of manure storage and distribution practices that place the environment at risk.

North American dairy producers need a viable alternative to the prevailing paradigm. The grass-based dairying systems more typical in the rest of the world may be the solution to at least some of these problems. Large scale feed import would still be problematic if manure nutrients are not returned to their place of origin. However, relying on grass-based mixtures to supply increased amounts of the energy and protein for the dairy ration could provide many opportunities for enhancing both profitability and sustainability of plant and animal agriculture.

References

Atwal, A.S., Hidiroglou, M. and Kramer, J.K.G. (1991) Effects of feed Protec and alpha-tocopherol on fatty acid composition and oxidative stability of cow's milk. *Journal of Dairy Science* 74, 140–145.

Bacon, S.C., Lanyon, L.E. and Schlauder, R.M., Jr (1990) Plant nutrient flow in managed pathways of an intensive dairy farm. *Agronomy Journal* 82, 755–761.

Ball, D.M., Pedersen, J.F. and Lacefield, G.D. (1993) The tall-fescue endophyte. *American Scientist* 81, 370–380.

Barnes, R.F., Miller, D.A. and Nelson, C.J. (eds) (1995) *Forages Volume I: An Introduction to Grassland Agriculture,* 5th edn. Iowa State University Press, Ames, Iowa, 516pp.

Berry, W. (1986) *The Unsettling of America: Culture and Agriculture.* Sierra Club Books, San Francisco, California, 228pp.

Boyle, G.E. (1993) Who produces the cheapest milk? *Tierzuchter* 45, 20–24.

Bromfeld, L. (1945) *Pleasant Valley.* Harper and Brothers, New York, 300pp.

Buxton, D.R. and Mertens, D.R. (1995) Quality-related characteristics of forages. In: Barnes, R.F., Miller, D.A. and Nelson, C.J. (eds) *Forages Volume II: The Science of Grassland Agriculture,* 5th edn. Iowa State University Press, Ames, Iowa, pp. 83–96.

Carter, M.R. and Kunelius, H.T. (1997) Quantifying the effect of grasses on soil physical quality of fine sandy loams in Prince Edward Island. In: *Proceedings XVIIIth International Grassland Congress, 8–19 June, 1997, Winnipeg and Saskatoon, Canada.* Congress Secretariat, Grasslands 2000, Calgary, Alberta, Canada, pp. 19/65–19/66.

Carter, M.R., Angers, D.A. and Kunelius, H.T. (1994) Soil structural form and stability, and organic matter under cool-season perennial grasses. *Soil Science Society of America Journal* 58, 1194–1199.

Cheeke, P.R. (1993) *Impacts of Livestock Production on Society, Diet/health and the Environment.* Interstate Publishers, Danville, Illinois, 241pp.

Cherney, J.H. and Cherney, D.J.R. (1997) Potassium accumulation in perennial cool-season grass forage. In: *Proceedings XVIIIth International Grassland Congress, 8–19 June, 1997, Winnipeg and Saskatoon, Canada.* Congress Secretariat, Grasslands 2000, Calgary, Alberta, Canada, pp. 17/21–17/22.

Cherney, J.H., Johnson, K.D., Volenec, J.J. and Greene, D.K. (1991) Biomass potential of selected grass and legume crops. *Energy Sources* 13, 283–292.

Clark, E.A. (1998) Landscape variables affecting livestock impacts on water quality in the humid temperate zone. *Canadian Journal of Plant Science* (in press).

Clark, E.A. and Poincelot, R.P. (eds) (1996) *The Contribution of Managed Grasslands to Sustainable Agriculture in the Great Lakes Basin.* Haworth Press, New York, 189pp.

Clubine, S.E. (1995) Managing forages to benefit wildlife. In: Barnes, R.F., Miller, D.A. and Nelson, C.J. (eds) *Forages Volume II: The Science of Grassland Agriculture,* 5th edn. Iowa State University Press, Ames, Iowa, pp. 263–275.

Coyne, M.S., Gilfillen, R.A., Rhodes, R.W. and Blevins, R.L. (1995) Soil and fecal coliform trapping by grass filter strips during simulated rain. *Journal of Soil and Water Conservation* 50, 405–408.

Doran, J.W. and Werner, M.R. (1990) Management of soil biology. In: Francis, C.A., Flora, C.B. and King, L.D. (eds) *Sustainable Agriculture in Temperate Zones.* Wiley Interscience, New York, pp. 205–230.

Duxbury, J.M., Harper, L.A. and Mosler, A.R. (1993) Contributions of agroecosystems to global climate change. In: Rolston, D.E., Harper, L.A., Mosler, A.R. and Duxbury, J.M. (eds) *Agricultural Ecosystems Effects on Trace Gases and Global Climate Change.* ASA Special Publication No. 55, American Society of Agronomy, Madison, Wisconsin, pp. 1–18.

Ferguson, R. (1991) *Compare the Share. Phase I. Canadian Farmers Need a Fair Share of the Consumer Food Dollar.* Discussion Paper, House of Commons, Ottawa, Canada, 12pp.

Fick, G.W. (1996) Plant:society interface – utilization of grassland resources in agri-
cultural diversification. In: Wedin, W.F. and Jones, J.P. (eds) *Proceedings of the
Workshop on Innovative Systems for Utilization of Forage/Grassland/Rangeland
Resources, 22–24 September, 1993, Airlie, Virginia.* University of Minnesota
Extension Service, St Paul, pp. 93–97.

Fick, G.W. and Cox, W.J. (1995) *The Agronomy of Dairy Farming in New York State.
SCAS Teaching Series No. T95-1.* Department of Soil, Crop and Atmospheric Sciences, Cornell University, Ithaca, New York, 8pp.

Fisher, M.J., Rao, I.M., Ayarza, M.A., Lascano, C.E., Sanz, J.I., Thomas, R.J. and Vera,
R.R. (1994) Carbon storage by introduced deep-rooted grasses in the South
American savannas. *Nature* 371, 236–238.

Flachowsky, E., Matthey, M., Graf, H., Ochrimenko, W.I., Beyersdorfer, S., Dorn, W.
and Flachowsky, G. (1993) Influence of season keeping and vitamin D3 supply
on vitamin A, D, and E concentration in milk as well as the 25-OH vitamin D3
concentration in plasma of dairy cows. *Monatshefte für Veterinärmedizin* 48,
197–202.

Flaherty, D.J. and Drelich, J., Jr (1997) The use of grasslands to improve water
quality in the New York City watershed. In: *Proceedings XVIIIth International
Grassland Congress, 8–19 June, 1997, Winnipeg and Saskatoon, Canada.* Congress Secretariat, Grasslands 2000, Calgary, Alberta, Canada, pp. 27/15–27/16.

Follett, R.F. and Wilkinson, S.R. (1995) Nutrient management of forages. In: Barnes,
R.F., Miller, D.A. and Nelson, C.J. (eds) *Forages Volume II: The Science of Grassland Agriculture,* 5th edn. Iowa State University Press, Ames, Iowa, pp. 55–82.

Fussell, G.E. (1965) *Farming Technique from Prehistoric to Modern Times.* Pergamon
Press, Oxford, 269pp.

Fussell, G.E. (1966) *The English Dairy Farmer 1500–1900.* Frank Cass, London,
357pp.

Gillespie, G.W., Jr (1995) Sustainable agriculture and prospects for rural community
development in the U.S. *Research in Rural Sociology and Development* 6,
167–191.

Goldberg, J.J., Wildman, E.E., Pankey, J.W., Kunkel, J.R., Howard, D.B. and Murphy,
B.M. (1992) The influence of intensively managed rotational grazing, traditional
continuous grazing, and confinement housing on bulk tank milk quality and
udder health. *Journal of Dairy Science* 75, 96–104.

Goss, M.J., Ogilvie, J.R., Beauchamp, E.G., Stonehouse, D.P., Miller, M.H. and Parris,
K. (1993) Current state of the art on manure/nutrient management. *Final Report
to Agriculture Canada* (Contract No. 01689-2-3920/01-XSE), Land and Resource
Science, University of Guelph, Guelph, Ontario.

Grigg, D.B. (1974) Dairying. In: Grigg, D.B. (ed.) *The Agricultural Systems of the
World – An Evolutionary Approach.* Cambridge University Press, London, pp.
187–209.

Hansen, J.L., Viands, D.R., Steffens, J.C. and Sniffen, C.J. (1992) Heritability and
improvement of protein and nitrogen concentrations in wilted alfalfa forage.
Crop Science 32, 879–883.

Hassanein, N. and Kloppenburg, J.R., Jr (1995) Where the grass grows again: knowledge exchange in the sustainable agriculture movement. *Rural Sociology* 60,
721–740.

Heichel, G.H. and Martin, N.P. (1980) Alfalfa. In: Pimentel, D. (ed.) *Handbook of
Energy Utilization in Agriculture.* CRC Press, Boca Raton, Florida, pp.155–161.

Hodgson, J. and Illius, A.W. (eds) (1996) *Ecology and Management of Grazing Systems*. CAB International, Wallingford, UK, 466pp.

Hoveland, C.S. (1993) Importance and economic significance of the *Acremonium* endophytes to performance of animals and grass plants. *Agriculture, Ecosystems and Environment* 44, 3–12.

Jackson, W. (1994) *Becoming Native to this Place*. University of Kentucky Press, Lexington, 121pp.

Jackson-Smith, D., Barham, D., Nevius, M. and Klemme, R. (1996) *Grazing in Dairyland. The Use of Performance Management Intensive Rotational Grazing Among Wisconsin Dairy Farms. Technical Report No. 5*. Agricultural Technology and Family Farm Institute, University of Wisconsin, Madison, Wisconsin, 58pp.

Johnson, K.A. and Johnson, D.E. (1995) Methane emissions from cattle. *Journal of Animal Science* 73, 2483–2492.

Jonker, J.S., Cherney, D.J.R., Fox, D.G., Chase, L.E., and Cherney, J.H. (1997) Orchardgrass utilization by dairy cattle. *Proceedings of the American Forage and Grassland Council* 6, 148–152.

Kaffka, S.R. (ed.) (1989) *Sustaining the Smaller Dairy Farm in the Northeast*. Sunny Valley Foundation, New Milford, Connecticut, 162pp.

Kauppi, L. (1990) Hydrology: water quality changes. In: Solomon, A.M. and Kauppi, L. (eds) *Toward Ecological Sustainability in Europe*. RR-90-6. International Institute for Applied Systems Analysis, Laxenburg, Austria, pp. 43–66.

Knoblauch, W.A., Milligan, R.A., Haslem, R.J. and van Lieshout, M.M. (1980) *An Economic Analysis of New York Field Crop Enterprises. Agricultural Extension Publication 80-6*. Department of Agricultural Economics, Cornell University Ithaca, New York, 47pp.

Knoblauch, W.A., Smith, S. and Putnam, L. (1996) Dairy-farm management. *New York Economic Handbook 1997*. Department of Agricultural, Resource, and Managerial Economics, Cornell University, Ithaca, New York, pp. 7/1–7/17.

Kohn, R.A. and Allen, M.S. (1995) Effect of plant maturity and preservation method on in vitro protein degradation of forages. *Journal of Dairy Science* 78, 1544–1551.

Landstrom, S., Lomakka, L. and Anderson, S. (1996) Harvest in spring improves yield and quality of reed canary grass as a bioenergy crop. *Biomass and Bioenergy* 11, 333–341.

Lanyon, L.E. (1995) Does nitrogen cycle?: Changes in the spatial dynamics of nitrogen with industrial nitrogen fixation. *Journal of Production Agriculture* 8, 70–78.

Larsen, R.E., Miner, J.R., Buckhouse, J.C. and Moore, J.A. (1994) Water-quality benefits of having cattle manure deposited away from streams. *Bioresource Technology* 48, 113–118.

Liaghat, A. and Prasher, S.O. (1996) A lysimeter study of grass cover and water table depth effects on pesticide residues in drainage water. *Transactions of the ASAE* 39, 1731–1738.

McAllister, T.A., Okine, E.K., Mathison, G.W. and Cheng, K.J. (1996) Dietary, environmental and microbiological aspects of methane production in ruminants. *Canadian Journal of Animal Science* 76, 231–243.

McSweeny, W.T. (1987) How to use forages profitably. In: *Future of Pennsylvania Agriculture and the Rural Community – A Regional Perspective*. Pennsylvania State College of Agriculture, College Park, pp. 79–80.

Mayland, H.F. and Cheeke, P.R. (1995) Forage-induced animal disorders. In: Barnes, R.F., Miller, D.A. and Nelson, C.J. (eds) *Forages Volume II: The Science of Grassland Agriculture*, 5th edn. Iowa State University Press, Ames, Iowa, pp. 121–135.

Meister, E. and Koch, B. (1997) Graded management intensity of grassland systems for enhancing biodiversity on a species and landscape scale. In: *Proceedings XVIIIth International Grassland Congress, 8–19 June, 1997, Winnipeg and Saskatoon, Canada*. Congress Secretariat, Grasslands 2000, Calgary, Alberta, Canada, pp. 12/13–12/14.

Minc, L.D. and Vandermeer, J. (1990) The origin and spread of agriculture. In: Carroll, C.R., Vandermeer, J.H. and Rosset, P.M. (eds) *Agroecology*. McGraw-Hill, New York, pp. 65–111.

Misselbrook, T.H., Laws, J.A. and Pain, B.F. (1996) Surface application and shallow injection of cattle slurry on grassland: nitrogen losses, herbage yields and nitrogen recoveries. *Grass and Forage Science* 51, 270–277.

Moore, K.C. and Gerrish, J.R. (1995) Economics of grazing systems versus row crop enterprises. *Proceedings of the American Forage and Grassland Council* 4, 112–116.

Mueller, R. (1996) Grazers whip confinees. *Pasture Talk* 2(1), 1, 10.

Murphy, W.M., Rice, J.R. and Dugdale, D.T. (1986) Dairy farm feeding and income effects of using Voisin grazing management of permanent pastures. *American Journal of Alternative Agriculture* 1, 147–152.

Nation, A. (1995) Irish competitiveness study full of bad news for US and Canada. *Stockman Grass Farmer* (January), 9–10.

Nelson, P.N., Cotsaris, E. and Oades, J.M. (1996) Nitrogen, phosphorus, and organic carbon in streams draining two grazed catchments. *Journal of Environmental Quality* 25, 1221–1229.

Olson, K.R. and Olson, G.W. (1985) Use of agronomic data and enterprise budgets in land assessment evaluations. *Journal of Soil and Water Conservation* 40, 455–457.

Oltjen, J.W. and Beckett, J.L. (1996) Role of ruminant livestock in sustainable agricultural systems. *Journal of Animal Science* 74, 1406–1409.

Parker, W.J., Muller, L.D. and Buckmaster, D.R. (1997) Management and economic implications of intensive grazing on dairy farms in the northeastern states. *Journal of Dairy Science* 75, 2587–2597.

Pearson, C.J. and Ison, R.L. (1987) *Agronomy of Grassland Systems*. Cambridge University Press, Cambridge, 169pp.

Perelman, M. and Shea, K.P. (1972) The big farm. *Environment* 14, 10–15.

Peters, R.L. and Lovejoy, T.E. (eds) (1992) *Global Warming and Biodiversity*. Yale University Press, New Haven, Connecticut, 386pp.

Pirtle, T.R. (1926a) The Netherlands. In: Pirtle, T.R. (ed.) *History of the Dairy Industry*. Mojonnier Brothers, Chicago, Illinois, pp. 283–309.

Pirtle, T.R. (1926b) Germany. In: Pirtle, T.R. (ed.) *History of the Dairy Industry*. Mojonnier Brothers, Chicago, Illinois, pp. 337–356.

Reaume, C.M., Joy, D.M., Etches, C., Lee, H., Whiteley, H.R. and Zelin, S. (1994) Bacterial contamination from land application of liquid manure. *Proceedings*

Canadian Society of Civil Engineering (Environmental Engineering Conference, Winnipeg) 3, 87–96.

Reid, J.T. (1970) The future role of ruminants in animal production. *Proceedings 3rd International Symposium on Physiology of Digestion and Metabolism in the Ruminant*. Oriel Press, Newcastle upon Tyne, pp. 1–22.

Reid, W.S. (1985) Soil erosion and crop productivity. In: Follett, R.F. and Stewart, B.A. (eds) *Regional Effects of Soil Erosion*. American Society of Agronomy, Madison, Wisconsin, pp. 235–239.

Russelle, M.P., Satter, L.D., Dhiman, T. and Kanneganti, V.R. (1997) Nitrogen cycling in pastures grazed by lactating dairy cows. In: *Proceedings XVIIIth International Grassland Congress, 8–19 June, 1997, Winnipeg and Saskatoon, Canada*. Congress Secretariat, Grasslands 2000, Calgary, Alberta, Canada, pp. 20/11–20/12.

Rust, J.W., Sheaffer, C.C., Eidman, V.R., Moon, R.D. and Mathison, R.D. (1995) Intensive rotational grazing for dairy cattle feeding. *American Journal of Alternative Agriculture* 10, 147–151.

Ryden, J.C., Ball, P.R. and Garwood, E.A. (1984) Nitrate leaching from grassland. *Nature* 311, 50–53.

Sanderson, M.A., Reed, R.L., McLaughlin, S.B., Wullschleger, S.D., Conger, B.V., Parrish, D.J., Wolf, D.D., Taliaferro, C., Hopkins, A.A. and Ocumpaugh, W.R. (1996) Swithchgrass as a sustainable energy crop. *Bioresource Technology* 56, 83–93.

Scharpenseel, H.W. and Becker-Heidmann, P. (1997) Carbon sequestration by grassland and woodland soils of different climate zones as revealed by (thin) layer wise carbon-14 dating. In: *Proceedings XVIIIth International Grassland Congress, 8–19 June, 1997, Winnipeg and Saskatoon, Canada*. Congress Secretariat, Grasslands 2000, Calgary, Alberta, Canada, pp. 9/3–9/4.

Shaish, A., Daugherty, A., O'Sullivan, F., Schonfeld, G. and Heinecke, J.W. (1995) Beta-carotene inhibits atherosclerosis in hypercholesterolemic rabbits. *Journal of Clinical Investigation* 96, 2075–2082.

Sharp, W.C., Schertz, D.L. and Carlson, J.R. (1995) Forages for conservation and soil stabilization. In: Barnes, R.F., Miller, D.A. and Nelson, C.J. (eds) *Forages Volume II: The Science of Grassland Agriculture,* 5th edn. Iowa State University Press, Ames, Iowa, pp. 243–262.

Sherratt, A. (1981) Plough and pastoralism: aspects of the secondary products revolution. In: Hodder, I., Isaac, G. and Hammond, N. (eds) *Pattern of the Past – Studies in Honor of David Clarke*. Cambridge University Press, Cambridge, pp. 261–305.

Soule, J.D. and Piper, J.K. (1992) *Farming in Nature's Image – An Ecological Approach to Agriculture*. Island Press, Washington, DC, 286pp.

Stout, W.L., Elwinger, G.F., Fales, S.L., Muller, L.D., Schnabel, R.R. and Priddy, W.E. (1997) Nitrogen leaching from dairy pastures in the northeast United States. In: *Proceedings XVIIIth International Grassland Congress, 8–19 June, 1997, Winnipeg and Saskatoon, Canada*. Congress Secretariat, Grasslands 2000, Calgary, Alberta, Canada, pp. 20/1–20/2.

Subcommittee on Global Change Research (1997) *Our Changing Planet. The FY 1998 U.S. Global Change Research Program*. Global Change Research Information Office, University Center, Michigan, 118pp.

Tebib, K., Rouanet, J.M. and Besancon, P. (1997) Antioxidant effects of dietary polymeric grape seed tannins in tissues of rats fed a high cholesterol-vitamin E-deficient diet. *Food Chemistry* 59, 135–141.

Tilman, D., Wedin, D. and Knops, J. (1996) Productivity and sustainability influenced by biodiversity in grassland ecosystems. *Nature* 379, 718–720.

Todd, M. (1972) *Everyday Life of the Barbarians*. G.P. Putnam's Sons, New York, 184pp.

Van Soest, P.J. (1973) The uniformity and nutritive availability of cellulose. *Federation Proceedings* 32, 1804–1808.

Van Soest, P.J. (1994) Feeding strategies, taxonomy, and evolution. In: Van Soest, P.J. (ed.) *Nutritional Ecology of the Ruminant*, 2nd edn. Cornell University Press, Ithaca, New York, pp. 22–39.

Vasey, D.E. (1992) Humid temperate lands. In: Vasey, D.E. (ed.) *An Ecological History of Agriculture – 10,000 b.c. – a.d. 10,000*. Iowa State University Press, Ames, Iowa, pp. 146–179.

Vogel, K.P., Haskins, F.A. and Gorz, H.J. (1987) Potential for hydrocyanic acid poisoning of livestock by indiangrass. *Journal of Range Management* 40, 506–509.

Weaver, J.E. (1968) *Prairie Plants and Their Environment*. University of Nebraska Press, Lincoln, Nebraska, 276pp.

Webster, J. (1993) *Understanding the Dairy Cow*. Blackwell Science, Oxford, 382pp.

Wilkerson, V.A., Casper, D.P. and Mertens, D.R. (1995) The prediction of methane production of Holstein cows by several equations. *Journal of Dairy Science* 78, 2402–2414.

Breeding Cool-season Grasses   2

M.D. Casler

Department of Agronomy, University of Wisconsin-Madison, 1574 Linden Drive, Madison, Wisconsin 53706-1597, USA

Introduction

Humans have been selecting forage crops for thousands of years, often without explicit knowledge of having done so. The effects of domestication are easily seen in grain crops, for which wild and weedy relatives have dramatically different phenotypes from the domesticated form. Forage grasses, in contrast, have changed little in phenotype from their wild states. Plants and seeds collected on plant exploration trips to the most remote and untrammelled parts of the world are usually phenotypically similar to representatives of the same species growing on dairy farms in similar climates. Nevertheless, some dramatic improvements have been made to cool-season forage grasses, making them more suitable and/or useful for dairy farming. This chapter will explore some of the most notable of these improvements and the techniques used to create them.

Domestication, Natural Variation and Unconscious Selection

Early agriculturists probably had little reason to collect seeds of grasses used principally as animal feed. Forage grasses occur naturally in meadows and open areas that contain high levels of intraspecific and interspecific variation. Population buffering and mild natural selection provided a relatively

©CAB INTERNATIONAL 1998. *Grass For Dairy Cattle*
(eds. J.H. Cherney and D.J.R. Cherney)

stable and reliable feed source for thousands of years, probably until human population pressures began to dictate changes in animal husbandry and feeding. Until animal agriculture developed the frequent use of hay harvesting and storage, forage grass evolution probably changed little with the exception of added grazing pressure and changes in climate. The practice of cutting hay is probably little more than 1000 years old (Casler *et al.*, 1996).

Italian ryegrass (*Lolium multiflorum* Lam.) and perennial ryegrass (*L. perenne* L.) are examples of domesticated forage crops. These species appear to be derived by opposite selection pressures from a 'huge hybrid swarm' that may be the ancestor of European ryegrass populations (Tyler *et al.*, 1987). Italian ryegrass apparently evolved in the Lombardy and Piedmont plains of Italy during the 12th century, as an adaptation to hay harvesting and occasional reseeding (Beddows, 1953). Conversely, perennial ryegrass appears to have evolved its lower growth habit, higher tiller density, and greater longevity in response to centuries of grazing pressure. As animal agriculture spread and intensified throughout Europe, wild grasses were subjected to increasing pressure to survive and reproduce in the presence of grazing animals. Perennial ryegrass is now rarely found in the wild, except in association with grazing animals (Charmet *et al.*, 1990). Apart from a few examples of improved seed retention or shattering resistance (Oram *et al.*, 1985; Casler *et al.*, 1996), there are no other examples of cool-season forage grasses that have been domesticated. Human influences on these two ryegrass species appear to have been applied over a longer time period and to a greater extent than for any other cool-season forage grass.

Most phenotypic variation present in cool-season forage grasses has arisen as a result of natural selection for adaptation. Grasses have evolved tolerance to a wide range of environmental stresses (including drought, soil acidity, salinity, ozone, heavy metals, low fertility and temperature extremes) and biotic stresses (including pathogens, insects and large herbivores). Variation in the magnitude and combination of these stresses has resulted in a plethora of land race populations or ecotypes of all cool-season forage grasses. Many of these populations have evolved peculiar traits that make them specifically adapted to micro-sites or climates. Early agriculturists were aware that local populations were generally superior to exotic populations (Harlan, 1992).

The first documentations of a fledgling grass seed industry were based on reports of field tests of land races, and premium seed prices for superior strains, in Great Britain during the late 18th century (Casler *et al.*, 1996). Formalized selection and breeding programmes began 100 years later in the USA, Great Britain and Sweden. As human populations grew and became more technologically skilled, the selection pressures applied to forage grasses became more varied and complex, and more dependent on human ingenuity, imagination and disturbance.

Efficiency and Effectiveness of Grass Breeding Programmes

Documented progress made by breeding cool-season forage grasses has significantly lagged behind that of grain crops during the 20th century. Increases in harvested dry matter yield have averaged 3.5% decade^{-1} for forage grasses compared to 12.5% decade^{-1} for grain crops (Humphreys, 1998). Humphreys gave four reasons for the lag observed in forage crops: (i) their longer breeding cycle, (ii) gains in harvest index of grain crops, (iii) greater exploitation of heterosis in grain crops, and (iv) the importance of other traits in forage crops. While each of these is a legitimate reason for reduced progress for dry matter yield in forage crops, each area has considerable potential for making improvements in the efficiency of forage grass breeding programmes.

Shortening the breeding cycle

Forage breeders have made significant strides toward reducing the time required to complete the selection process and recombine selected plants to form the new generation. Burton (1974) described a set of restrictions to phenotypic selection that he termed Recurrent Restricted Phenotypic Selection. With some additional restrictions (Burton, 1982), he was able to coax a 1-year recurrent selection cycle time out of Pensacola bahiagrass (*Paspalum notatum* var. *saure* Parodi) by collecting data, making selections, and producing seed on selected plants during the establishment year. Furthermore, he acheived selection among both female and male gametes by excising reproductive culms, placing them in water-filled bottles in a laboratory, and shaking them daily during anthesis. Six cycles of selection increased forage yield by 6.3% cycle^{-1}, far surpassing the average gains for both forage and grain crops reported by Humphreys (1998).

The vernalization requirement and the juvenility period of most cool-season forage grasses is an impediment to rapid recombination of selected plants. Methods have been developed to advance one generation in less than 6 months in reed canarygrass, *Phalaris arundinacea* L. (Heichel *et al.*, 1980) and orchardgrass, *Dactylis glomerata* L. (Ikegaya *et al.*, 1985). This would allow theoretical completion of two breeding cycles per year for seedling traits.

While formal grass breeding efforts have existed for over 100 years, most cultivar registrations suggest that it typically requires a minimum of 10 years, but often 15–20 years, to produce a new cultivar. This has severely limited the number of recombinations and the cumulated selection pressure that has been applied during the past 100 years. Improvements in the efficiency of selection programmes, such as those made by Burton, represent one of the greatest opportunities to increase genetic progress.

Increasing 'harvest index' of forage grasses

Forage grasses have no classical harvest index, as defined for grain crops, except in their seed production phase. In forage swards, the equivalent of the grain crop harvest index would be the ratio of shoot dry matter to total plant dry matter (shoots plus roots). While it might be tempting to select for a higher shoot : root ratio, this would upset the functional evolutionary balance between shoots and roots. Kemp and Culvenor (1994) suggest dramatic increases in the shoot : root ratio should be avoided, unless accompanied by other changes such as an increased number of root tips or an increased water and nutrient uptake capacity per unit of root biomass, i.e. a smaller, but more efficient root system.

Forage grasses generally have highly plastic root and shoot development, combined with a relatively short lifespan of individual roots and shoots (Langer, 1963; Eason and Newman, 1990). These inherent characteristics bring about 'a constant re-adjustment in the spatial distribution of the leaf canopy and the actively absorbing part of the root system', allowing them to be effective competitors in productive environments (Grime, 1988). The root system may be the most important component of a grass plant's ability to compete against its neighbours (Donald, 1958; Hofman and Ennik, 1980). In tall fescue (*Festuca arundinacea* Schreb.), genotypes with larger primary root diameters, but fewer primary roots, are better able to penetrate hardpans and can extract water from twice the depth as genotypes with smaller but more roots (Torbert *et al.*, 1990). Conversely, when rooting depth is not inhibited, the smaller-root-diameter genotypes had greater Mg uptake, shoot dry matter, water-use efficiency, and lance nematode (*Hoplolaimus galeatus* (Cobb.) Sher.) tolerance (Table 2.1).

Alternatively, shoot dry matter may be partitioned into functional units that appear amenable to modification by selection and breeding. For example, Kemp and Culvenor (1994) define sward growth rate as:

$$\text{(plants ha}^{-1}) \times \text{(tillers plant}^{-1}) \times \text{(leaves tiller}^{-1}) \times \text{(leaf growth rate).}$$

Although there is compensation among these four components of sward growth rate, all are amenable to genetic modification, and changes in any one component are not necessarily fully compensated by opposite changes in the others. For example, tall fescue populations selected for high leaf area expansion rate (LAER) had higher mid- and late-summer forage yield under infrequent cutting than populations selected for low LAER (Sleper and Nelson, 1989). Reproductive growth did not differ between high- and low-LAER populations.

Pasture dry matter can be partitioned into stem, live leaf, and dead tissue components (Hodgson, 1985), where live leaf tissue is preferentially selected by most grazing animals. Dead tissue may be reduced by the use of more disease and insect resistant cultivars or by genetic modifications to canopy architecture, reducing shading and/or early senescence of lower

Table 2.1. Rooting characteristics, magnesium uptake, forage yield, lance nematode tolerance, and water-use efficiency of two tall fescue genotypes differing in mean primary root size.

Reference/measurement variable	Tall fescue genotype	
	Small roots	Large roots
Torbert *et al.* (1985)		
Mean root diameter (mm)	0.72	0.90
Mean root dry weight (g plant^{-1})	0.78	0.43
Mean root volume (cm^3)	14.2	6.8
Magnesium uptake (μM g^{-1} day^{-1})	15.3	13.9
Elkins *et al.* (1979)		
Median root diameter (mm)	1.0	2.2
Number of nematodes (g root^{-1})	22	45
Forage yield, without nematodes (g plant^{-1})	1.28	0.97
Forage yield, with nematodes (g plant^{-1})	1.17	0.48
Yield suppression due to nematodes (%)	9	50
Relative water-use efficiency (%)	100	71

leaves. Stem tissue may be reduced by breeding of sparse-flowering pasture cultivars. Sparse-flowering orchardgrass plants in the northeastern USA had normal flowering and seed production in a northwestern USA seed production region (Hovin *et al.*, 1966).

Exploitation of heterosis

Heterosis is responsible for a large portion of the genetic gains made on grain crops such as maize (*Zea mays* L.), sorghum (*Sorghum bicolor* L.), and pearl millet (*Pennisetum americanum* (L.) Leeke). It has been significantly utilized in cultivars of only one forage grass, the warm-season species bermudagrass. The development of a system to propagate single clones vegetatively on vast hectarages allows bermudagrass breeding programmes to utilize all sources of genetic variation in cultivar development. Although heterosis *per se* has had little impact on cool-season forage grass production, numerous genetic mechanisms for exploiting heterosis have undergone recent development and refinement. They offer considerable hope for incorporation into applied breeding programmes.

All current cool-season forage grass cultivars are either synthetic populations or strain crosses among multiple complementary populations. In species with disomic inheritance, either diploids or allopolyploids, a small amount of heterosis is usually observed in the first generation following synthesis, after which there is a small, but progressive decline in performance with each generation of seed multiplication (Bingham *et al.*, 1994). Conversely, with polysomic inheritance of autopolyploids, which characterizes most cool-season forage grasses, there is progressive heterosis to the second generation with non-inbred parents or the third generation with inbred parents (Bingham *et al.*, 1994). Furthermore, preferential pairing of chromosomes in autotetraploids, a common phenomenon that is genetically controlled, maintains much more heterozygosity than is possible in diploids (Breese *et al.*, 1981).

Exploitation of heterosis in autopolyploid crops is largely a function of complementary gene interactions between blocks of chromosomes derived from different parents (Bingham *et al.*, 1994). Greater genetic diversity among parents leads to greater exploitation of heterosis for forage yield in their progeny (Humphreys, 1991) due to greater allelic diversity within the progeny population. Chromosome blocks selected in different parents complement each other within the progeny, an example in which 'the whole is greater than the sum of the parts'. Indeed, selection for increased seed yield and forage traits in orchardgrass was unsuccessful in four populations, when each was selected at a single location. Conversely, when subsets of populations selected at four different locations were strain-crossed to create a pooled progeny population, two cycles of multi-location selection led to a 7–8% increase in forage yield (Casler *et al.*, 1997a) combined with a 23–32% increase in seed yield (Barker *et al.*, 1997). Selection at diverse locations apparently led to accumulation of favourable genes for both forage and seed yield at all locations. Allelic diversity among the locations led to complementary gene interactions within the strain-cross progeny.

Self-incompatability systems exist within most cool-season forage grasses, typically controlled by the two-locus SZ system. Hayward (1988) proposed a system to produce 83% hybrid seed, utilizing paired inbred populations that are internally self-incompatable, but externally cross-compatable, based on their SZ alleles. The system has yet to be tested, partly due to the great time and expense of developing the appropriate inbred populations and testing them for combining ability and seed production potential.

Many forage species, including ryegrasses and orchardgrass, are capable of producing gametes (pollen or eggs, or both) with the sporophytic chromosome number (Mariani and Tavoletti, 1992). Production of $2n$ gametes is under genetic control, allowing it to be transferred from one genotype to another. Production of $2n$ pollen in perennial ryegrass, via fusion of adjacent polar nuclei, is an effective means of transferring approximately 80% of the heterozygosity of a diploid genotype to the tetraploid level (Chen *et al.*, 1997). In orchardgrass, when combined with selection at both the diploid and tetraploid levels, $2n$ eggs led to significant heterosis of progeny compared with their tetraploid male parent (Casler and Hugessen, 1988).

Apomixis is a method of reproducing an unlimited numbers of copies of a favourable genotype via asexual seed formation. Because it holds such an advantage of fixing favourable genotypes without the drawbacks of many hybrid production systems it represents, more than any other single trait, the 'holy grail' of many breeding programmes in grain, forage, and turf species. Apomixis is usually under simple genetic control and methods have been developed to transfer genes from wild relatives to cultivated species using traditional or molecular methods (Hanna, 1995). Although apomixis has been discovered in many species, there is only one prominent example of a cool-season forage grass: Kentucky bluegrass, *Poa pratensis* L. (Hanna, 1995). However, Hanna and Bashaw (1987) suggest that apomixis should be discovered in more crop species if we were to be more conscious, diligent and methodical in our searches for sources of apomixis.

Importance of traits other than forage yield

Forage yield *per se* has received relatively little emphasis in cool-season forage grass breeding. There are very few references to direct selection for forage yield (Casler *et al.*, 1996). I believe there are five principal reasons for this: (i) the difficulty and expense of using sward plots compared with spaced plants, (ii) the reduced additive genetic variance among families encountered in sward-plot tests, (iii) the additional cycle time required for sward-plot tests, (iv) recent emphasis on forage quality traits more directly related to animal performance, and (v) the ease with which other traits can be manipulated by selection among spaced plants.

Spaced plantings vs. sward plots
The vast majority of grass breeding takes place on an individual-plant basis, usually under relatively non-competitive conditions. Spaced plants allow maximum phenotypic expression, selection within and among germplasm sources, and observation of plant traits not evident under sward conditions. While forage yield of spaced-plants is heritable, the correlation between spaced plants and sward plots for forage yield is highly variable, but often low (Casler *et al.*, 1996). Although grass breeders have been successful at increasing sward-plot forage yield by selection among spaced plants, these gains have generally been due to correlated responses of forage yield to selection for other traits, not due to selection for forage yield *per se*.

Many cultivar registrations cite direct selection for forage yield on sward plots in the later stages of the cultivar development programme (Casler *et al.*, 1997a). However, selection pressures are usually very mild because of the small number of families typically evaluated. Without definitive comparisons, it is impossible to weigh the relative benefits of direct selection for forage yield on sward plots vs. indirect selection on spaced plants.

Additive genetic variance

The use of sward-plots for estimating forage yield requires vegetative prop-agation of clonal ramets, which is very time-consuming, or a seed genera-tion, such as polycross or topcross families. Because genetic variation among polycross or topcross families utilizes only 25% of additive genetic variance and less of additive forms of inter-locus interaction (epistatic) vari-ance, family testing is seldom used on a large scale (Breese and Hayward, 1972).

Utilization of all additive genetic variance requires selection within fam-ilies. Within-family selection can be facilitated by selecting tillers from sward plots or by planting spaced plants of selected families. Selecting surviving tillers from sward plots is subject to genetic changes associated with natural selection. A literature review of natural selection in pasture grasses shows a general response toward a more densely tillering, prostrate, shorter plant type, but highly variable, and usually unpredictable, genetic changes in forage yield and maturity (Falkner, 1996). Reduced seed production of sur-viving plants, observed by Charles (1961; 1964) and possibly caused by lack of selection pressure for seed production, would be a death knell for many cultivars. The use of supplemental spaced plantings would require either additional years per cycle or additional plantings (Vogel and Pedersen, 1993).

Van Dijk (1979) overcame these drawbacks by selecting perennial ryegrass spaced-plants in an overseeded timothy (*Phleum pratense* L.) sward. In this case, among- and within-family selection gave a 4–25% higher selection response for forage yield compared with among-family selection alone. Theoretically, addition of within-family selection should increase selection response by 3.6–5.4 times (Vogel and Pedersen, 1993). The dis-crepancy between actual gains in the Van Dijk study (1979) and theoretical gains may relate to the fact that selection was practised for vigour of overseeded spaced-plants and response was measured in monoculture swards. As discussed under the heritability section, this suggests a low, but positive, genetic correlation between these two traits. Overseeded spaced-plant nurseries have become a common practice for grass breeders (Wilkins, 1991).

Cycle time

Progeny-test selection in sward plots, the most effective measure of forage yield, requires a longer cycle time than phenotypic or mass selection. Many forage breeders are not willing to devote additional time to a selection pro-cedure that requires more time and labour, and is of unknown benefit. A survey of 51 orchardgrass cultivar registrations between 1961 and 1996 showed 51% to have employed some form of progeny test, but with only mild selection pressure (26–67% of families selected) (Casler *et al.*, 1997a). Although Vogel and Pedersen (1993) have suggested some potentially useful modifications to traditional family selection methods, there remains little empirical evidence comparing family selection to phenotypic selection. Both gave similar gains per year for seed yield of intermediate wheatgrass, *Thinopyrum intermedium* (Host) Bark. & D.R. Dewey (Knowles, 1977).

Potential impact on animal performance

Other traits are often more important for dairy production systems than forage yield *per se*. For example, Clark and Wilson (1993) predicted that a 5% increase in digestibility of pasture, combined with a 5% decrease in forage yield, would still result in increased profit for dairy farming, despite the loss of milk production per hectare. Similarly, the highest-yielding perennial ryegrass cultivar in a group of four gave the lowest liveweight gain of lambs (Munro *et al.*, 1992). Numerous studies have shown that digestibility, fibre concentration, and crude protein concentration can be modified rather easily by selecting among spaced plants (Buxton and Casler, 1993; Casler *et al.*, 1996; Casler, 1998). Furthermore, genetic gains made in forage nutritive value of spaced plants invariably hold up when populations are tested under sward conditions, for laboratory analysis or animal performance (see later section).

Heritability of other traits and correlation with yield

Breeding programmes have generally focused on solving specific problems such as increasing disease resistance (Casler and Pederson, 1996) or identifying specific components of forage yield that can be easily modified. In theory, if a yield component has a high genetic correlation with forage yield and a higher heritability than forage yield, selection for the yield component should be as effective or more effective than selection for forage yield *per se* (Falconer, 1989). Because forage yield is generally considered to have low heritability, most breeders opt for indirect improvements.

Genetic progress has been documented for a range of potential forage yield components, including: leaf growth rate (Sleper and Nelson, 1989), survival under competition (Van Dijk, 1979; Novy *et al.*, 1995), and tolerances to environmental stresses such as cold, drought, heavy metals, and salt (Casler *et al.*, 1996). These are traits for which seedling or spaced-plant performance is highly correlated with sward-plot performance. Traits such as these, that relate to agricultural fitness and survival, contribute to long-term persistence, stress tolerance and ability to produce dry matter over an extended life of a stand and/or a range of environments. Van Wijk *et al.* (1993) logically argue that many such improvements have led to genetic increases in forage yield potential of cool-season forage grasses.

Genetic progress for forage yield may be slowed if there is a low or unpredictable genetic correlation between the putative yield component and forage yield *per se*. For example, selection for cold tolerance of Italian ryegrass seedlings led to large genetic gains in cold tolerance of seedlings, but to genetic losses of adult-plant field persistence (Hides, 1979). This illustrates the complexity of persistence, that there are many external factors potentially causing loss of persistence and/or forage yield. Relaxation of selection pressure for field adaptation can have rapid and dire consequences. Conversely, field selection for survival of perennial ryegrass, hybrid ryegrass (*L. hybridum* Hausskn.), and festulolium (*Festulolium braunii* K.A.) has been effective in increasing persistence of selected populations in mixed swards (Novy *et al.*, 1995) and has been associated with increased freezing tolerance in festulolium (Casler *et al.*, 1997b).

Most disease resistance breeding programmes widely and successfully rely on seedling evaluations. Genetic increases in resistance generally provide protection of forage yield potential under disease outbreaks, without loss of forage yield under disease-free conditions (Casler and Pederson, 1996). This suggests that disease resistance carries no physiological cost and may actually have a small positive genetic correlation with forage yield.

Grass Breeding Now and in the Future

Efforts to improve cool-season forage grasses can be grouped into two broad categories: (i) adaptation and production and (ii) forage nutritional value. Obviously the first category is paramount; if a species or cultivar is unadapted to a region, the quality of its forage is of no consequence. Given marginal adaptation or short-term persistence of a cultivar with superior nutritional value, economics will determine the cost-benefits of the trade-off between perenniality and production vs. improved nutritional value. The marginal cold tolerance of perennial ryegrass in the northcentral USA is an example of this decision rule. Adaptation is typically measured in terms of harvestable forage dry matter. Because forage crops are seldom allowed to undergo sexual reproduction under forage production conditions, measures of plant or tiller survival are the agricultural analogues to Darwinian fitness (number of surviving progeny) and are considered here as a component of agricultural fitness.

Selection criteria in cool-season grass breeding

Adaptation, production and agricultural fitness
The comparison of genetic gains for forage grasses vs. grain crops (Humphreys, 1998), combined with the morphological similarity between natural and 'domesticated' populations of forage grass species (Harlan, 1992), suggest that forage grass breeding is still in its infancy. This conclusion is further validated by the heavy reliance of grass breeders on natural variation and natural selection. Most cool-season forage grasses have wide regions of adaptation and have demonstrated a wealth of climatic and

geographic variation for quantitative traits. Although many thousands of collections have been made by many germplasm organizations, continued collection is still considered necessary and is often made a higher priority than preservation of existing collections (Harlan, 1983; Brown *et al.*, 1997). Some breeding programmes rely almost exclusively on natural collections as the basis for all genetic variability (e.g. Charmet *et al.*, 1990; Vogel and Pedersen, 1993).

Natural genetic polymorphisms exist for a wide range of stresses, allowing forage grasses to be productive and agriculturally useful under a wide range of edaphic and climatic conditions. Adaptation of forage grasses to stressful environments is often achieved by screening a wide array of introduced germplasm. The most superior germplasm in the stress environment is likely to derive from similar environments in other parts of the globe. For example, germplasm to improve the persistence of timothy in marginal sub-arctic regions is being derived from naturalized populations in the most severe of those environments (Schjelderup *et al.*, 1994). Similarly, the most freezing tolerant perennial ryegrass germplasm has been collected from locations with low mean winter temperatures (Humphreys and Eagles, 1988). Perennial ryegrass collections from some of these sites are making a rapid contribution to breeding programmes (Jones and Humphreys, 1993).

Development of heavy metal tolerance in grass cultivars derives principally from the evolution of tolerance in pastures or on mine spoils (Casler *et al.*, 1996). Many natural populations of forage grasses have genes for heavy metal tolerance maintained as natural genetic polymorphisms in extremely low frequencies. They were not discovered until accumulated human disturbance led to selection pressures that specifically favoured the survival of tolerant genotypes over intolerant genotypes. Although the most common examples of evolved tolerance are from mine spoils and smelter sites, Zn solubilization from rainwater on electricity pylons has led to evolved tolerance to Zn in contaminated pasture soils (Fig. 2.1). Other examples of evolved tolerances that may be useful in forage grass cultivars are ozone, NO_2, SO_2 and salt (Casler and Pederson, 1996; Casler *et al.*, 1996).

Inheritance studies generally show such tolerances to be monogenic (McNair, 1991), suggesting they can be transferred easily to other germplasm. An alcohol dehydrogenase gene of *Bromus mollis* L. appears to control both germination rate under cold conditions and flooding tolerance, with different alleles favouring each stress tolerance (Brown *et al.*, 1976). Apart from this example, there has been little success at identifying molecular markers associated with specific stress tolerances or adaptation traits of forage grasses. The only other notable exception is the identification of a chromosome segment that controls 80% of the genetic variation for timing of reproductive maturity in a perennial ryegrass cross (Hayward *et al.*, 1994).

Most cool-season forage grasses are used in mixed swards where a certain degree of compatability is necessary for multiple species to coexist

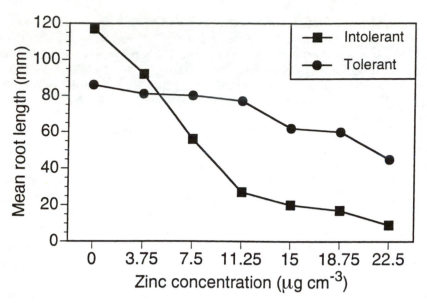

Fig. 2.1. Response of two tufted hairgrass (*Deschampsia caespitosa* L.) populations, one Zn tolerant and one intolerant, to increasing levels of Zn in solution culture (Coulaud and McNeilly, 1992).

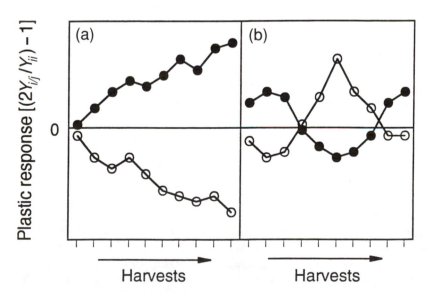

Fig. 2.2. Plastic responses of forage yield for components of binary forage mixtures: (a) competition in which one species exerts immediate dominance driving the other toward elimination and (b) competition in which each species fluctuates in exhibiting temporary dominance allowing coexistence of both mixture components (adapted from Zannone *et al.*, 1986).

within a pasture or hayfield. Darwin (1859) suggested that an individual's neighbours are the most important part of its environment. Excessive competitive ability leads to dominance of a single species, while inability to compete leads to loss of species (Fig. 2.2a). The dynamics of most forage mixtures, particularly between grasses and legumes, are regulated by competition (Zannone *et al.*, 1986). However, if mixture components are properly balanced with respect to competitive abilities, mixtures can offer higher mean forage yield and greater buffering against temporal and spatial heterogeneity in the environment (Zannone *et al.*, 1986).

There is considerable debate as to the most efficient means of developing components of forage mixtures. On one hand, accumulated results from a large number of studies of several grass species suggest a strong positive correlation between performance in monoculture and mixed stands (Zannone *et al.*, 1986). These authors strongly suggest that mixture components matched according to vigour based on monoculture performance will provide a mixture with plastic responses capable of buffering temporal environmental variability (Fig. 2.2b). Key to this conclusion was a long-term evaluation of mixtures, sufficient to allow fluctuations in the temporal dominance of individual components to be observed.

There is also good evidence for an alternative mechanism regulating mixture performance: coevolution. The approach has been to collect plants from old pastures of perennial ryegrass and white clover (*Trifolium repens* L.), and plant clonal ramets or seedlings in grass–clover mixtures of coexisting pairs or non-coexisting pairs (Evans *et al.*, 1989). Coexisting pairs generally have higher mean forage yield than non-coexisting pairs (Table 2.2), suggesting an evolved compatability favouring enhanced dry matter production. In a review by Turkington (1996), additional citations support this phenomenon, including evidence that (i) coexisting pairs may evolve a 'tolerance' toward each other, (ii) close neighbours within a pasture can evolve a greater compatability toward each other than distant neighbours, and (iii)

Table 2.2. Dry matter yield of white clover–perennial ryegrass mixtures based on coexisting pairs from long-term (40+ years) pastures or non-coexisting pairs (perennial ryegrass cultivars). Data from Evans *et al.* (1989).

White clover origin	Perennial ryegrass origin		
	Coexisting (kg ha^{-1})	Non-coexisting 1 (kg ha^{-1})	Non-coexisting 2 (kg ha^{-1})
Switzerland	5749	5019	4313
Switzerland	7549	6974	6431
Italy	6901	6664	6425
France	6893	6986	6490
Mean	6773	6411	5915

this phenomenon can be found in other grass species, such as annual blue-grass (*Poa annua* L.) and poverty oatgrass (*Danthonia spicata* (L.) Beauv.).

For both of the above phenomena, there are results that suggest a broad generalization to all mixture systems is not possible (Eagles, 1983; see review by Turkington, 1996). Other factors may regulate coexistence and the relationship between monoculture and mixed stand performance. Chanway *et al.* (1989) found that yield advantages of coexisting white clover–perennial ryegrass mixtures were due to the effects of *Rhizobium* strain and ryegrass genotype, not clover genotype, despite a lack of physical association between *Rhizobium* and perennial ryegrass. Progress may be made toward developing superior and stable forage mixtures by either method, but the preponderance of evidence suggests that neither system is universal and each mixture system must be evaluated in its appropriate target environment before we can determine the appropriate mechanism for additional or ultimate genetic improvement. In lieu of waiting 40+ years for natural selection to take its course, Hill (1994) has proposed a selection scheme designed to maximize compatability between mixture partners.

Despite their heavy reliance on natural variation, forage grass breeders have shown many notable successes at improving grasses by methodical selection. Forage yield (Fig. 2.3), seed yield, competitive ability, cold or freezing tolerance, leaf growth rate, drought tolerance, disease resistance

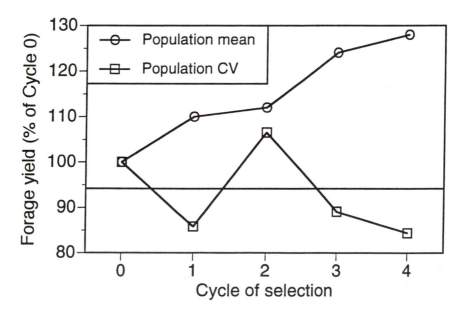

Fig. 2.3.   Forage yield mean and coefficient of variability response to four cycles of selection for high forage yield in rye, *Secale cereale* L. (Bruckner *et al.*, 1991).

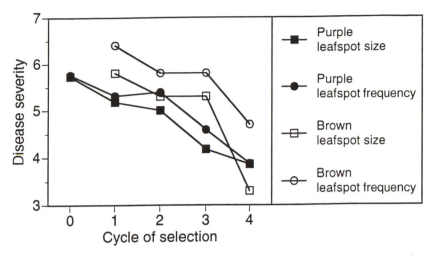

Fig. 2.4. Responses to four cycles of selection for increased resistance to purple leafspot (*Stagonospora arenaria* Sacc.) in orchardgrass (Zeiders *et al.*, 1984) and brown leafspot (*Pyrenophora bromi* (Died.) Drechs.) in smooth bromegrass, *Bromus inermis* Leyss. (Berg *et al.*, 1986).

(Fig. 2.4), and the length of the growing season have all been improved by methodical selection and breeding. Many additional examples are provided in reviews by Casler *et al.* (1996), Casler and Pederson (1996), Humphreys (1998), Johnson and Asay (1993), and Larson (1994).

Forage nutritional value
Efforts to breed grasses for improved forage nutritional value fall into two categories. Reactive strategies are aimed at correcting problems such as chemical toxicities, mineral imbalances, or physical characteristics. Proactive strategies are aimed at increasing energy and/or protein intake and availability to the ruminant.

One of the more remarkable stories in breeding cool-season forage grasses has been the development of improved cultivars in two species of the genus *Phalaris* – reed canarygrass (Marten *et al.*, 1976) and phalaris, *P. aquatica* L. (Oram *et al.*, 1985). In reed canarygrass, the presence of tryptamine or β-carboline alkaloids was associated with a 13-fold increase in the incidence of diarrhoea of sheep. While variation in gramine concentration did not cause differential animal health, a 68% reduction in gramine concentration was associated with an average of 185% greater lamb gains and 136% greater steer gains. Recent cultivars of reed canarygrass all have reduced gramine concentration, combined with no tryptamine/β-carboline alkaloids.

Several alkaloids are similarly responsible for reduced animal performance and health following grazing of perennial ryegrass or tall fescue.

These alkaloids are produced by endophytic fungi that have evolved to form a mutualistic association with their host plants (Clay, 1990). The fungi derive nutrients and water from their host, while the host is partially protected from ruminant and insect herbivory, via the anti-palatability and possibly toxicological effects of the fungal alkaloids. The endophyte is seed transmitted. Removal of the endophyte from tall fescue seed stock, relatively easily accomplished by heat and humidity, has led to changes in plant breeding objectives. Breeders may now focus more attention toward proactive strategies, such as increased digestibility, physiological efficiency and mineral uptake (Pedersen and Sleper, 1988). Conversely, endophyte removal may also force a reactive effort to improve stress tolerances that may have been conveyed by the endophyte. For example, lack of genetic sources of resistance to Argentine stem weevil (*Listronotus bonariensis* Kuschel) in perennial ryegrass prevents removal of the endophyte, which confers adequate levels of resistance (Prestidge and Gallagher, 1988). Identification of more agriculturally favourable endophyte biotypes (Christensen *et al.*, 1991) or genetically transformed genotypes of the endophyte (Murray *et al.*, 1992) may allow for a combination of the endophyte's beneficial effects without its harmful effects.

Mineral composition of grasses seldom matches the exact needs of dairy cows, particularly those that are on ration-balancing programmes that follow nutritional guidelines. A selected perennial ryegrass population with 56% more Mg than a check cultivar had greater intake of Mg, Ca and P by grazing ewes (Moseley and Baker, 1991). Similarly, in tall fescue, a high-Mg cultivar gave 12% higher blood serum Mg levels than 'KY-31' (D.A. Sleper, Missouri, 1997, personal communication). Many grasses are chronically low in P concentration for lactating dairy cows (e.g. Berg and Hill, 1983; Casler and Reich, 1987). Breeding for higher P concentration would reduce or alleviate the need for P supplementation in the ration, reducing input costs and increasing profit. Furthermore, high-P grass cultivars would simultaneously help to draw down excess soluble P reserves in pasture soils, reducing the potential for P contamination of ground and surface water.

The development of an *in vitro* dry matter digestibility technique revolutionized efforts to improve forage grass nutritional value. Increased *in vitro* digestibility, due to selection and breeding, has been documented in smooth bromegrass, orchardgrass, and perennial ryegrass, as well as numerous other forage species (Buxton and Casler, 1993). In Italian ryegrass, a 6% increase in *in vitro* organic matter digestibility, combined with a 5% increase in forage availability, led to a 3% increase in daily milk production and a 6% increase in milk production per hectare in a replicated grazing study (Wilman *et al.*, 1992).

Generally, forage yield is not affected by selection for increased *in vitro* digestibility (Casler *et al.*, 1996). Because seed production (and the price of seed) does not appear to be affected by increases in digestibility, these genetic improvements represent increases in animal production without accompanying increases in production costs, i.e. pure profit compared with pre-existing cultivars. As long as a cultivar is well adapted to a region, increases in forage nutritional value are one of the most economical means for a grower to increase animal production.

Improvements in forage nutritional value are generally consistent across environments, managements and harvesting systems (Wilman *et al.*, 1992; Casler, 1998). Forage nutritional value can be improved without modifying the timing of reproductive maturity. Such improvements are generally due to biochemical modifications to the plant, rather than changes in morphological traits, such as leaf : stem ratio. Most reports show that increased digestibility is due to reduced lignin concentration, modified lignin composition, and/or increased water soluble carbohydrate (WSC) concentration (Humphreys, 1998; Casler, 1998). Genetic changes in WSC concentration within 'S23' perennial ryegrass were closely associated with changes in allele frequency of a single genetic marker, Pgi/2 (Hayward *et al.*, 1994).

There has been recent interest in breeding forage grasses for reduced fibre concentration, an indicator of voluntary intake in ruminants (Van Soest, 1994). These efforts have been successful in smooth bromegrass and reed canarygrass, but have been accompanied by correlated reductions for forage yield (Surprenant *et al.*, 1988; Casler, 1998). Recent data suggest, however, that forage yield reductions may have resulted from inbreeding rather than a positive genetic correlation between fibre concentration and forage yield *per se* (Casler, 1998).

Recent advances in plant transformation techniques suggest that enzymes in the lignin biosynthetic pathway can be down-regulated with cloned antisense genes (Boudet *et al.*, 1995). These techniques are too recent to have been applied to cool-season forage grasses. Furthermore, their effects on non-graminaceous species are variable, giving unpredictable changes in lignin concentration or composition, sometimes with severe modifications to plant morphology. Nevertheless, plant transformation has the potential to produce dramatic changes in forage quality which, when incorporated into a grass breeding programme, may lead to novel and useful germplasm.

Crude protein (CP) concentration can be rather easily increased by breeding, but genetic increases in CP concentration are usually associated with decreases in forage yield and are negated by increasing rates of N fertilizer (Fig. 2.5). Furthermore, protein ingested by ruminants is often used inefficiently. Most soluble protein is degraded to ammonia in the rumen, much of which is absorbed into the bloodstream and eventually excreted as urea. Non-degradable protein passes through the rumen where it is more efficiently utilized in the lower digestive tract. Because native proteins differ dramatically in their rumen degradability, plant transformation may be an effective mechanism for inserting genes coding for non-degradable proteins into forage grasses (McNabb *et al.*, 1993). This method has been demonstrated, in principle, as a means of increasing the concentration of sulphur-rich proteins in forage legumes (Khan *et al.*, 1996). Genetic variation

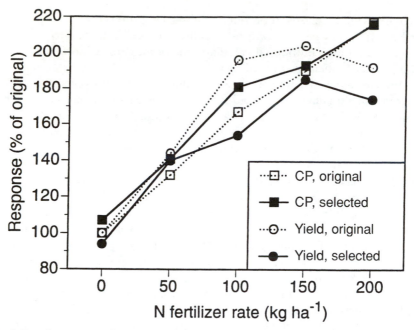

Fig. 2.5. Responses of two perennial ryegrass populations, one selected from the other for increased crude protein (CP) concentration, to increasing rates of N fertilizer (Arcioni *et al.*, 1983).

for ruminal protein degradation parameters is present within lucerne, *Medicago sativa* L. (Broderick and Buxton, 1991) and may also exist within some grass species.

Limits to future genetic progress

There is no evidence that we are approaching any limits to the progress that can be achieved by breeding cool-season forage grasses. Genetic variation within and among populations is still extremely high, showing no signs of decreasing. While there may be evidence for founder effects in some grain crops, drastically decreasing genetic variation within populations or land races (Mayr, 1976), there is no evidence for such genetic bottlenecks in forage grasses. Selection responses are generally linear, showing no signs of plateaus or reductions in genetic variance (Aldrich, 1987; Bruckner *et al.*, 1991). Logic also supports this argument, by the high likelihood that fewer than 20 selection cycles have been completed in any species since grass breeding efforts began in the late 19th century.

Wide hybridization within species, between species, and between genera has been used to create new genetic variability and expand the already wide boundaries of many species (Asay *et al.*, 1979; Asay and Jensen, 1996; Humphreys, 1998). Examples of successful gene transfers or recombinations include: early- × late-maturing ryegrasses or perennial × Italian ryegrass hybrids to improve seasonal yield distribution, transfer of drought tolerance from *Festuca* to *Lolium*, transfer of high nutritional value from *Lolium* to *Festuca*, and transfer of salinity tolerance and persistence from *Elytrigia* to *Pseudoroegneria*. There is also evidence that even more diverse grass genomes can be united by intergeneric hybridization, such as *Agropyron* × *Bromus* (Heszky *et al.*, 1993) and *Dactylis* × *Lolium/Festuca* (Oertl *et al.*, 1996). Several cultivars have resulted from some of these programmes, and generally represent transfers of a small number of genes from one species to another (Humphreys, 1989). As with most of the progress made in cool-season forage grass breeding, human imagination and ingenuity, combined with available funding, are the only real limitations to our ability to contribute to improved dairy production systems.

References

Aldrich, D.T.A. (1987) Developments and procedures in the assessment of grass varieties at NIAB 1950-87. *Journal of the National Institute of Agricultural Botany* 17, 313–327.

Arcioni, S., Veronesi, F., Mariotti, D. and Falcinelli, M. (1983) Evaluation of the possibility of improving protein yield in *Lolium perenne* L. *Zeitschrift für Pflanzenzüchtung* 91, 203–210.

Asay, K.H. and Jensen, K.B. (1996) Wheatgrasses. In: Moser, L.E., Buxton, D.R. and Casler, M.D. (eds) *Cool-Season Forage Grasses*. American Society of Agronomy, Madison, Wisconsin, pp. 691–724.

Asay, K.H., Frakes, R.V. and Buckner, R.C. (1979) Breeding and cultivars. In: Buckner, R.C. and Bush, L.P. (eds) *Tall Fescue*. American Society of Agronomy, Madison, Wisconsin, pp. 111–139.

Barker, R.E., Casler, M.D., Carlson, I.T., Berg, C.C., Sleper, D.A. and Young III, W.C. (1997) Convergent-divergent selection for seed yield and forage traits in orchardgrass. II. Seed yield response in Oregon. *Crop Science* 37, 1054–1059.

Beddows, A.R. (1953) The ryegrasses in British agriculture: a survey. *Welsh Plant Breeding Station, Bulletin Series H*, No. 17, 81pp.

Berg, C.C., and Hill, R.R., Jr (1983) Quantitative inheritance and correlations among forage yield and quality components in timothy. *Crop Science* 23, 380–384.

Berg, C.C., Sherwood, R.T. and Zeiders, K.E. (1986) Recurrent phenotypic selection for resistance to brown leaf spot in smooth bromegrass. *Crop Science* 26, 533–536.

Bingham, E.T., Groose, R.W., Woodfield, D.R., and Kidwell, K.K. (1994) Complementary gene interactions in alfalfa are greater in autotetraploids than diploids. *Crop Science* 34, 823–829.

Boudet, A.M., Lapierre, C. and Grima-Pettenati, J. (1995) Biochemistry and molecular biology of lignification. *New Phytologist* 129, 203–236.

Breese, E.L. and Hayward, M.D. (1972) The genetic basis of present breeding methods in forage crops. *Euphytica* 21, 324–336.

Breese, E.L., Lewis, E.J. and Evans, G.M. (1981) Interspecies hybrids and ploidy. *Philosophical Transactions of the Royal Society of London* B 292, 487–497.

Broderick, G.A. and Buxton, D.R. (1991) Genetic variation in alfalfa for ruminal protein degradability. *Canadian Journal of Plant Science* 71, 755–760.

Brown, A.H.D., Marshall, D.R. and Munday, J. (1976) Adaptedness of variants at an alcohol dehydrogenase locus in *Bromus mollis* L. (soft bromegrass). *Australian Journal of Biological Sciences* 29, 389–396.

Brown, A.H.D., Brubaker, C.L. and Grace, J.P. (1997) Regeneration of germplasm samples: wild versus cultivated plant species. *Crop Science* 37, 7–13.

Bruckner, P.L., Raymer, P.L. and Burton, G.W. (1991) Recurrent phenotypic selection for forage yield in rye. *Euphytica* 54, 11–17.

Burton, G.W. (1974) Recurrent restricted phenotypic selection increases forage yields of pensacola bahiagrass. *Crop Science* 14, 831–834.

Burton, G.W. (1982) Improved recurrent restricted phenotypic selection increases bahiagrass forage yields. *Crop Science* 22, 1058–1061.

Buxton, D.R., and Casler, M.D. (1993) Environmental and genetic effects on cell wall composition and degradability. In: Jung, H.G., Buxton, D.R., Hatfield, R.D. and Ralph, J. (eds) *Forage Cell Wall Structure and Digestibility*. ASA-CSSA-SSSA, Madison, Wisconsin, pp. 685–714.

Casler, M.D. (1998) Breeding for improved forage quality: potentials and problems. In: *Proceedings XVIIIth International Grassland Congress, 8–19 June, 1997, Winnipeg and Saskatoon, Canada*. Congress Secretariat, Grasslands 2000, Calgary, Alberta, Canada (in press).

Casler, M.D. and Hugessen, P.M. (1988) Performance of tetraploid progeny derived from 2x–4x intersubspecific crosses in *Dactylis glomerata* L. *Genome* 30, 591–596.

Casler, M.D. and Pederson, G.A. (1996) Host resistance and tolerance and its deployment. In: Chakraborty, S., Leath, K.T., Skipp, R.A., Pederson, G.A., Bray, R.A., Latch, G.C.M. and Nutter, F.W., Jr (eds) *Pasture and Forage Crop Pathology*. American Society of Agronomy, Crop Science Society of America, and Soil Science Society of America, Madison, Wisconsin, pp. 475–507.

Casler, M.D. and Reich, J.M. (1987) Genetic variability for mineral element concentrations in smooth bromegrass related to dairy cattle nutritional requirements. In: Gabelman, H.W. and Loughman, B.C. (eds) *Genetic Aspects of Plant Mineral Nutrition*. Martinus Nijhoff, Dordrecht, pp. 569–577.

Casler, M.D., Pedersen, J.F., Eizenga, G.C. and Stratton, S.D. (1996) Germplasm and cultivar development. In: Moser, L.E., Buxton, D.R. and Casler, M.D. (eds) *Cool-Season Forage Grasses*. American Society of Agronomy, Madison, Wisconsin, pp. 413–469.

Casler, M.D., Berg, C.C., Carlson, I.T. and Sleper, D.A. (1997a) Convergent-divergent selection for seed yield and forage traits in orchardgrass. III. Correlated responses for forage traits. *Crop Science* 37, 1059–1065.

Casler, M.D., Pitts, P.G., Bilkey, P.C. and Rose-Fricker, C.A. (1997b) Selection for field survival increases freezing tolerance in festulolium. In: *Proceedings XVIIIth International Grassland Congress, 8–19 June, 1997, Winnipeg and Saskatoon, Canada*. Congress Secretariat, Grasslands 2000, Calgary, Alberta, Canada, pp. 4/25–4/26.

Chanway, C.P., Holl, F.B. and Turkington, R. (1989) Effect of *Rhizobium legumino-sarum* biovar *trifolii* genotype on specificity between *Trifolium repens* and *Lolium perenne*. *Journal of Ecology* 77, 1150–1160.

Charles, A.H. (1961) Differential survival of cultivars of *Lolium, Dactylis,* and *Phleum*. *Journal of the British Grassland Society* 16, 69–75.

Charles, A.H. (1964) Differential survival of plant types in swards. *Journal of the British Grassland Society* 19, 198–204.

Charmet, G., Balfourier, F. and Bion, A. (1990) Agronomic evaluation of a collection of French perennial ryegrass populations: multivariate classification using genotype × environment interactions. *Agronomie* 10, 807–823.

Chen, C., Sleper, D.A., Chao, S., Johal, G.S. and West, C.P. (1997) RFLP detection of 2n pollen formation by first and second division restitution in perennial ryegrass. *Crop Science* 37, 76–80.

Christensen, M.J., Latch, G.C.M. and Tapper, B.A. (1991) Variation within isolates of *Acremonium* endophytes from perennial rye-grasses. *Mycological Research* 95, 918–923.

Clark, D.A. and Wilson, J.R. (1993) Implications of improvements in nutritive value on plant performance and grassland management. In: Baker, M.J. (ed.) *Grasslands for Our World*. SIR Publishing, Wellington, New Zealand, pp. 165–171.

Clay, K. (1990) Fungal endophytes of grasses. *Annual Review of Ecology and Systemics* 21, 275–297.

Coulaud, J., and McNeilly, T. (1992) Zinc tolerance in populations of *Deschampsia cespitosa* (Gramineae) beneath electricity pylons. *Plant Systemics and Evolution* 179, 175–185.

Darwin, C. (1859) *The Origin of Species by Means of Natural Selection,* 6th edn. John Murray, London, 502pp.

Donald, C.M. (1958) The interaction of competition for light and for nutrients. *Australian Journal of Agricultural Research* 9, 421–432.

Eagles, C.F. (1983) Relationship between competitive ability and yielding ability in mixtures and monocultures of populations of *Dactylis glomerata* L. *Grass and Forage Science* 38, 21–24.

Eason, W.R. and Newman, E.I. (1990) Rapid cycling of nitrogen and phosphorus from dying roots of *Lolium perenne*. *Oecologia* 82, 432–436.

Elkins, C.B., Haaland, R.L., Rodriguez-Kabana, R. and Hoveland, C.S. (1979) Plant-parasitic nematode effects on water use and nutrient uptake of a small- and a large-rooted tall fescue genotype. *Agronomy Journal* 71, 497–500.

Evans, D.R., Hill, J., Williams, T.A. and Rhodes, I. (1989) Coexistence and the productivity of white clover-perennial ryegrass mixtures. *Theoretical and Applied Genetics* 77, 65–70.

Falconer, D.S. (1989) *Introduction to Quantitative Genetics,* 3rd edn. Longman Group, UK Ltd, Harlow, Essex, 438pp.

Falkner, L.K. (1996) Effects of natural and laboratory selection on nutritive value and palatability of smooth bromegrass. PhD thesis, University of Wisconsin, Madison, Wisconsin.

Grime, J.P. (1988) The C-S-R model of primary plant strategies – origins, implications and tests. In: Gottlieb, L.D. and Jain, S.K. (eds) *Plant Evolutionary Biology*. Chapman and Hall, New York, pp. 371–393.

Hanna, W.W. (1995) Use of apomixis in cultivar development. *Advances in Agronomy* 54, 333–350.

Hanna, W.W. and Bashaw, E.C. (1987) Apomixis: Its identification and use in plant breeding. *Crop Science* 27, 1136–1139.

Harlan, J.R. (1983) The scope for collection and improvement of forage plants. In: McIvor, J.G. and Bray, R.A. (eds) *Genetic Resources of Forage Plants*. CSIRO, East Melbourne, Australia, pp. 3–14.

Harlan, J.R. (1992) Origins and processes of domestication. In: Chapman, G.P. (ed.) *Grass Evolution and Domestication*. Cambridge University Press, New York, pp. 159–175.

Hayward, M.D. (1988) Exploitation of the incompatability mechanism for the production of F_1 hybrid forage grasses. *Euphytica* 39, 33–37.

Hayward, M.D., McAdam, N.J., Jones, J.G., Evans, C., Evans, G.M., Forster, J.W., Ustin, A., Hossain, K.G., Quader, B., Stammers, M. and Will, J.K. (1994) Genetic markers and the selection of quantitative traits in forage grasses. *Euphytica* 77, 269–275.

Heichel, G.H., Hovin, A.W. and Henjum, K.I. (1980) Seedling age and cold treatment effects on induction of panicle production in reed canarygrass. *Crop Science* 20, 683–687.

Heszky, L.E., Janovszky, J., Gyulai, G., Kiss, E. and Hangyel, L.T. (1993). Application of the somaclone method in fertility restoration of the partially sterile hybrid of *Agropyron repens × Bromus inermis*: callus initiation and plant regeneration. In: Baker, M.J. (ed.) *Proceedings of the XVII International Grassland Congress, 8–21 February, 1993, Hamilton, NZ and Rockhampton, Australia*. New Zealand Grassland Association, Palmerston North, pp. 1048–1049.

Hides, D.H. (1979) Winter hardiness in *Lolium multiflorum* Lam. III. Selection for improved cold tolerance and its effect on agronomic performance. *Grass and Forage Science* 34, 119–124.

Hill, J. (1994) Breeding components for mixture performance. *Euphytica* 92, 135–138.

Hodgson, J. (1985) The control of herbage intake in the grazing ruminant. *Proceedings of the Nutrition Society* 44, 339–346.

Hofman, T.B. and Ennik, G.C. (1980) Investigation into plant characters affecting the competitive ability of perennial ryegrass (*Lolium perenne* L.). *Netherlands Journal of Agricultural Science* 28, 97–109.

Hovin, A.W., Rincker, C.M. and Wood, G.M. (1966) Breeding nonflowering orchardgrass, *Dactylis glomerata* L. *Crop Science* 6, 239–241.

Humphreys, M.O. (1991) A genetic approach to the multivariate differentiation of perennial ryegrass (*Lolium perenne* L.) populations. *Heredity* 66, 437–443.

Humphreys, M.O. (1998) The contribution of conventional plant breeding to forage crop improvement. In: *Proceedings XVIIIth International Grassland Congress, 8–19 June, 1997, Winnipeg and Saskatoon, Canada*. Congress Secretariat, Grasslands 2000, Calgary, Alberta, Canada (in press).

Humphreys, M.O. and Eagles, C.F. (1988) Assessment of perennial ryegrass (*Lolium perenne* L.) for breeding. I. Freezing tolerance. *Euphytica* 38, 75–84.

Humphreys, M.W. (1989) The controlled introgression of *Festuca arundinacea* genes into *Lolium multiflorum*. *Euphytica* 42, 105–116.

Ikegawa, F., Kawabata, S. and Sato, S. (1985) A method of accelerating generation-cycles in orchardgrass In: *Proceedings XV International Grassland Congress, 24–31 August, Kyoto, Japan*. Japanese Society of Grassland Science, Nishi-nasumo, Tochigi-ken, Japan, pp. 211–213.

Johnson, D.A. and Asay, K.H. (1993) Viewpoint: Selection for improved drought response in cool-season grasses. *Journal of Range Management* 46, 194–202.

Jones, M.L. and Humphreys, M.O. (1993) Progress in breeding interspecific hybrid ryegrasses. *Grass and Forage Science* 48, 18–25.

Kemp, D.R. and Culvenor, R.A. (1994) Improving the grazing and drought tolerance of temperate perennial grasses. *New Zealand Journal of Agricultural Research* 37, 365–378.

Khan, M.R.I., Ceriotti, A., Tabe, L., Aryan, A., McNabb, W., Moore, A., Craig, S., Spencer, D. and Higgins, T.J.V. (1996) Accumulation of a sulfur-rich seed albumin from sunflower in the leaves of transgenic subterranean clover (*Trifolium subterraneum* L.). *Transgenic Research* 5, 179–185.

Knowles, R.P. (1977) Recurrent mass selection for improved seed yields in intermediate wheatgrass. *Crop Science* 17, 51–54.

Langer, R.H.M. (1963) Tillering in herbage grasses. *Herbage Abstracts* 33, 141–148.

Larson, A. (1994) Breeding winter hardy grasses. *Euphytica* 77, 231–237.

McNabb, W.C., Higgins, C. Tabe, L. and Higgins, T.J.V. (1993) Transfer of genes encoding proteins with high nutritional value into pasture legumes. In: Baker, M.J. (ed.) *Proceedings of the XVII International Grassland Congress, 8–21 February, 1993, Hamilton, NZ and Rockhampton, Australia.* New Zealand Grassland Association, Palmerston North, pp. 1085–1086.

McNair, M.R. (1991) Why the evolution of resistance to anthropogenic toxins normally involves major gene changes: the limits to artificial selection. *Genetica* 84, 213–219.

Mariani, A. and Tavoletti, S. (eds) (1992) *Gametes with the Somatic Chromosome Number in the Evolution and Breeding of Polyploid Polysomic Species: Achievements and Perspectives.* Tipolitografia Porziuncola, Assisi, Italy, 103pp.

Marten, G.C., Jordan, R.M and Hovin, A.W. (1976) Biological significance of reed canarygrass alkaloids and associated palatability variation to grazing sheep and cattle. *Agronomy Journal* 68, 909–914.

Mayr, E. (1976) *Evolution and the Diversity of Life.* Harvard University Press, Cambridge, Massachusetts, 721pp.

Moseley, G. and Baker, D.H. (1991) The efficacy of a high magnesium grass cultivar in controlling hypomagnasaemia in grazing animals. *Grass and Forage Science* 46, 375–380.

Munro, J.M.M., Davies, D.A., Evans, W.B. and Scurlock, R.V. (1992) Animal production evaluation of herbage varieties. 1. Comparison of Aurora with Frances, Talbot, and Melle perennial ryegrass when grown alone and with clover. *Grass and Forage Science* 47, 259–273.

Murray, F.R., Latch, G.C.M. and Scott, D.B. (1992) Surrogate transformation of perennial ryegrass, *Lolium perenne*, using genetically modified *Acremonium* endophyte. *Molecular and General Genetics* 233, 1–9.

Novy, E.M., Casler, M.D. and Hill, R.R., Jr (1995) Selection for persistence of tetraploid ryegrasses and festulolium in mixture with perennial legumes. *Crop Science* 35, 1046–1051.

Oertl, C., Fuchs, J. and Matzk, F. (1996) Successful hybridization between *Lolium* and *Dactylis*. *Plant Breeding* 115, 101–105.

Oram, R.N., Schroeder, H.E. and Culvenor, R.A. (1985) Domestication of *Phalaris aquatica* as a pasture grass. In: *Proceedings XV International Grassland*

Congress, 24–31 August, 1985, Kyoto, Japan. Japanese Society of Grassland Science, Nishi-nasumo, Tochigi-ken, Japan, pp. 220–221.

Pedersen, J.F. and Sleper, D.A. (1988) Considerations in breeding endophyte-free tall fescue forage cultivars. *Journal of Production Agriculture* 1, 127–132.

Prestidge, R.A. and Gallagher, R.T. (1988) Endophyte fungus confers resistance to ryegrass: Argentine stem weevil larval studies. *Ecological Entomology* 13, 429–435.

Schjelderup, I., Aastveit, A.H. and Aastveit, K. (1994) Winter hardiness in marginal populations of timothy. *Euphytica* 77, 193–198.

Sleper, D.A. and Nelson, C.J. (1989) Productivity of selected high and low leaf area expansion *Festuca arundinacea* strains. In: Desroches, R. (ed.) *Proceedings of the XVIth International Grassland Congress, 4–11 October, 1989, Nice, France*. French Grassland Society, Versailles Cedex, France, pp. 379–380.

Surprenant, J., Barnes, D.K., Busch, R.H. and Marten, G.C. (1988) Bidirectional selection for neutral detergent fiber and yield in reed canarygrass. *Canadian Journal of Plant Science* 68, 705–712.

Torbert, H.A., Edwards, J.H. and Pedersen, J.F. (1985) Growth and magnesium uptake of tall fescue clones with varying root diameters. *Journal of Plant Nutrition* 8, 731–749.

Torbert, H.A., Edwards, J.H. and Pedersen, J.F. (1990) Fescues with large roots are drought tolerant. *Applied Agricultural Research* 5, 181–187.

Turkington, R. (1996) Intergenotypic interactions in plant mixtures. *Euphytica* 92, 105–119.

Tyler, B.F., Chorlton, K.H. and Thomas, I.D. (1987). Collection and field-sampling techniques for forages. In: Tyler, B.F. (ed.) *Collection, Characterization and Utilization of Genetic Resources of Temperate Forage Grass and Clover*. IBPGR Training Courses: Lecture series 1. International Board of Plant Genetic Resources, Rome, pp. 3–10.

Van Dijk, G.E. (1979) Breeding *Lolium perenne* L. for yield and persistence under heavy nitrogen and infrequent cutting. In: Ceccarelli, S. (ed.) *Breeding for Stress Conditions*, EUCARPIA Fodder Crops Sec. 4–6 Sept. 1979. Perugia, Italy, pp. 131–135.

Van Soest, P.J. (1994) Feeding strategies, taxonomy, and evolution. In: Van Soest, P.J. (ed.) *Nutritional Ecology of the Ruminant,* 2nd edn. Cornell University Press, Ithaca, New York, pp. 22–39.

van Wijk, A.J.P., Boonman, J.G. and Rumball, W. (1993) Achievements and perspectives in the breeding of forage grasses and legumes. In: Baker, M.J. (ed.) *Proceedings of the XVII International Grassland Congress, 8–21 February, 1993, Hamilton, NZ and Rockhampton, Australia*. New Zealand Grassland Association, Palmerston North, pp. 379–383.

Vogel, K.P. and Pedersen, J.F. (1993) Breeding systems for cross-pollinated perennial grasses. *Plant Breeding Reviews* 11, 251–274.

Wilman, D., Walters, R.J.K., Baker, D.H. and Williams, S.P. (1992) Comparison of two varieties of Italian ryegrass (*Lolium multiflorum*) for milk production, when fed as silage and when grazed. *Journal of Agricultural Science*, Cambridge 118, 37–46.

Wilkins, P.W. (1991) Breeding perennial ryegrass for agriculture. *Euphytica* 52, 201–214.

Zannone, L., Rotili, P., Paoletti, R. and Scotti, C. (1986) Experimental studies of grass–legume associations. *Agronomie* 6, 931–940.

Zeiders, K.E., Berg, C.C. and Sherwood, R.T. (1984) Effect of recurrent phenotypic selection on resistance to purple leaf spot in orchardgrass. *Crop Science* 24, 182–185.

Breeding Tropical and Subtropical Grasses

3

J.B. Hacker[1] and L. Jank[2]

[1]ATFGRC, CSIRO Tropical Agriculture, 306 Carmody Road, St Lucia, Queensland 4069, Australia
[2]CNGGC/EMBRAPA, CxP 154, 79002-970 Campo Grande, Mato Grosso do Sul, Brazil

Introduction

Regions of the subtropics and tropics where commercial dairying is practised are frequently where the climate is cooler, associated with higher altitude, and/or where soils are more fertile. In subtropical regions, dairying is generally based on a balance of cool-season and warm-season forages, the latter being invariably lower in quality, whereas in tropical regions there is a total reliance on warm-season grasses. In developing countries in the tropics, there is an increasing demand for animal products, including milk. Many such countries lack montane districts and dairying necessarily needs to be developed in districts unsuited to cool-season grasses. In these countries, there is a reliance on warm-season grasses for dairying, supplemented with crop by-products.

Forage conservation – in the form of hay or silage, is less common in the tropics than in temperate regions. There is no requirement to house cattle indoors during the cool or dry season, and haymaking during summer is difficult or can be impossible.

High forage protein concentration and high digestibility are desirable in forages for dairying, but are difficult to achieve in warm-season forages, as they vary with age of the sward and protein concentration is affected by plant nutrition. This is also true of cool-season forages, although they tend, as a group, to be intrinsically higher in both protein and digestibility than warm-season grasses (Minson, 1990). The low protein concentrations of tropical forages are associated with high growth rates and there is little opportunity to increase protein concentration by breeding while, at the same time, maintaining their high dry matter productivity. In developed countries, protein concentration in the diet of dairy cattle is increased through applying N fertilizer to the grass. Given adequate N, tropical grasses can carry higher stocking rates than temperate grasses, which to a considerable extent compensates for low per head milk production.

Tropical grasses are about 13% lower in digestibility than temperate grasses at the same growth stages. Their intrinsically lower digestibility is in part due to the characteristic leaf anatomy of panicoid C_4 (tropical) grasses, and in part associated with climate, since high temperatures increase cell-wall fibre concentration (Wilson, 1982; Chamberlain, 1989). In most developed countries, provision of supplements and/or inclusion of a legume in the sward make up for deficiencies in the grass.

Selection Criteria

In general, warm-season grasses utilized for dairying have been selected for overall productivity, acceptability to ruminant livestock and adaptation, rather than any specific attribute that relates to dairying. Prime selection criteria are overall yield, seasonal distribution of yield and leafiness, all attributes that equally apply to grasses for beef cattle. An appropriate propagation method is also essential, whether the selected cultivar is for beef or dairy cattle. In many developing countries, vegetative propagation is a viable option, and elephant grass (*Pennisetum purpureum* Schumach.) and its hybrids and bermudagrass (*Cynodon dactylon* (L.) Pers.) are invariably planted vegetatively. For most species, though, seed production is imperative. Problems associated with seed production have been a serious limitation to developing pastures in the low-latitude humid tropics.

Temperate forages have already undergone long processes of domestication with many years of breeding and selection, while very little effort has been devoted to the improvement of tropical forages, and most cultivars available commercially are wild forms, which have not been derived through breeding programmes. In all available germplasm of tropical species studied to date, there is considerable variation in quality traits (Hacker, 1982). Improvement in quality through breeding is possible, as demonstrated for the bermudagrass cultivar Coastcross-1 (Burton, 1972), which has much higher dry matter digestibility (DMD) than the Coastal progenitor. Usually quality traits and mineral composition of forages are not considered in breeding programmes.

Tropical Grasses Used for Dairy Pastures

Australia

The major proportion of the dairy industry in Australia is in the temperate regions of southern States, notably in Victoria and southern New South Wales. In these regions, C_3 grasses are the grasses generally used for dairying. Intake by dairy cows is considerably greater for ryegrass–white clover pastures than for pastures dominated by the warm-season *Paspalum dilatatum* Poiret (Stockdale, 1985) and cows produce more milk at equivalent dry matter intakes, and maintain better condition (King and Stockdale, 1984). In the tropics, dairying is largely on the Atherton Tableland, where the climate is comparatively cool, soils are fertile, and a range of tropical grasses such as *Panicum maximum* Jacq. and *Brachiaria decumbens* Stapf. are grown. In the subtropics (i.e. north from central NSW) dairying relies on a balance of C_3 (cool-season) and C_4 (warm-season) grasses, depending on climate, altitude, soils and availability of irrigation. The cool-season grasses – ryegrasses, oats and, uncommonly, prairie grass (*Bromus unioloides* Kunth) are grown in valleys where irrigation is available; oats are also often grown in dryland situations. The warm-season grasses are generally grown on the slopes, almost invariably as nitrogen-fertilized grass-only pastures. In the southern subtropics, kikuyu grass (*Pennisetum clandestinum* Hochst. ex Chiov.) is widely naturalized where soils are volcanic and more fertile and, with the development of seeding varieties, new areas are being sown. Other warm-season grasses used by the Australian dairy industry are *Panicum maximum* var. *trichoglume* Robijns (cv. Petrie), *P. maximum* (cvv. Hamil, Gatton, Riversdale), Rhodes grass (*Chloris gayana* Kunth cv. Callide), setaria (*Setaria sphacelata* (Schumach.) Stapf & Hubbard ex M.B. Moss cvv. Narok, Solander), signal grass (*Brachiaria decumbens*) and Dallis grass (*P. dilatatum*). *P. maximum, P. clandestinum* and *Digitaria eriantha* Steudel (pangola grass) were considered to be the best grasses for dairying in northern Australia in the 1960s, with 8–9 kg cow^{-1} day^{-1} being the maximum milk production that can be achieved over long periods (Stobbs, 1971). The former two species are still highly rated as grasses for dairying; pangola grass is less popular because it needs to be vegetatively planted and is susceptible to diseases.

Some resistance to use of nitrogen fertilizers is developing in the dairy industry, partly because of cost and partly for environmental reasons (B.G. Cook, Australia, 1997, personal communication). The two legumes *Vigna parkeri* Baker and *Arachis pintoi* Krap. & Greg., both tolerant of heavy grazing, have been released for use in intensive dairy systems.

Central and South America

In tropical America, milk production from tropical pastures can reach 10–12 l cow^{-1} and > 10,000 l ha^{-1} if adapted species are used and fertilized, and supplements are considered to be unnecessary when forages are young (Tergas, 1983). Dairy production in Central America is strongly based on stargrass (*Cynodon nlemfuensis* Vanderyst) pastures. Such is the case in Ecuador, Puerto Rico, Southern Mexico, Belize and Jamaica. In Cuba, no differences in milk yields were obtained from *D. eriantha* (two genotypes), Coastal bermudagrass, *C. nlemfuensis, Brachiaria ruziziensis* Germain & Evrard, *Echinochloa polystachya* Rojas, *B. mutica* (Forsk.) Stapf and *P. maximum* (Perez-Infante and Gonzalez, 1985). In Puerto Rico and Jamaica, *Brachiaria* spp. and guineagrass (*Panicum maximum*) are used to a limited extent. *P. maximum* cv. Tanzânia-1 is increasingly being used in Equador, because of the high fertility soils. In Honduras and the Virgin Islands, pangola (*Digitaria eriantha*) and guineagrass are mostly used. Guineagrass is predominant in most of the Caribbean. In the higher altitudes of Guatemala emphasis is given to kikuyu grass, *Pennistum clandestinum*.

In general, in South American countries, much *Brachiaria* is used, especially *B. brizantha* and *B. decumbens*. On the more fertile soils, *Panicum maximum* is frequent, and *B. humidicola* ((Rendle) Schweick) permits the utilization of wetter soils. Temperate species are used in higher altitude pastures. In Guyana, milk production is based on *Brachiaria humidicola* (Rendle) Schweick, *B. decumbens* and *B. mutica* pastures, but also on pangola, bahiagrass and common bermudagrass. *Echinochloa pyramidalis* Hitchcock & Chase is used in Guyana and in Surinam to a greater extent. In Brazil, dairy production is increasing from pastures of *B. decumbens* and *B. brizantha*, and *P. maximum* cvv. Tanzânia-1, Tobiata and Mombaça. More and more farmers are seeing the benefits of rotational grazing of these pastures. In smaller areas, elephant grass (*Pennisetum purpureum*) is frequently used either for grazing or cut-and-carry. In Southeast Brazil, *Digitaria eriantha* (pangola), *D. swazilandensis* Stent., *C. nlemfuensis* and *Pennisetum purpureum* are recommended for dairy pastures (Aronovich *et al.*, 1989). When dairy cattle graze these species and are provided with 2 kg supplements and 20 kg day^{-1} sugar cane with 1% urea, expected milk yields are 10 kg cow^{-1} day^{-1} instead of the 3 kg day^{-1} generally obtained. However, unsupplemented dairy cattle grazing nitrogen fertilized grass pastures gave 7–10 kg cow^{-1} day^{-1}.

Brazil exports seed of *B. brizantha, B. decumbens, B. humidicola* and *P. maximum* cv. Tanzânia-1 to 21 countries in Central and South America for both beef and dairy production. Some countries, such as Costa Rica and Colombia, do not permit the import of non-scarified seed, so *P. maximum* cultivars are not sold to these countries.

Central and Southern Africa

Although dairy products have traditionally been an important part of the diet of many African communities, little forage grass is sown. The majority of the tropical grasses used in agriculture are of African origin and much of the natural grassland, providing it has not been subjected to overgrazing, provides good pasture, although lacking in the quality legume component.

In South Africa, ryegrasses and kikuyu grass are favoured by dairy producers, augmented by concentrates. In addition, a considerable proportion of dairy production is based on temperate cereals – particularly oats and triticale. Maize silage is fed in winter, particularly in the high altitude regions, where it is sown at high densities and harvested when the heads are immature, wilted, and ensiled. Other species used for hay making are *Eragrostis curvula* Nees and *Digitaria eriantha*. About 200,000 ha of the latter species were established in South Africa between 1990 and 1994 (Terblanche *et al.,* 1996). In warmer areas, kikuyu grass is grown for dairy cattle. This strongly stoloniferous species can tolerate heavy grazing and is best adapted to a fertile soil. In South Africa it is generally grown with irrigation and heavily fertilized. One of the few tropical grasses with a reputation for high quality, Nile grass (*Acroceras macrum* Stapf), has been the subject of a breeding programme and the cultivar Cedara has been released (Rhind and Goodenough, 1976; Theron and Arnott, 1979). However, it has not found wide acceptance, largely due to poor seed production and viability.

Southern and Southeast Asia

Dairy products have not traditionally been a significant component of the diet in Southeast Asian countries, except in the Indian sub-continent. In India, dairy cattle are not usually provided with sown pastures. Where pastures are sown, *Cenchrus ciliaris* L. and *C. setigerus* Vahl are the more widely used species. Milk production from *C. ciliaris* pastures is more than that from *Sehima nervosum* Stapf pastures because the former is higher yielding (Upadhyay *et al.,* 1980). *Pennisetum purpureum* (and a hybrid between this species and *P. glaucum* R. Br.) is also used in some districts. In Kerala, in the south of the country, *Brachiaria ruziziensis* is used as a feed for dairy cattle (Krishnan, 1998). This species is sown along contour banks on hillsides, for erosion control. *B. ruziziensis* has been selected not for any specific benefit as a feed for dairying, but because it seeds reliably in a climate where seed production is frequently difficult, and because of its soil conservation qualities.

In several Southeast Asian countries increasing living standards have resulted in increasing demand by the population for dairy products. In most countries in the region, ruminants largely rely on forage and browse along roadsides and on common land, and rice straw during the dry season.

However, in the last 15 years, Thailand has developed a sophisticated pasture seed industry, with much of the seed being grown and sold by smallholders, either to each other or to Government enterprises (Satjipanon, 1995). The mainstay of this industry is the grass ruzi (*Brachiaria ruziziensis*), which was developed in the early 1980s, primarily for beef production. There has now been a switch in emphasis to dairying but ruzi is still the dominant grass used. In recent years the government has been promoting the further development of dairying, to reduce reliance on imports, and flood-prone land previously under rice is being developed for this purpose. *Paspalum atratum* Swallen shows considerable promise as a dairy species and also produces good seed crops, except in the south of the country. Promotion of dairying is also occurring in north Thailand, through the building of an ultra heat treated (UHT; long-life) milk processing plant. Low rainfall, a pronounced dry season and the low water table mean irrigation in north Thailand is out of the question. Farmers maintain cattle over the dry season on rice straw and concentrates. Although a species related to ruzi (*B. decumbens* cv. Basilisk) has some capacity to grow during the early dry season, it produces little or no seed in north Thailand and imported seed, from Australia, is expensive and not always reliable. Recent studies with a range of accessions of *B. decumbens* and the closely related *B. brizantha* suggest that it should be possible to select one with commercially acceptable seed yields in the region. In Central Thailand *B. mutica* was the favoured grass for dairying. More recently, Thai purple guinea (*Panicum maximum* cv. Purple) is being promoted for dairy pastures.

In Malaysia there is a policy to increase size of dairy farms, reducing emphasis on smallholders. For dairying, the main grasses used are *Pennisetum purpureum* and *P. maximum*. *B. decumbens* is also used, with quality maintained through rotational grazing and growth periods of 2–3 weeks (Aminah and Chen, 1991). The importance of frequent grazing to retain forage quality is coming to be recognized, and common by-products, such as sweet corn stover, which may be ensiled as a quality diet for the dry season, have been identified. Although a considerable amount of research has been carried out to identify adapted forages for Malaysia, comparatively small areas are sown to forage. One of the main problems is seed production in the humid tropical environment (Chen *et al.*, 1995). Some species currently being developed appear to produce adequate yields of seed, at least under experimental conditions (Chen *et al.*, 1995).

There is little dairying in the Philippines. The grass used is commonly *P. purpureum* cv. 'Florida'.

Southern USA

In the Southern USA, dairy production is based on bermudagrass, stargrass and rhizoma peanut (*Arachis glabrata*), as permanent pastures, and pearl millet (*Pennisetum glaucum*) and sorghum × sudangrass as annual pastures. The bermudagrass cultivars mainly used are common, 'Coastal' and 'Tifton 85' released by Burton *et al.* (1993). This latter is a cross between *C. dactylon* and *C. nlemfuensis* and is more productive than other cultivars. The pearl millet is usually cv. Tifleaf-2, released in Georgia, USA. Stargrass is not very cold tolerant and is used in South Florida in highly fertilized, frequently defoliated pastures. Some systems in Florida are based on annual warm-season grasses in summer and annual cool-season species, such as annual ryegrass (*Lolium multiflorum* Lamk.) and white clover (*Trifolium repens* L.) in winter. Another system is to use perennial warm-season grasses overseeded with annual ryegrass (*Lolium multiflorum*) in winter.

Other temperate regions

Some summer-growing grass species are grown in the northern parts of North Island, New Zealand. Naturalized subtropical grasses form a substantial component of dairy pastures in some North Island regions, and there is an expectation that they may increase with global warming (Bell and Keene, 1996), causing concerns over weed potential. The predominant summer-growing pasture species is kikuyu grass (*Pennisetum clandestinum*), although Dallis grass (*Paspalum dilatatum*) is also grown. Both grasses have a considerable degree of frost tolerance and can retain green leaf following sub-zero temperatures (Hacker *et al.*, 1974). Some research has been directed towards identifying other subtropical species which are more productive during the brief summer drought period in the northern part of the island (Rumball, 1991), but little seed of other species is sown. However, the New Zealand dairy industry is largely reliant on cool-season grasses, especially *Lolium* spp.

In southern Japan, both temperate and tropical forages are grown, with tropical forages grown mostly as annuals, except in the southern island of Okinawa. Dairy cattle fed on silage made from warm-season *C. gayana* and supplemented with concentrates produce less milk than those fed on cool-season *Lolium multiflorum* (Miaki *et al.*, 1990a).

Breeding and Selection of Warm-season Grasses Used in Dairying

Improvement of temperate forages by breeding and selection has been very intense in the past 50–75 years and has resulted in the release of numerous improved cultivars. The same cannot be said, however, of tropical forage species. While some temperate pasture grasses have already been genetically transformed with 'foreign' genes (Hodges *et al.*, 1993), tropical ones still depend on the availability of germplasm collections and traditional plant breeding techniques. One reason for the relatively small and unrepresentative collections of tropical forage germplasm is the African origin of most tropical forage grasses, and difficulties and expense in assembling comprehensive collections. Another reason is the stage of development of most tropical countries, which cannot afford to mount collecting expeditions. Excluding Australia and the international Consultative Group on International Agricultural Research (CGIAR) Centres (Centro Internacional Agricultura Tropical, Colombia (CIAT); International Livestock Research Institute, Ethiopia (ILRI)), the countries interested in developing tropical grass cultivars, have never organized collection expeditions to Africa.

With few exceptions, all the warm-season grasses used for dairying are naturally occurring ecotypes. Before the 1950s, cultivars were generally selected from the limited collections of germplasm then available. More recently, substantial collections of the more promising genera have been assembled at various centres, including CIAT; ILRI; Empresa Brasiliera de Pesquisa Agropecuária/Centro Nacional de Pesquisa de Recursos Genéticos e Biotecnologia, Brazil (EMBRAPA/CENARGEN) and Australian Tropical Forages Genetic Resource Centre, Australia (ATFGRC). Screening of these large collections is resulting in the identification of new ecotypes potentially suited to dairying, for example in *Brachiaria* spp. (Keller-Grein *et al.*, 1996) and *Panicum maximum* (Savidan *et al.*, 1990).

Apomixis is the reproductive mode of at least 60% of the tropical forage grasses. The discovery of a sexual mutant in the normally apomictic buffel grass (*Cenchrus ciliaris*) led to development and release of the new cultivars Nueces and Llano (Bashaw, 1980). In other tropical grasses, there have been significant inputs only in the past one or two decades. Only in a few species has apomixis been manipulated, and sexual representatives which permit genetic recombination been discovered. So, if accessions 160% more productive than local commercial cultivars can be selected from a well-represented germplasm, as in *Panicum maximum* (Jank *et al.*, 1989), this could also be possible in other species. The lack of available germplasm, and lack of available sexual forms in the apomictic species *Melinis minutiflora* Beauv., have been responsible for this species being neglected by selectors. This species used to be, not so long ago, the basis of the dairy industry in some regions of Brazil.

Brachiaria spp.

The availability of the African grass *Brachiaria decumbens* has permitted the utilization of many regions of savannas of South America, especially the Cerrados of Brazil. It is persistent, productive, and adapted to the infertile acid soils of the region. The cultivar Basilisk was imported to South America from Australia in the early 1970s, having been selected from a small collection of germplasm originating from Africa.

The productive *B. brizantha* (A. Rich.) Stapf cv. Marandu was released in Brazil in 1984. It is not so well adapted to acid soils as *B. decumbens*, but has rapidly spread throughout the country and other Latin American countries. It performs very well both in extensive management and in rotational high-input systems. *B. humidicola* is extensively utilized on moist soils; it is, however, of low nutritive value. It is recommended in preference to other species for milk and beef production in the Brazilian province of Rondonia (Goncalves and Oliveira, 1982). CIAT recently released *B. dictyoneura* cv. Llanero, which is not so well adapted to wet soils as *B. humidicola*, but is of better quality and is more productive. All these cultivars are selections from naturally occurring ecotypes.

The recent collection expedition to East Africa (Keller-Grein *et al.*, 1996), has provided the basis for major *Brachiaria* hybridization programmes in Colombia (CIAT) and Brazil (CNPGC/EMBRAPA) (Miles and Valle, 1996). Sexual diploid forms of the usually tetraploid apomictic *B. decumbens* have been discovered among the accessions (Valle *et al.*, 1989), and of *B. brizantha* (Valle and Savidan, 1996), but will only be available for breeding after the successful doubling of chromosome numbers. Current breeding programmes are based on the sexual diploid species *B. ruziziensis*. Belgian scientists have synthesized autotetraploid *B. ruziziensis*, which has been hybridized with *B. decumbens* and *B. brizantha*, producing fertile progeny. Breeding seeks to combine the grazing tolerance and adaptation to acid soils of *B. decumbens* with the spittle-bug (Homoptera:Cercopidae) resistance and productivity of *B. brizantha* and quality of *B. ruziziensis*. Breeding of *B. humidicola* and *B. dictyoneura* depend on the identification of compatible sexual forms (Miles and Valle, 1996).

Chloris gayana

Chloris gayana (Rhodes grass) cultivars have been developed in Australia, Kenya, Zimbabwe and Japan. The species was first domesticated in South Africa in 1895 (Bogdan, 1977) and was first tested in Australia in 1902–3 (Oram, 1990). The more widely grown cultivars are Pioneer, Katambora and Callide, all of which are naturally occurring ecotypes, and all are sexual, in common with the species as a whole. Unlike the other two cultivars, which are diploids, Callide is a tetraploid and is later flowering, more robust and more leafy. Holstein-Friesian cows grazing nitrogen fertilized Callide pastures can produce 3500 kg milk year^{-1}, when supplemented with silage, forage oats and concentrates (Cowan *et al.*, 1985).

Katambora is the most popular cultivar in Japan, where it is grown as an annual, or as a perennial in Okinawa. A new cultivar, cv. Hatsunatsu, has been bred, which produces better early growth than Katambora and has good regrowth (Nakagawa *et al.*, 1993).

Cynodon spp.

The genus *Cynodon* is very important for beef and dairy production in the Southern United States and in many tropical and subtropical countries of Central and South America. Two species of *Cynodon* are utilized in dairy pastures. *C. dactylon* (bermudagrass) is a pan-tropical species which is extremely variable and has been the subject of intensive breeding studies since the 1960s by G.W. Burton and colleagues, in Tifton, Georgia (USA). It is stoloniferous and rhizomatous, and, as far as we are aware, all cultivars are planted vegetatively. Many cultivars have been released. Coastcross-1 was released in 1972 and is derived from a cross between a clone previously commercialized – cv. Coastal – and an accession from Kenya. It is more digestible than Coastal (Burton, 1972) and is widely planted in the Caribbean region where its vulnerability to winter injury is not a disadvantage. A more cold-tolerant hybrid, cv. Tifton 44, was registered in 1978 (Burton and Monson, 1978), but is not grown to any marked extent in the tropics.

Other vegetatively propagated hybrid clones are also available, including cv. Callie. In trials in Cuba, this cultivar has yielded more milk than cv. Coastcross-1 (Lamela *et al.*, 1984). Two sister lines were bred from the cold-tolerant Tifton 44 and cv. Callie and released; Tifton 78 (Burton and Monson, 1988) and Florakirk (Mislevy, 1995). This latter was released in Florida, because Tifton 78 did not perform well in this state (Sollenberger *et al.*, 1995b). Florakirk is more cold-tolerant and is reported to have withstood temperatures of −13°C. It is also more adapted to wetter sites than other bermudagrasses (Sollenberger *et al.*, 1995a). Lately, Tifton 85 was released (Burton *et al.*, 1993). It is a hybrid between *C. dactylon* (an accession from South Africa) and *C. nlemfuensis* (cv. Tifton 68). It is the most productive of all bermudagrasses so far released. It produced 26% more dry matter, with 11% higher digestibility than Coastal bermudagrass (Burton *et al.*, 1993). Experiments with beef cattle have shown Tifton 85 to result in higher live-weight gains per hectare than Tifton 78 (Hill *et al.*, 1993) and Florakirk (Sollenberger *et al.*, 1995b).

The other species of *Cynodon* used in dairy pastures are *C. nlemfuensis* (stargrass) and *C. plectostachyus* (K. Schum.) Pilg., both of which are also planted vegetatively. Stargrasses Florico and Florona (released in Florida) are higher in digestibility than the bermudagrasses (Sollenberger *et al.*, 1995a). The former species has been hybridized with other members of the genus, whereas *C. plectostachyus* is genetically isolated (de Wet and Harlan, 1970). In Mexico, dairy cows grazing either species may produce milk yields exceeding 4000 kg ha^{-1} year^{-1} (Fernandez-Baca *et al.*, 1986). In Cuba, in one trial, no difference in milk production was found between Coastcross-1, *C. nlemfuensis* and pangola grass (Jerez *et al.*, 1984), although in another trial, in which concentrates were added to the cows' diet, better results were obtained from *C. nlemfuensis* than Coastcross-1 or pangola (Jerez *et al.*, 1989), or between cv. Coastcross 67 and common guinea (*P. maximum*) (Jerez *et al.*, 1988).

Digitaria spp.

The species of *Digitaria* commonly used for grazing are in section *Eriantbae* of the genus and of tropical or southern African distribution. Most would now be classified in *Digitaria eriantha* (including pangola grass, *D. decumbens*) or *D. milanjiana* (Rendle) Stapf. Most are utilized as much or more for grazing beef cattle as for dairy cattle. Pangola is a totally sterile, triploid clone, collected by J. Oakes, USDA, near Nelspruit (Henrard, 1950), although its name is a corruption of Pongola (River), which is substantially further south. Attempts to re-collect pangola have failed, as the collection site was destroyed when a bridge was built (A. Kruger and R. Ellis, South Africa, 1997, personal communication). Pangola was introduced to the USA in 1930 and has since been widely planted in the tropics and subtropics of both hemispheres. Other clones, such as Taiwan and Transvala digitgrasses (Boyd *et al.*, 1973; Kretschmer *et al.*, 1978) have been selected for agronomic attributes and disease and insect resistance, and released, but none is widely planted. There is no stoloniferous cultivar of *D. eriantha* commercially propagated by seed.

D. eriantha also includes tussock-forming morphotypes previously known as *D. smutsii* Stent. The cultivar Irene is widely sown in central and northern high-altitude regions of South Africa. A tussock-forming morphotype introduced to Australia as 'northern strain' has been released as cv. Premier, although it is not grown in areas where dairying is practised. In South Africa a breeding programme aimed at production of a homogeneous cultivar with improved seed quality and leafiness has resulted in cv. Tip Top, which is currently (1997) being multiplied for commercial release (Terblanche *et al.*, 1996; C.J. Terblanche, Lynn East, South Africa, 1997, personal communication).

D. milanjiana is a species which is very similar to *D. eriantha*, and is also very variable. It is of more tropical distribution. The vegetatively propagated cv. Mardi was selected in Malaysia and limited areas are planted. In Australia, cv. Jarra, an ecotype originating from Malawi, was released as a vegetatively planted cultivar, primarily for control of nematodes in banana orchards (Anon., 1993). However, it has since been shown to produce commercially viable quantities of seed and is sown in pastures largely for beef cattle. Another ecotype of *D. milanjiana* has more recently been released in

Australia as cv. Strickland, and is adapted to more sub-humid districts where beef cattle are the main enterprise.

A breeding study in *D. milanjiana* focused on producing genotypes with contrasting leaf *in vitro* digestibility (Hacker, 1986). Differences (1.9–4.5 digestibility units) were still expressed when plants were grown in swards (Masaoka *et al.*, 1991) and fed to sheep in pens (Minson and Hacker, 1986). In a grazing trial with unsupplemented dairy cattle, milk yields from the high digestibility selection were 16.9 kg cow^{-1} day^{-1}, compared with 16.0 kg cow^{-1} day^{-1} from pangola (Lowe *et al.*, 1991). However, stolon production from the *D. milanjiana* selection was not adequate for commercial vegetative planting.

Panicum maximum

Panicum maximum is a species which is generally adapted to higher fertility situations. It was probably introduced to India in 1793, to South America also in the 18th century (Bogdan, 1977) and to Australia in the late 19th century. Many *Panicum maximum* cultivars are reported in the literature, amongst the most widely utilized being cvv. Likoni, Gatton, Makueni, Petrie, Riversdale and Tanzânia-1. All of these are based on wild ecotypes. Although the species includes sexual, apomictic and sexual/apomictic genotypes (Savidan, 1983), most cultivars are naturally occurring ecotypes which have been selected as having agronomic merit.

Selection and breeding programmes are under way in Brazil (Savidan *et al.*, 1989). The cultivars Tanzânia-1 and Mombaça were released in 1991 and 1993, respectively; both were selected from naturally occurring ecotypes. In comparative trials, they were shown to be, respectively, 80% and 130% more productive (leaf dry matter yield) than the commercial cv. Coloniao (Savidan *et al.*, 1990). They are increasingly being used for dairy cattle in high-input rotational grazing systems. Carrying capacities of 12 animal units (AU) ha^{-1} have been reported for these cultivars (M. Corsi, Brazil, 1996, personal communication). Many South and Central American countries have already imported seed from Brazil. In Equador, Tanzânia-1 received the name 'Saboya'.

P. maximum is also being bred in Japan. Wild germplasm was collected in East Africa (Hojito and Horibata, 1982) and sexual plants were isolated from this collection (Nakajima *et al.*, 1979). In Japan, tropical grasses are usually used as annuals as persistence is poor over the cold winters, except in the south of the country. The cultivar Natsukaze was bred from an open-pollinated sexual accession from Africa and is widely adapted as an annual forage for hay production (Sato *et al.*, 1993). Another cultivar, cv. Natsuyutaka, is an ecotype that was selected from a collection of accessions from Africa and is adapted to the southwest islands of Japan, where it persists as a perennial (Shimizu *et al.*, 1993).

The more widely-sown modern cultivars in Australia – Gatton, Petrie, Riversdale, Hamil – all trace their origins to the 1930s or earlier (Oram, 1990). Cultivar Likoni was introduced to Cuba in 1971 and is recommended for milk production (Hernandez and Caceres, 1983). In a trial in Cuba, milk production from cv. Likoni silage was 22% greater than from stargrass and 40% greater than from king grass (*P. purpureum* × *P. glaucum*) silage (Esperance and Diaz, 1985). Similarly, daily milk production in another trial in Cuba was 8% higher from *P. maximum* cv. SIH-127 than from Coast-cross-1 bermudagrass (Lamela *et al.*, 1984). In Puerto Rico, milk production from fertilized *P. maximum* was 39% higher than that from pangola grass and 29% more than that from *Pennisetum purpureum* (Caro-Costas and Vicente-Chandler, 1969).

Paspalum dilatatum, P. notatum, P. atratum

Paspalum is a large grass genus with its centre of diversity in South America. It has been extensively collected in Brazil and adjacent countries in recent years. It is being evaluated in various research centres in Brazil, but no major breeding programmes are under way. The programme in Georgia, USA, has led to the release of *P. notatum* Flueggé cv. Tifton 9, derived through recurrent restricted phenotypic selection within cv. Pensacola (Burton, 1989). Tifton 9 is reported as having greater seedling vigour than Pensacola, being more palatable and having 47% higher yield. The higher yields, however, are due to increased allocation of dry matter to harvestable foliage and possibly greater production of non-root biomass (Pedreira and Brown, 1996).

Dallis grass (*P. dilatatum*) is an important grass throughout the warmer regions of the world. Breeding, however, is limited to cytogenetical studies and relationship to other species, since common Dallis grass is an obligate apomictic pentaploid. The female parent commonly used in the breeding programmes is a sexual tetraploid. These studies are being conducted in Brazil, Argentina (Quarin and Caponio, 1995) and USA (Burson, 1995).

Evaluation of small germplasm collections and cytological and reproductive studies are showing *P. atratum* to be a promising forage species in Brazil (Quarín *et al.*, 1997). It is also showing considerable promise in the Philippines, Thailand and subtropical Australia (B.G. Cook, Australia, 1996, personal communication). Some variability is apparent in this species but it is unlikely that a breeding programme is warranted until the merits and deficiencies of accessions currently being promoted have been identified. *P. guenoarum* Arech and *P. nicorae* Parodi, also in the *plicatulum* group of the genus, have desirable palatability and/or cool season production, and could be commercially developed in the future.

Pennisetum clandestinum

Kikuyu grass *(Pennisetum clandestinum)* is a native of higher altitudes in East Africa. It is one of the most important grasses for dairying in the subtropics and montane tropics. Although different morphotypes are recognized in East Africa (Bogdan, 1977), the main attempts at breeding and selection have been in Australasia. Kikuyu grass was first introduced to Australia in 1919 (Mears, 1970) and to New Zealand, about the same time (Rumball and Lambert, 1985).

For many years, seed production was a problem, partly because it is set close to the ground and partly because some genotypes are male sterile, the male sterility controlled by a single dominant gene (Piggot and Morgan, 1986). Propagation was therefore largely by vegetative means, until the 1970 release in Australia of cv. Whittet, an ecotype introduced from Kenya, which also was higher yielding than local ecotypes (Oram, 1990). At the same time, management practices were developed to increase seed production and harvest the seed. Breakwell was developed soon after the release of Whittet, from an hermaphrodite ecotype naturalized in NSW. Cultivar Crofts is a clone which was selected for cold tolerance and is better adapted to higher latitudes in NSW than is Whittet, but is susceptible to the disease kikuyu yellows. Cultivar Noonan was selected, from a population of open-pollinated Breakwell and Whittet, for its resistance to kikuyu yellows, and was registered in 1983. Only Whittet is widely used.

In New Zealand, most local ecotypes are apparently male sterile, and some high-altitude Kenyan introductions are more vigorous as well as being fully fertile (Rumball and Lambert, 1985). However, these authors do not consider that replacing kikuyu grass pastures with alternative kikuyu grass genotypes is warranted.

Current studies in Australia are aimed at improving the digestibility of kikuyu grass through mutagenesis (Luckett *et al.*, 1996). Field screening of 40,000 treated plants resulted in the identification of 212 off-types, some of which had comparatively high *in vitro* dry matter digestibility values.

Pennisetum purpureum and hybrids

Pennisetum purpureum (elephant or napier grass) is a robust perennial native to tropical Africa but now widespread throughout the humid tropics. It produces little or no seed and is vegetatively propagated, either for grazing or for 'cut-and-carry'. Elephant grass is widely used for beef and dairy production, both by smallholders and large-scale farmers. In Brazil, Olivo *et al.* (1992) found no difference in milk production between *P. purpureum* and *Setaria sphacelata* var. *sericea* (Stapf) W.D. Clayton. There are many cultivars, with local names. Dwarf forms of the species also exist, a recent dwarf cultivar being cv. Mott (Sollenberger *et al.*, 1988). Mott has a high leaf production, high overall yield and is of high digestibility (Sollenberger and Jones, 1989). Silage made from Mott is considered to have high potential for dairy cows; when substituted for maize silage, it resulted in only a small decrease in milk yield (Ruiz *et al.*, 1992).

A *P. purpureum* × *P. glaucum* hybrid which is widely planted in the humid tropics is known as king grass. In Cuba, milk production from cows fed king grass silage was less than from *Panicum maximum* or *Cynodon plectostachyus* silage (Esperance and Diaz, 1985). Both elephant grass and king grass are best adapted to fertile soils and in the more widespread acidic soils of the humid tropics, other species, such as *Brachiaria decumbens* and *B. brizantha*, are preferred.

Setaria sphacelata

Setaria sphacelata has received attention from grass selectors and breeders since the 1950s. The diploid cultivar Nandi was selected in Kenya in the 1950s and subsequent selection produced Nandi Mark II and Nandi Mark III (Bogdan, 1977). Nandi was introduced to Australia in 1961. In Australia, evaluation of a collection of introductions of the species showed that high-altitude ecotypes from the Aberdares range in Kenya retain green leaf after light to moderate frosts. This was considered to be a valuable attribute. Selection within one of these accessions (a tetraploid), for winter yield and winter greenness, produced the cultivar Narok, released in 1969.

Another tetraploid which was grown in South Africa and released in Australia in 1962 is cv. Kazungula. This unselected ecotype originates from Zambia. It is better adapted to climates with high temperatures and drought than are the cultivars currently (1997) available and also is a comparatively heavy and reliable seed producer. Early experience with Narok showed that it produced much less seed than other cultivars and a breeding programme successfully incorporated much of the seed producing ability of accessions of the Kazungula type with the winter-greenness of Narok (Hacker, 1991, 1994). The new cultivar, Solander, was released in Australia in 1985. Both Narok and Solander are grown in dairy pastures, and also Nandi on the Atherton Tablelands of North Queensland. Kazungula, a cultivar which tends to become very stemmy in summer, unless very carefully managed, is better suited to beef production, although, as seed is comparatively cheap, it may be sown on some dairy pastures for perceived economic reasons.

A deficiency of this species is the accumulation of oxalate, particularly when pastures are heavily fertilized with nitrogen or potassium (Jones and Ford, 1972). High levels of oxalate can be lethal to cattle, although if they are introduced to pastures with high oxalate concentrations progressively, they adapt to it. Cultivars differ in their propensity to accumulate oxalate, but no accession has been found which lacks the compound (Hacker, 1974).

Annual warm-season forages

These include forage sorghum, maize and pearl millet (*Pennisetum glaucum*). In some countries *Echinochloa frumentacea* (Roxburgh) Link is grown as a minor warm season forage crop.

Forage sorghums are generally F_1 hybrids based on Sudan grass (*S.* × *drummondii* (Steud.) Millsp. & Chase) and grain sorghum (*S. bicolor* (L.) Moench). Hybrids are produced by private industry and are based on male sterile lines. Annual forage sorghum may be grazed or ensiled. Forage pearl millet is grown in Australia and USA. Early cultivars, such as Tamworth and Katherine (Oram, 1990) were land races introduced from Africa, where they were grown for grain. More recently, male sterile lines have been identified and F_1 hybrids developed by private industry, including Supermill, Feedmill and Nutrifeed.

Maize is used as a silage in temperate regions and a considerable amount of research has been carried out to investigate opportunities for improving its nutritive value (e.g. Dolstra and Miedema, 1986). Quality in maize is largely associated with the ear component, and development of the ear is sensitive to moisture stress. Thus maize is not a preferred option in subhumid regions where drought may occur at flowering (Minson *et al.*, 1993), unless irrigation is available.

Dairy cows fed on sorghum, maize and pearl millet silage gave similar milk yields in trials in Japan (Miaki *et al.*, 1990b). When grazed, Sudan grass pastures are reported to give a similar milk yield to maize (Acosta *et al.*, 1984; Mieres *et al.*, 1988). Sorghum may also be chopped and fed to milk buffaloes (Pathak and Pal, 1983).

Conclusions

Warm-season forages have been developed to suit most tropical and sub-tropical climates (excluding arid climates) and most soil types. In most situations where there is a requirement for dairy production, there is now a suite of adapted grasses available. However, milk production is less than from the higher quality cool-season grasses and there is little likelihood that the high milk yields obtained from temperate grasses will be duplicated in the tropics.

However, tropical forage grass breeding is only in its infancy. Germ-plasm collections for most species are extremely limited and there are very few pasture grass breeders working in the tropics, as compared with temperate regions. Modes of reproduction are not yet fully understood. Much research is still necessary before all the tools are available for efficient tropical grass breeding.

A survey of recent literature indicates that expected milk yields from warm-season grasses are in the order of 8–12 kg cow^{-1} day^{-1}. Aronovich *et al.* (1989) contrast the 3 kg cow^{-1} day^{-1} generally obtained in Brazil with their estimate of 10 kg cow^{-1} day^{-1} possible with adapted sown forages and supplements of sugarcane and urea. In the short term, the best opportunity for increasing milk production in developing countries is more likely to be through promoting the adoption of existing adapted cultivars and appropriate management practices, rather than continued searching for higher quality tropical grasses. A further factor which deserves attention is the incorporation of a legume in the dairy cow's diet. In tropical grass–legume pastures in South America, a 15–20% increase in milk production has been reported in response to the legume (Serpa, 1983; Lascano and Avila, 1991), and inclusion of the shrub legume *Leucaena leucocephala* in the diet of dairy cattle increases milk production by 2–33% (Jones, 1994). In this context, the potential of grazing tolerant tropical legumes such as *Arachis pintoi* is being explored.

Acknowledgements

We are indebted to A. Kruger (South Africa), C.C. Wong (Malaysia), C. Phaikaew and B. Hare (Thailand), Francisco Gabunada Jr (Philippines), W.J. Fulkerson and T. Cowan (Australia), W. Williams (New Zealand), L. Sollenberger (USA), G. and L. dos Santos and M. Corsi (Brazil), B. Macoon (Guyana), E. Valencia (Belize) and M. Adjei (Virgin Islands) for freely providing information which we have included in this chapter.

References

Acosta, Y.M., Mieres, J.M. and Durán, H. (1984) Comparación de Sudangras y maíz para la producción de leche bajo pastoreo. *Investigaciones Agronomicas, Centro de Investigaciones Agricolas 'Alberto Boerger'* 5, 75–78.

Aminah, A. and Chen, C.P. (1991) Future prospects for fodder and pasture production. *FAO Animal Production and Health Paper* No. 86, 127–141.

Anon. (1993) New herbage plant cultivars. A. Grasses 23. Digitaria (b) *Digitaria milanjiana* (Rendle) Stapf. (finger grass) cv. Jarra. *Tropical Grasslands* 27, 55–57.

Aronovich, S., Dusi, G.A. and Valle, L.C.S. (1989) Tropical pastures for milk production in southeastern region of Brasil. In: Desroches, R. (ed.) *Proceedings of the XVIth International Grassland Congress, 4–11 October, 1989, Nice, France.* French Grassland Society, Versailles Cedex, France, pp. 1153–1154.

Bashaw, E.C. (1980) Registration of Nueces and Llano buffelgrass. *Crop Science* 20, 112.

Bell, C.C. and Keene, B. (1996) Subtropical grasses in Bay of Plenty dairy pastures. *Proceedings of the New Zealand Grassland Association* 57, 55–58.

Bogdan, A.V. (1977) *Tropical Pasture and Fodder Plants*. Longman, London, 475 pp.

Boyd, F.T., Schank, S.C., Smith, R.L., Hodges, E.M., West, S.H., Kretschmer, A.E., Jr, Brolmann, J.B. and Moore, J.E. (1973) Transvala digitgrass. *Circular S-222. Florida Agricultural Experiment Stations IFAS*, University of Florida, 16pp.

Burson, B.L. (1995) Genome relationship and reproductive behavior of intraspecific *Paspalum dilatatum* hybrids: yellow-anthered × Uruguaian. *International Journal of Plant Sciences* 156, 326–331.

Burton, G.W. (1972). Registration of Coastcross-1 Bermudagrass. *Crop Science* 12, 125.

Burton, G.W. (1989) Registration of 'Tifton 9' Pensacola bahiagrass. *Crop Science* 29, 1326.

Burton, G.W. and Monson, W.G. (1978) Registration of 'Tifton 44' bermudagrass. *Crop Science* 18, 911.

Burton, G.W. and Monson, W.G. (1988) Registration of 'Tifton 78' bermudagrass. *Crop Science* 28, 187–188.

Burton, G.W., Gates, R.N. and Hill, G.M. (1993) Registration of 'Tifton 85' bermudagrass. *Crop Science* 33, 644.

Caro-Costas, R. and Vicente-Chandler, J. (1969) Milk production with all-grass rations from steep intensively managed tropical pastures. *Journal of Agricultural Research of Puerto Rico* 53, 251–258.

Chamberlain, A. (1989) Available feedstuffs and their use in milk production; pastures and forage crops. In: Payne, W.J.A. (ed.) *Milk Production in the Tropics*. Longman, Harlow, UK, pp. 75–85.

Chen, C.P., Aminah, A. and Khairuddin, G. (1995) Forage Seeds Project in Malaysia: Activities, results and conclusions. In: Stür, W.W., Cameron, A.G. and Hacker, J.B. (eds) *Proceedings of the Third Meeting of the Southeast Asian Regional Forage Seeds Project, Samarinda, Indonesia*, CIAT Working Document No. 143, pp. 20–32.

Cowan, R.T., Upton, P.C., Lowe, K.F. and Bowdler, T.M. (1985) Effect of level of fertiliser on milk output from a Callide Rhodes grass pasture. The Challenge: efficient dairy production. *Proceedings of the Conference organised by the Australian and New Zealand Societies of Animal Production, March 25–28, 1985, Albury-Wodonga*, Australia, Australian Society of Animal Production, pp. 45–47.

de Wet, J.M.J. and Harlan, J.R. (1970) Biosystematics of *Cynodon* L.C. Rich (Gramineae). *Taxon* 19, 565–569.

Dolstra, O. and Miedema, P. (eds) (1986) *Breeding of Silage Maize. Proceedings of the 13th Congress of the Maize and Sorghum Section of EUCARPIA*. Pudoc, Wageningen, 191pp.

Esperance, M. and Diaz, D. (1985) Valor nutritivo y producción de leche en los ensilajes sin miel de guinea likoni, pasto estrella y king grass. *Pastos y Forages* 8, 297–305.

Fernandez-Baca, S., Lucia, R. de and Jara, L.C. (1986) Mexico; milk and beef production from tropical pastures – an experience in the humid tropics. *World Animal Review* 58, 2–12.

Goncalves, C.A. and Oliveira, J.R. da C. (1982) Formacao, recuperacao e manejo de pastagens em Rondonia. Informacoes practicas. *Circular Tecnica, Unidade de Execucao de Pesquisa de Ambito Estadual de Porto Velho* No 1, 22pp.

Hacker, J.B. (1974) Variation in oxalate, major cations, and dry matter digestibility of 47 introductions of the tropical grass setaria. *Tropical Grasslands* 8, 145–154.

Hacker, J.B. (1982) Selecting and breeding better quality grasses. In: Hacker, J.B. (ed.) *Nutritional Limits to Animal Production from Pastures*. Commonwealth Agricultural Bureaux, Farnham Royal, UK, pp. 305–326.

Hacker, J.B. (1986) Selecting for nutritive value in *Digitaria milanjiana* 1. Breeding of contrasting full-sib clones differing in leaf digestibility. *Australian Journal of Experimental Agriculture* 26, 543–549.

Hacker, J.B. (1991) Evaluation of bred populations and cultivars of *Setaria sphacelata. Tropical Grasslands* 25, 245–252.

Hacker, J.B. (1994) Seed production and its components in bred populations and cultivars of winter-green *Setaria sphacelata* at two levels of applied nitrogen fertiliser. *Australian Journal of Experimental Agriculture* 34, 153–160.

Hacker, J.B., Forde, B.J. and Gow, J.M. (1974) Simulated frosting of tropical grasses. *Australian Journal of Agricultural Research* 25, 45–57.

Henrard, J.Th. (1950) *Monograph of the Genus* Digitaria. University of Leiden, Leiden, The Netherlands, 999pp.

Hernandez, M. and Caceres, O. (1983) Guinea Likoni. *Pastos y Forajes* 6, 1–16.

Hill, G.M., Gates, R.N. and Burton, G.W. (1993) Forage quality and steer performance from Tifton 85 and Tifton 78 bermudagrass pastures. *Journal of Animal Science* 71, 3219–3225.

Hodges, T.K., Rathore, K.S. and Peng, J. (1993) Advances in genetic transformation of plants In: Baker, M.J. (ed.) *Proceedings of the XVII International Grassland Congress, 8–21 February, 1993, Hamilton, NZ and Rockhampton, Australia*. New Zealand Grassland Association, Palmerston North, pp. 1013–1023.

Hojito, S. and Horibata, T. (1982) Plant exploration, collection and introduction from Africa. In: *Nekken shiryo 58. Tropical Agriculture Research Center,* Japan, 1–20.

Jank, L., Savidan, Y.H., Costa, J.C.G. and Valle, C.B., do. (1989) Pasture diversification through selection of new *Panicum maximum* cultivars in Brazil. In: Desroches, R. (ed.) *Proceedings of the XVIth International Grassland Congress, 4–11 October, 1989, Nice, France*. French Grassland Society, Versailles Cedex, France, pp. 275–276.

Jerez, I., Rodriguez, V. and Rivero, J.L. (1984) Milk production from three tropical pastures: Coast cross Bermudagrass No. 1 (*Cynodon dactylon*), improved star grass (*Cynodon nlemfuensis*) and pangola grass (*Digitaria decumbens*) during the rainy season. *Cuban Journal of Agricultural Science* 18, 253–260.

Jerez, I., Perez, M. and Rivero, J.L. (1988) A comparative study of Coast Cross 67 bermudagrass (*Cynodon dactylon*) and common guinea grass (*Panicum maximum*) with or without supplementation on milk yield and composition. *Cuban Journal of Agricultural Science* 22, 143–148.

Jerez, I., Menchaca, M.A., Jordan, H., Garcia, T.R., Gonzalez, L., Sosa, M., Rivero, J.L. and De la Rosa, E. (1989) Contribution to the study of tropical pasture management for intensive milk production. In: Desroches, R. (ed.) *Proceedings of the XVIth International Grassland Congress, 4–11 October, 1989, Nice, France*. French Grassland Society, Versailles Cedex, France, pp. 1145–1146.

Jones, R.J. and Ford, C.W. (1972) Some factors affecting the oxalate content of the tropical grass *Setaria sphacelata. Australian Journal of Experimental Agriculture and Animal Husbandry* 12, 400–406.

Jones, R.M. (1994) The role of leucaena in improving the productivity of grazing cattle. In: Gutteridge, R.C. and Shelton, H.M. (eds) *Forage Tree Legumes in Tropical Agriculture*. CAB International, Wallingford, UK, pp. 232–244.

Keller-Grein, G., Maass, B.L. and Hanson, J. (1996) Natural variation in *Brachiaria* and existing germplasm collections. In: Miles, J.W., Maass, B.L. and Valle, C.B., do (eds) *Brachiaria: Biology, Agronomy and Improvement*. CIAT, Colombia, pp. 16–42.

King, K.R. and Stockdale, C.R. (1984) Effects of pasture type and grazing management in autumn on performance of dairy cows in late lactation and on subsequent pasture productivity. *Australian Journal of Experimental Agriculture and Animal Husbandry* 24, 312–321.

Kretschmer, A.E., Jr, Allen, R.J., Jr and Hodges, E.M. (1978) Taiwan digitgrass. *Circular S-258. Florida Agricultural Experiment Stations, IFAS*, University of Florida, 12pp.

Krishnan, K. (1998) The Kerala experience with forage seed production and supply system. *Proceedings of International Training Course for Trainers entitled Forage Agronomy, Seed Production and Seed Supply Systems held at Khon Kaen and Pakchong; October–November 1996*. CIAT, Colombia (in press).

Lamela, L., Pereira, E. and Silva, O. (1984) Evaluación comparativa de pastos para la producción de leche. 1. Bermuda Cruzada-1, Bermuda Callie y Guinea SIH-127. *Pastos y Forrajes* 7, 395–408.

Lascano, C.E. and Avila, P. (1991) Potencial de producción de leche en pasturas solas y asociadas con leguminosas adaptadas a suelos ácidos. *Pasturas Tropicales* 13, 2–10.

Lowe, K.K., Moss, R.J., Cowan, R.T., Minson, D.J. and Hacker, J.B. (1991) Selecting for nutritive value in *Digitaria milanjiana* 4. Milk production from an elite genotype compared with *Digitaria eriantha* ssp. *pentzii* (pangola grass). *Australian Journal of Experimental Agriculture* 31, 603–608.

Luckett, D., Kaiser, A. and Virgona, A. (1996) Innovative breeding of high digestibility kikuyu cultivars to increase milk production. *Final Report to the Dairy Research and Development Corporation, Project DAN 063*. DRDC, Melbourne.

Masaoka, Y., Wilson, J.R. and Hacker, J.B. (1991) Selecting for nutritive value in *Digitaria milanjiana* 3. Relation of chemical composition and morphological and anatomical characteristics to the difference in digestibility of divergently selected full sibs, and comparison with *D. eriantha* ssp. *pentzii* (pangola grass). *Australian Journal of Experimental Agriculture* 31, 631–638.

Mears, P.T. (1970) Kikuyu – (*Pennisetum clandestinum*) as a pasture – a review. *Tropical Grasslands* 4, 139–152.

Miaki, T., Tanaka, S., Kawamura, O., Adania, K., Ohashi, T., Yamauchi, K., Haga, S., Hamakawa, H., Katayama, H. and Misumi, M. (1990a) Comparison of feeding value between Rhodesgrass silage and Italian ryegrass silage for lactating cows. *Bulletin of the Faculty of Agriculture, Miyazaki University* 37, 139–143.

Miaki, T., Tanaka, S., Kawamura, O., Ohashi, T., Yamauchi, K., Haga, S., Hamakawa, H., Katayama, H. and Misumi, M. (1990b) A comparison of the feeding value of sorghum, pearl millet and maize silages on the basis of milk production. *Bulletin of the Faculty of Agriculture, Miyazaki University* 36, 361–366.

Mieres, J., Acosta, Y. and Durán, H. (1988) Comparación de Sudangrás y maíz en dos estados fisiológicas contrastantes para la producción de leche bajo pastoreo.

Investigaciones Agronomicas, Centro de Investigaciones Agricolas 'Alberto Boerger', Uruguay 6, 68–71.

Miles, J.W. and Valle, C.B., do (1996) Manipulation of apomixis in *Brachiaria* breeding. In: Miles, J.W., Maass, B.L. and Valle, C.B., do (eds) *Brachiaria: Biology, Agronomy and Improvement.* CIAT, Colombia, pp. 164–177.

Minson, D.J. (1990) *Forage in Ruminant Nutrition.* Academic Press, San Diego, 483pp.

Minson, D.J. and Hacker, J.B. (1986) Selecting for nutritive value in *Digitaria milanjiana* 2. Intake and digestibility of divergently selected full sibs compared with *D. decumbens. Australian Journal of Experimental Agriculture* 26, 551–556.

Minson, D.J., Cowan, T. and Havilah, E. (1993) Northern dairy feedbase 2001. 1. Summer pasture and crops. Proceedings of a Workshop held at Cedar Lake, Queensland, March 29–31, 1993. *Tropical Grasslands* 27, 131–149.

Mislevy, P. (ed.) (1995) Florakirk bermudagrass. *Circular S 394. Florida Agricultural Experiment Stations, IFAS*, University of Florida, 5pp.

Nakagawa, H., Shimizu, N. and Sato, H. (1993) A new registered cultivar 'Hatsunatsu' rhodesgrass (*Chloris gayana* Kunth). *Bulletin of the Kyushu National Agricultural Experiment Station* 27, 399–416.

Nakajima, K., Komatsu, T., Mochizuki, N. and Suzuki, S. (1979) Isolation of diploid and tetraploid sexual plants in guineagrass (*Panicum maximum* Jacq.). *Japanese Journal of Breeding* 93(2), 228–238.

Olivo, C.J., Moreira, J.C., Barreto, I.L., Diefenbach, J., Ruviaro, C.F. and Sanchez, L.M.B. (1992) Utilizacao de pastagens de capim-elefante e capim setaria como base da alimentacao de vacas em lactacao, durante o verao. *Revista da Sociedade Brasiliera de Zootecnia* 21, 347–352.

Oram, R.N. (1990) *Register of Australian Herbage Plant Cultivars*, 3rd edn. CSIRO, Melbourne, 304pp.

Pathak, J.N. and Pal, R.N. (1983) Effect of chopping green forage on milk production and comparison in Murrah buffaloes. II Kharif fodder (sorghum). *Haryana Agricultural University Journal of Research* 13, 362–365.

Pedreira, C.G.S. and Brown, R.H. (1996) Yield of selected and unselected bahiagrass populations at two cutting heights. *Crop Science* 36, 134–137.

Perez-Infante, F. and Gonzalez, F. (1985) Performance of different pasture species with grazing cows. *Cuban Journal of Agricultural Science* 19, 249–256.

Piggot, G.J. and Morgan, H.M. (1986) A genetic basis for male sterility in kikuyu grass. *Tropical Grasslands* 20, 34–36.

Quarin, C.L. and Caponio, I. (1995) Cytogenetics and reproduction of *Paspalum dasypleurum* and its hybrids with *P. urvillei* and *P. dilatatum* ssp. *flavescens. International Journal of Plant Science* 156, 232–235.

Quarín, C.L., Valls, J.F.M. and Urbani, M.I. (1997) Cytological and reproductive behaviour of *Paspalum atratum*, a promising forage grass for the tropics. *Tropical Grasslands* 31, 114–116.

Rhind, J.M.L. and Goodenough, D.C.W. (1976) The assessment and breeding of *Acroceras macrum* Stapf. *Proceedings of the Grassland Society of Southern Africa* 11, 115–117.

Ruiz, T.M., Sanchez, W.K., Staples, C.R. and Sollenberger, L.E. (1992) Comparison of 'Mott' dwarf elephantgrass silage and corn silage for lactating dairy cows. *Journal of Dairy Science* 75, 533–543.

Rumball, P.J. (1991) Performance of several subtropical grasses in Northland hill pastures. *New Zealand Journal of Agricultural Research* 34, 375–382.

Rumball, P.J. and Lambert, J.P. (1985) Local ecotypes compared with Kenyan introductions. In: Piggot, G.J. (ed.) *Proceedings of a 'Kikuyu Grass Forum' held at Whangarei, 1 November 1985*. Gallagher Electronics, Hamilton, New Zealand, pp. 42–44.

Satjipanon, C. (1995) Pasture seed production in Thailand. In: Wong, C.C. and Ly, L.V. (eds) *Enhancing sustainable livestock-crop production in smallholder farming systems. Proceedings of the 4th Meeting of the Forage Regional Working Group on Grazing and Feed Resources of Southeast Asia*, FAO, Rome, pp. 73–81.

Sato, H., Shimizu, N., Nakagawa, H., Nakajima, K. and Ochi, M. (1993) A new registered cultivar 'Natsukaze' guineagrass (*Panicum maximum* Jacq.). *Bulletin of the Kyushu National Agricultural Experiment Station* 27, 417–437.

Savidan, Y.H. (1983) Genetics and utilization of apomixis for the improvement of guineagrass (*Panicum maximum* Jacq.). In: Smith, J.A. and Hays, V.W. (eds) *Proceedings of the XIVth International Grassland Congress, 15–24 June, Lexington, Kentucky*. Westview Press, Boulder, Colorado, pp. 182–184.

Savidan, Y.H., Jank, L., Costa, J.C.G. and Valle, C.B., do. (1989) Breeding *Panicum maximum* in Brazil: 1. Genetic resources, modes of reproduction and breeding procedures. *Euphytica* 41, 107–112.

Savidan, Y.H., Jank, L. and Costa, J.C.G. (1990) *Registro de 25 acessos selecionades de* Panicum maximum. EMBRAPA-CNPGC Document 44. EMBRAPA, Campo Grande, Brazil, 68pp.

Serpa, A. (1983) Subsidios para um sistema de producao de leite em pastagens. *Circular-Tecnica, PESAGRO-RIO* No. 6, 28pp.

Shimizu, N., Nakagawa, H., Sato, H. and Nakajima, K. (1993) A new registered cultivar 'Natsuyutaka' guineagrass (*Panicum maximum* Jacq.). *Bulletin of the Kyushu National Agricultural Experiment Station* 27, 439–458.

Sollenberger, L.E. and Jones, C.S. (1989) Beef production from nitrogen-fertilized Mott dwarf elephantgrass and Pensacola bahiagrass pastures. *Tropical Grasslands* 23, 129–134.

Sollenberger, L.E., Prine, G.M., Ocumpaugh, W.R., Hanna, W.W., Jones, C.S., Jr, Schank, S.C. and Kalmbacher, R.S. (1988) 'Mott' dwarf elephantgrass: A high quality forage for the subtropics and tropics. *Circular S-356. Florida Agricultural Experiment Stations IFAS*, University of Florida, 18pp.

Sollenberger, L.E., Chambliss, C.G., Wright, D.L. and Staples, C.R. (1995a) Warm-season forages and pasture management for dairy cows. In: Carbia, V.M. (ed.) *Proceedings of the International Conference on Livestock in the Tropics*. University of Florida, Gainesville, pp. 46–57.

Sollenberger, L.E., Pedreira, C.G., Mislevy, P. and Andrade, I.F. (1995b) New *Cynodon* forages for the subtropics and tropics. In: Carbia, V.M. (ed.) *Proceedings of the International Conference on Livestock in the Tropics*. University of Florida, Gainesville, pp. 22–26.

Stobbs, T.H. (1971) Quality of pasture and forage crops for dairy production in the tropical regions of Australia. 1. Review of the literature. *Tropical Grasslands* 5, 159–170.

Stockdale, C.R. (1985) Some factors affecting the consumption of irrigated pastures grazed by lactating dairy cows. *The Challenge: efficient dairy production*.

Proceedings of the Conference organised by the Australian and New Zealand Societies of Animal Production, March 25–28 1985. Albury-Wodonga, Australia. Australian Society of Animal Production.

Terblanche, C.J., Smith, A., Smith, M.F. and Greyling, R. (1996) The improvement of *Digitaria eriantha* (Smuts finger grass). In: *Proceedings Abstracts of the Grassland Grassland Society of South Africa Congress 31*, p. 45.

Tergas, L.E. (1983) Utilizaciónde pastos mejorados en sistemas de producción de leche América tropical. *Agro, Dominican Republic* 12, 29–33.

Theron, E.P. and Arnott, J.K. (1979) Notes on the performance of *Acroceras macrum* Stapf. cv. Cedara selected in Natal. *Proceedings of the Grassland Society of Southern Africa* 14, 23–25.

Upadhyay, V.S., Rai, P. and Patil, B.D. (1980) Comparison of two tropical grass species for growth and milk production in dairy cattle under grazing conditions. *Indian Journal of Range Management* 1, 129–135.

Valle, C.B., do and Savidan, Y.H. (1996) Genetics, cytogenetics, and reproductive biology of *Brachiaria*. In: Miles, J.W., Maass, B.L. and Valle, C.B., do (eds) *Brachiaria: Biology, Agronomy and Improvement*. CIAT, Colombia, pp. 147–163.

Valle, C.B. do, Savidan, Y.H. and Jank, L. (1989) Apomixis and sexuality in *Brachiaria decumbens* Stapf. In: Desroches, R. (ed.) *Proceedings of the XVIth International Grassland Congress, 4–11 October 1989, Nice, France.* French Grassland Society, Versailles Cedex, France, pp. 407–408.

Wilson, J.R. (1982) Environmental and nutritional factors affecting herbage quality. In: Hacker, J.B. (ed.) *Nutritional Limits to Animal Production from Pastures.* Commonwealth Agricultural Bureaux, Farnham Royal, UK, pp. 111–131.

Sward Characteristics and Management Effects on Cool-season Grass Forage Quality

4

C.C. Sheaffer[1], P. Seguin[2] and G.J. Cuomo[3]

[1,2]*Department of Agronomy and Plant Genetics, University of Minnesota, 1991 Buford Circle, 411 Borlaug Hall, St Paul, Minnesota 55108, USA*
[3]*West Central Experiment Station, State Highway B26, Morris, Minnesota 56267, USA*

Introduction

Forage quality is best evaluated by measuring dairy performance. However, agronomically, we often describe forage quality in terms of nutritive value, intake potential, and antiquality components (Marten, 1985). Plant species have inherent differences in chemical composition and forage quality that are influenced by sward characteristics, maturity, management, and the environment. We will emphasize important management strategies such as defoliation frequency, stockpiling, and use of grass–legume mixtures that influence forage quality and sward characteristics of cool-season grasses. Finally, dairy cattle disorders encountered with cool-season grasses are discussed.

Grass Traits

Grass development

Forage grasses and varieties often differ in morphology and ontogeny. All perennial and annual species flower in the spring in response to long days and the inflorescence is borne on an upright stem (culm). Subsequent regrowth of perennial cool-season species such as smooth bromegrass (*Bromus inermis* Leyss.), reed canarygrass (*Phalaris arundinacea* L.), and quackgrass (*Elytrigia repens* (L.) Nevski), which have a vernalization requirement, is stemmed but vegetative (i.e. tillers are elongated but remain vegetative). In contrast, timothy (*Phleum pratense* L.) has no vernalization requirement and thus can flower at each regrowth; whereas regrowth of orchardgrass (*Dactylis glomerata* L.), perennial ryegrass (*Lolium perenne* L.) and tall fescue (*Festuca arundinaceae* Schreb) is stemless and vegetative. To quantify morphological development, Moore and Moser (1995) described five primary growth stages for perennial forage grasses: (i) germination, (ii) vegetative, (iii) elongation, (iv) reproductive and (v) seed (caryopsis) ripening. Because there can be significant variation for plant and tiller maturity within a cultivar, they recommended a weighted mean stage system for describing a population. Quantification of grass morphological stage is necessary for descriptive and predictive purposes relative to forage quality.

Annual and perennial cool-season grasses differ in seasonal growth pattern. Annual grasses have a single growth cycle with vegetative growth, internode elongation, maturity and death. The traditional annual cool-season grasses used for forage production are mostly wheat (*Triticum aestivum* L.), oat (*Avena sativa* L.) and annual ryegrass (*Lolium multiflorum* Lam.). Annual grasses can be planted in the autumn and used as winter forage in milder climates and can be used as companion crops for spring establishment of perennial forage species or as a summer forage in temperate climates.

Variability in morphology due to environment, species and time of year has an obvious effect on the proportion of leaves and stems and the quality of forage available to livestock. Sward management and especially grazing will have profound effects on grass development (Hume and Brock, 1997) and thus on forage quality. Jones *et al.* (1982) reported that management strategies (i.e. mowing vs. continuous grazing), affected sward development characteristics such as canopy structure, tiller density and leaf turnover. The effects of grazing on plant morphology have recently been reviewed by Matches (1992).

Plant canopy components

Grass forage consists of varying proportions of leaves (blade and sheath), stems, and inflorescence depending on the growth stage. Leaves consist of blades with a pronounced mid-rib and sheath, which surrounds the stem and is attached at nodes. Leaf blades by virtue of their role in photosynthesis, consist of a high proportion of relatively thin walled mesophyll and epidermal cells. These cells contain relatively higher N, and lower proportions of highly lignified secondary cell walls than stem xylem cells (Wilson, 1994).

These characteristics generally result in leaf blades having higher forage quality than stems (Table 4.1). Typically, the crude protein (CP) concentration of perennial grass leaves is 100% greater than that of stems, neutral detergent fibre (NDF) concentration is 20% lower, and digestibility is 10% greater (Terry and Tilley, 1964; Mowat *et al.*, 1965; Hockensmith *et al.*, 1997). Similar relationships were reported for annual species by Cherney and Marten (1982b) who found that leaf blades of barley (*Hordeum vulgare* L.) averaged 780 g kg^{-1} digestibility while stems averaged 570 g kg^{-1}. However, Mullahey *et al.* (1992) reported that smooth bromegrass leaves and stems had similar concentrations of rumen escape proteins.

Leaf portions also differ in quality. Cherney and Marten (1982b) reported that leaf sheath digestibility averaged 650 g kg^{-1} making its composition more similar to that of the stem than leaf. The similarity of leaf sheath and stem digestibility is due to their greater fibre content compared with leaf blades (Hacker and Minson, 1981). Wilman *et al.* (1996a) also reported that leaf blades had greater digestibility than sheaths. In addition, Wilman *et al.* (1996a) reported greater *in vitro* true digestibility (IVTD) of the middle and basal leaf blade portion of annual and perennial ryegrass compared with the leaf tip portion (896 and 894 vs. 884 g kg^{-1}). As for the quality of inflorescences, Hume (1991) observed that prairie grass (*Bromus willdenowii* Kunth.) and annual ryegrass inflorescences have a relatively high N content; however, their greater cell wall content was associated with a relative low digestibility (75–62 g kg^{-1}) compared with other plant parts such as the leaf blade (82–78 g kg^{-1}). However, an increasing proportion of highly digestible grain in the inflorescence can improve total forage nutritive value of cereals harvested for forage (e.g. Cherney and Marten, 1982a).

The distribution of leaves and stems in the canopy affects the quality of forage available for animals. Usually, the top portion of the grass canopy is leafier and of greater quality than lower portions, which also have a greater proportion of low quality senescing leaves. When the top (> 20 cm) portion of a smooth bromegrass sward was compared with the bottom (< 20 cm) portion, it had lower NDF (637 vs. 704 g kg^{-1}), and higher *in vitro* digestible dry matter (IVDDM) (569 vs. 482 g kg^{-1}) and CP (137 vs. 68 g kg^{-1}) concentrations (Miller, 1986). Buxton and Marten (1989) also reported that young top leaves in the canopy of various grass swards were about 150 g kg^{-1} more digestible than bottom leaves. Likewise, stem tops of reed canarygrass were 90 g kg^{-1} more digestible than stem bottoms; however, stem bottoms of tall fescue and orchardgrass were more digestible than stem tops probably because stem bases are a primary site of carbohydrate storage for these species.

Table 4.1. Cool-season grass leaf and stem forage quality.

Grass species	Maturity/management	NDF (g kg⁻¹ DM) Leaf	NDF Stem[1]	CP (g kg⁻¹ DM) Leaf	CP Stem[1]	Digestibility[2] (g kg⁻¹ DM) Leaf	Dig. Stem[1]	Location	Authors
Kentucky bluegrass	Average of six stages	588	727	130	77	643	536	Minnesota, USA	Hockensmith et al. (1997)
Orchardgrass		537	661	132	58	704	666		
Reed canarygrass		551	714	183	100	697	613		
Smooth bromegrass		565	704	157	83	715	625		
Tall fescue		571	685	142	77	677	604		
Timothy		557	671	130	70	685	618		
Bromegrass	Cut weekly for 12 weeks	ND	ND	124	55	678	639	Ontario, Canada	Mowat et al. (1965)
Orchardgrass		ND	ND	108	52	631	548		
Timothy		ND	ND	108	62	660	636		
Orchardgrass	50% ear emergence	ND	ND	ND	ND	728	664	England, UK	Terry and Tilley (1964)
Perennial ryegrass		ND	ND	ND	ND	812	757		
Timothy		ND	ND	ND	ND	786	697		
Oat	Average of six stages	ND	ND	ND	ND	780	610	Minnesota, USA	Cherney and Marten (1982b)
Barley		ND	ND	ND	ND	730	545		

[1] Includes leaf sheath.
[2] Expressed as IVDDM (*in vitro* digestible dry matter).
ND, not determined; NDF, neutral detergent fibre; CP, crude protein.

Plant maturity and forage quality

Plant maturity is the most important factor affecting forage quality of grasses over a range of environments. Changes are most dramatic for grasses with culmed reproductive or vegetative tillers which have a decreasing leaf : stem ratio during maturation and less with grasses like perennial ryegrass, tall fescue and orchardgrass which are leafy during regrowth. For example, Buxton and Marten (1989) reported a decline in digestibility of 4 and 6 g kg^{-1} day^{-1} for plants with moderate and numerous reproductive tillers, respectively; while Cherney *et al.* (1993) reported a decrease in IVDDM of 8 g kg^{-1} day^{-1} for five perennial cool-season grasses. Also, the rate of fibre digestion declined linearly for the same five species. Changes in quality (Fig. 4.1) are associated with increases in lignin and cell wall deposition and a decrease in minerals, CP, and digestible cell solubles like starches

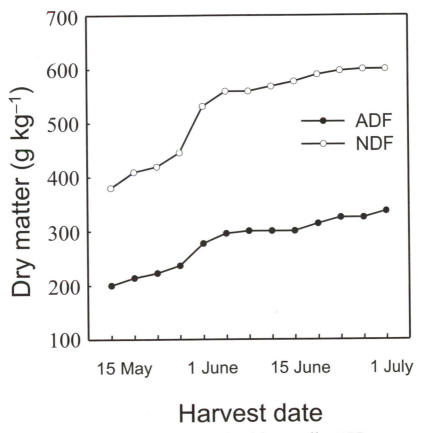

Fig. 4.1. Neutral detergent fibre (NDF) and acid detergent fibre (ADF) concentrations averaged over five cool-season grass species as affected by harvest date (adapted from Cherney *et al.*, 1993).

(Åman and Lindgren, 1983). Wilman *et al.* (1994) reported a decrease in N concentration by 50%, and P and K by 67% with increased maturity of perennial ryegrass leaf blades; however, there was a fivefold and twofold increase in Ca and Mg concentrations, respectively.

The decrease in forage quality with maturity is associated with a decline in quality of both leaves and stems and an increase in the proportion of stems in the canopy. Whole plant forage quality of mature grasses is influenced to a greater extent by stem than leaf chemical composition (Durand and Surprenant, 1993). Sanderson and Wedin (1989) reported that DM digestibility of herbage, leaf blades and stems of timothy and smooth bromegrass declined linearly with maturity. Leaf quality, however, declined at a slower rate with maturity than stem quality. Another example of this relationship was provided by Hockensmith *et al.* (1997) who reported that leaf blade digestibility of reed canarygrass averaged 697 g kg^{-1} and declined 25 g kg^{-1} week^{-1} from vegetative to flowering stages while stem digestibility averaged 613 g kg^{-1} and declined 58 g kg^{-1} week^{-1}. Terry and Tilley (1964) also reported reduced leaf digestibility of orchardgrass, perennial ryegrass, timothy and tall fescue with increased maturity, but digestibility changed at a slower rate for leaf blades than for leaf sheaths and stems.

Species and cultivar differences in forage quality

Cool-season grasses vary in quality when harvested at a given stage of maturity (Table 4.2). Annual and perennial ryegrasses frequently have had superior forage quality in comparative studies with other cool-season grasses including reed canarygrass (Fairey, 1985; Wilman *et al.*, 1996a). Cherney *et al.* (1993) observed differences in forage quality between five cool-season grasses. The average NDF and acid detergent fibre (ADF) concentrations were lowest for creeping foxtail (*Alopecurus arundinaceus* Poir.) while lignin concentration was greatest for timothy. Species also differed in forage IVDDM and CP concentrations but differences among species were not consistent over sites. Among commonly used grasses, reed canarygrass has consistently been among those with the highest concentration of CP when N has been applied in fertilizer, manure or sewage effluent (Sheaffer *et al.*, 1990a).

Some differences in forage quality between species are due to differences in developmental rate and plant morphology. Hockensmith *et al.* (1997) reported that Kentucky bluegrass (*Poa pratensis* L.) leaf concentration declined at the slowest rate, while leaf concentration of timothy and tall fescue declined at the fastest rate. Kentucky bluegrass was leafier than tall growing grasses but the leaf concentration was not correlated with herbage digestibility or CP concentration. Differences in forage quality between grass species might also arise from differences in their tissue composition and anatomy. Relative content of highly indigestible tissues, such as vascular tissue, as well as their organization, vary between species (Akin, 1989), and might result in differences in forage quality (Wilman *et al.*, 1996b). Grabber and Allinson (1992) reported that the digestion of reed canarygrass leaf blades is limited relative to that of tetraploid hybrid ryegrass (*Lolium hybridum* Hausskn.) by anatomical characteristics. In contrast, Bruckner and Hanna (1990) reported that differences in digestibility of various forage small grain cereals were not associated with microanatomical characters. Forage quality differences among species might also arise from differences in the timing of rates of deposition of cell wall components (Bidlack and Buxton, 1992).

Grass cultivars also differ in forage quality as shown for timothy in Table 4.3. Differences in forage quality among cultivars are often due to genotypic differences in rate of development and thus in morphogenic and structural characteristics (Hides *et al.*, 1983; Bélanger, 1996). For timothy, early-maturing cultivars have a lower *in vitro* dry matter digestibility (IVDMD) and a greater NDF concentration compared with late-maturing cultivars (Bélanger and McQueen, 1996). These early-maturing cultivars were reported to have a greater leaf extension rate (Bélanger, 1996). However, when early- and late-maturing timothy cultivars were harvested at a given growth stage, Míka (1983) reported opposite results: i.e. early-maturing cultivars had a greater digestibility then late-maturing ones. For small grains, Bolsen *et al.* (1976) reported large differences in intake and performance of steers fed different wheat cultivars. Steers fed silage made from 'Arthur' wheat gained 27.5% faster and consumed 18% more dry matter than those fed silage made from 'Parker' wheat.

Palatability (selection when a choice is given) is a complex animal response influenced by plant, animal, and environmental factors. Marten (1985) identified antiquality components, plant morphology, canopy height and maturity as several important plant factors influencing palatability of forages. Perennial ryegrass is know as a very palatable grass, probably because of its high leaf carbohydrate concentrations (Wilman *et al.*, 1996a). Sheaffer *et al.* (1990a) reported that on a scale of 1 (complete consumption) to 10 (complete rejection), smooth bromegrass rated 2.0, low alkaloid reed canarygrass rated 3.0, orchardgrass rated 4.4, and high alkaloid reed canarygrass rated 5.5. Palatability problems are also known in tall fescue due to the presence of alkaloids produced by endophytic fungi (Bush *et al.*, 1971). Wilman *et al.* (1996b) suggested that differences in physical characteristics (e.g. veining and leaf area) of grasses might explain differences in palatability. The impact of most palatability differences among species can be minimized if animal stocking rates are adjusted so as to force consumption of forages. However, palatability differences can result in overgrazing of more palatable species if animals are given a choice (Hodgson, 1985), and thus can represent a problem in mixed swards.

Variability in forage quality traits of many cool-season grasses is sufficient to have successful selection for improved forage quality. As reviewed

Table 4.2. Forage quality of cool-season grass species.

| Grass species | | Maturity/ management | Variables (g kg⁻¹ DM) | | | | | |
Common name	Scientific name		NDF	ADF	CP	Digest.	Location	Author(s)
Kentucky bluegrass	*Poa pratensis* L.	Vegetative	557	ND	173	697[1]	Minnesota, USA	Hockensmith et al. (1997)
Orchardgrass	*Dactylis glomerata* L.		486	ND	177	774		
Reed canarygrass	*Phalaris arundinaceae* L.		528	ND	204	758		
Smooth bromegrass	*Bromus inermis* Leyss.		537	ND	181	766		
Tall fescue	*Festuca arundinacea* Schreb.		524	ND	176	740		
Timothy	*Phleum pratense* L.		514	ND	166	761		
Quackgrass	*Elytrigia repens* (L.) Nevski	Average of three stages	635	ND	150	605[1]	Minnesota, USA	Sheaffer *et al.* (1990b)
Reed canarygrass	*Phalaris arundinaceae* L.		615	ND	160	599		
Smooth bromegrass	*Bromus inermis* Leyss.		614	ND	151	622		
Annual ryegrass	*Lolium multiflorum* Lam.	5 weeks' interval	382	245	ND	891[2]	Wales, UK	Wilman *et al.* (1996a)
Meadow fescue	*Festuca pratensis* Huds.		446	223	ND	888		
Perennial ryegrass	*Lolium perenne* L.		395	226	ND	897		
Tall fescue	*Festuca arundinacea* Schreb.		452	252	ND	873		
Barley	*Hordeum vulgare* L.	Average of six stages	476[3]	327	ND	714[1]	Minnesota, USA	Cherney and Marten (1982a)
Oats	*Avena sativa* L.		496	361	ND	653		
Triticale	× *Triticosecale* Wittmack		489	348	ND	690		
Wheat	*Triticum aestivum* L.		495	361	ND	661		

Parry oat grass	*Festuca scabrella* Scribn.	Average of two and four cuts	ND	370	128	536[1]	Alberta, Canada	Willms and Beauchemin (1991)
Rough fescue	*Danthonia parryi* Torr.		ND	406	102	519	Alberta, Canada	
Meadow fescue	*Festuca pratensis* Hud.	Three and four cuts	578–548	271–238	ND	ND	Prince Edward Island, Canada	Kunelius (1990)
Orchardgrass	*Dactylis glomerata* L.		613–585	296–262	ND	ND		
Tall fescue	*Festuca arundinacea* Schreb.		579–572	279–250	ND	ND		
Timothy	*Phleum pratense* L.		593–565	284–254	ND	ND		

[1] Expressed as IVDDM (*in vitro* digestible dry matter).
[2] Expressed as IVTD (*in vitro* true digestibility).
[3] Expressed as CWC (cell wall constituents).
ND, not determined; NDF, neutral detergent fibre; ADF, acid detergent fibre; CP, crude protein.

Table 4.3. Differences in forage quality among cool-season grasses cultivars.

Grass species	Cultivar	Management	Variables (g kg^{-1})				Location	Author(s)
			NDF	ADF	CP	Digestibility[1]		
Timothy	Champ	Average of two cuts	622	346	97	ND	Québec, Canada	Surprenant *et al.* (1993)
	Itasca		628	348	99	ND		
	Ott-px5		611	339	108	ND		
	Tiller		608	338	96	ND		
	Timfor		624	347	100	ND		
Timothy	Champ	Average of three and four cuts	582	269	ND	ND	Prince Edward Island, Canada	Kunelius (1990)
	Farol		572	264	ND	ND		
Orchardgrass	Frode		591	275	ND	ND		
	Prairial		603	277	ND	ND		
Orchardgrass	Napier	Average of three and five cuts	578	ND	ND	ND	Iowa, USA	Buxton (1990)
	Orion		560	ND	ND	ND		
Smooth bromegrass	Barton		592	ND	ND	ND		
	Rebound		565	ND	ND	ND		
Tall fescue	LLER[2]		580	ND	ND	ND		
	HLER		571	ND	ND	ND		
Orchardgrass	Saborto	Average of four cuts	ND	ND	ND	603	British Columbia, Canada	Fairey (1985)
	Tenderbite		ND	ND	ND	587		
Perennial ryegrass	Cropper		ND	ND	ND	658		
	Barlatra		ND	ND	ND	681		
Reed canarygrass	Castor		ND	ND	ND	630		
	NRG 721		ND	ND	ND	634		

[1]Expressed as IVDDM (*in vitro* digestible dry matter); [2]See Buxton (1990) for details on these germplasms. ND, not determined; NDF, neutral detergent fibre; ADF, acid detergent fibre; CP, crude protein.

by Vogel and Sleper (1994) and Casler (see Chapter 2), both qualitative and quantitative heritable traits related to improved forage quality have been identified and used in breeding programmes in the last 25 years; resulting in important forage quality improvements.

Harvest Management

Defoliation frequency

Because of the relationship between grass maturity and forage quality, harvest regimes with more frequent harvests during the growing season that remove forage at less mature growth stages often result in greater forage quality than less frequent harvest regimes. For example, Sheaffer *et al.* (1990b) reported that increasing seasonal cutting frequency of quackgrass, reed canarygrass, and smooth bromegrass from two-cuts (reproductive stage) to four-cuts (vegetative stage) increased forage CP and IVDDM concentration and decreased NDF concentration. Values for the three-cut system were intermediate. This relationship between forage quality and defoliation frequency was confirmed for diverse cool-season perennial species by Marten and Hovin (1980) who reported that a three-cut system, although intermediate between a two- and four-cut system in forage quality, had the greatest nutrient yields, mostly because of higher forage dry matter yields. Likewise, results reported by Kunelius (1990) suggest that quality of annual ryegrass hybrids is greater under a four-cut than a three-cut system, with lower NDF and ADF concentrations under four than three cuts (546 and 249 vs. 561 and 271 g kg^{-1}). Willms and Beauchemin (1991) reported that for rough fescue (*Festuca scabrella* Torr.) and Parry oat grass (*Danthonia parryi* Scribn.) increasing defoliation frequency from one to 16 cuts during a 16-week period reduced acid detergent-insoluble nitrogen (ADIN), ADF, and lignin, and increased CP and IVDDM concentrations of both grasses. Turner *et al.* (1996) also reported that average daily gain of steers was greater with increased defoliation/grazing frequency due to an increase in the quality of the forage consumed.

In selection of a growth stage for harvest, it is important to consider its impact on both yield and quality. Since forage yield usually increases until reproductive growth stages, harvest regimes with frequent defoliation aimed at producing a high quality forage will often reduce seasonal forage yields. A simplified version of this relationship is presented in Fig. 4.2. In addition, for some grasses, timing of harvest can have a significant effect on persistence. The defoliation regime that will optimize both forage quality, forage yield and persistence, varies with species and environmental conditions.

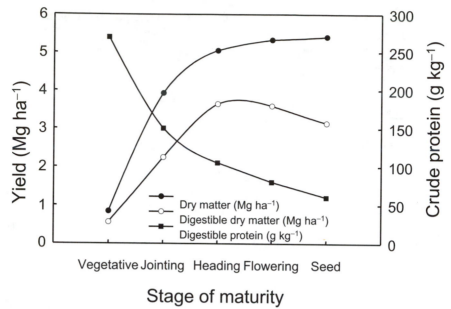

Fig. 4.2. Generalized changes in yield and quality of reed canarygrass with maturity (adapted from Decker *et al.*, 1967).

Stockpiling

Stockpiling forage has been described as 'the practice of allowing forage to accumulate in the field until it is needed for grazing' (Mays and Washko, 1960). Stockpiling forage for autumn and winter use extends the grazing season in the milder climates of the upper-South and mid-Atlantic states. The primary stockpiled forage in those areas has been tall fescue because relative to other cool-season grasses it produces more growth and a high quality forage following freezing temperatures (Matches, 1979). However, the outcome of stockpiling is greatly influenced by weather conditions and in northern areas cold temperatures and snowfall in mid-autumn may limit the usefulness of stockpiling. In addition, extreme winters may threaten tall fescue persistence.

Initiating stockpiling earlier in the growing season will result in greater forage yield but greater deterioration of forage in the field (Van Keuren, 1972; Matches and Tevis, 1973; Cuomo *et al.*, 1997). Matches and Tevis (1973), in Missouri, reported stockpiled forage mass of tall fescue ranging from 10 Mg ha^{-1} to 775 kg ha^{-1} depending on length of growth or regrowth period and environment. They also reported IVDMD in October that ranged from 579 g kg^{-1} for tall fescue accumulated from May compared with 684 g kg^{-1} for tall fescue accumulated from August. Crude protein for the same treatments averaged 83 and 101 g kg^{-1}.

Stockpiling has been done with species other than tall fescue, and there has been interest in extending the grazing season with stockpiled forage in areas where tall fescue is not adapted. Van Keuren (1972) in Ohio compared tall fescue, orchardgrass and Kentucky bluegrass stockpiled for the entire growing season, or stockpiled beginning in June or August until November. Stockpiled forage mass of tall fescue and Kentucky bluegrass were similar when stockpiled for the entire growing season (5.36 Mg ha^{-1}) or when stockpiled from June (3.66 Mg ha^{-1}). Orchardgrass forage mass was 1.5 and 1.3 Mg ha^{-1} less, respectively, under the same stockpile management systems. Wedin *et al.* (1966) in Iowa concluded that tall fescue maintained higher forage soluble carbohydrates and digestibility when stockpiled than reed canarygrass, although reed canarygrass also offered some advantages, primarily in greater crude protein and winter hardiness, as a stockpile crop in Iowa. Nitrogen fertilization increases forage mass, duration of leaf retention and forage quality (especially CP) of stockpiled forage (Taylor and Templeton, 1976; Archer and Decker, 1977; Balasko, 1977). However, high rates of N fertilization reduced stands of tall fescue and orchardgrass in Iowa (Wedin *et al.*, 1966).

Losses in forage mass and nutritive value of cool-season grasses occur during autumn and winter when temperatures prevent new growth (Van Keuren, 1972; Matches and Tevis, 1973; Taylor and Templeton, 1976; Archer and Decker, 1977; Ocumpaugh and Matches, 1977; Cuomo *et al.*, 1997). These losses are primarily associated with leaf death and degradation. For example, Archer and Decker (1977), in Maryland, reported that during autumn and early winter the proportion of dead leaves in tall fescue and orchardgrass pastures increased from 20 to 46% while the IVDMD of the total herbage declined about 4 units for each 10% increase in dead leaf content. Cuomo *et al.* (1997) reported dry matter losses of 67% for smooth bromegrass. Ocumpaugh and Matches (1977), in Missouri, reported that after autumn growth stopped dry matter losses for tall fescue were 25 kg ha^{-1} week^{-1}. Additionally, they reported losses of 10 g kg^{-1} week^{-1} in IVDMD, 1 g kg^{-1} week^{-1} in CP, and 1 g kg^{-1} week^{-1} in K.

Growth regulators

Growth regulators such as amidochlor {*N*-[(acetylamino)methyl] 2-chloro-*N*-(2,6-diethylphenyl) acetimide}, imazethapyr [5-ethyl-2-(4-isopropyl-4-methyl-5-oxo-2-imidazolin-2-yl) nicotic acid], and mefluidide {*N*-[2,4-dimethyl-5-(trifluoromethyl) sulfonyl amino]phenyl]acetamide} have been used in forage production systems to maintain a high proportion of leaves in the canopy, with the objective of increasing the quality of harvested forages. When applied to grasses when vegetative in spring, stem elongation and seedhead formation will be reduced and/or delayed. Sheaffer and Marten (1986) reported decreased NDF and increased CP with mefluidide application to several perennial forage grasses. However, there was year-to-year variation in its effectiveness. Fales and Hoover (1990) reported improved timothy forage quality with imazethapyr applications. For the first cut, NDF was reduced by 150 g kg^{-1}, and IVDDM and CP were increased by 80 and 27 g kg^{-1}, respectively. The quality of subsequent cuts were similar to untreated controls. However, yields of the first harvest were reduced by 50%. Use of growth regulators is infrequent in commercial dairy operations due to the cost of application and impact on yield, such that their profitability is questionable.

Other management related factors

Other management decisions affecting grass forage quality include defoliation height, which influences the amount of low quality stems harvested (Willms and Beauchemin, 1991) and the type of mowing/grazing management system used (i.e. continuous vs. rotational) (Undersander and Paine, 1997). These affect the forage quality by altering plant development and morphology (Jones *et al.*, 1982), but also affect plant populations and population dynamics in mixtures (Edwards *et al.*, 1996). The choice of grazing systems will have profound implications on the quality of forage consumed by grazing animals (see Chapter 12). Knowledge of grass development has led to the elaboration of grazing systems such as 'first and last grazing', which have been promoted to provide forage with variable levels of forage quality to groups of grazing cattle. As a result, Blaser *et al.* (1986) reported that first grazing milk cows produced 55% more milk than last grazing milk cows because at low grazing pressures cattle can also obtain a high quality diet by selecting leaves before stems and young plant parts before old plant parts. The tendency for grazing animals to select green leaf is well documented, as is the influence of the distribution of leaves within the sward canopy upon intake per bite (Hodgson, 1985).

Sward establishment strategies, such as the use of companion crops, seeding date, method and rate (Smouter *et al.*, 1996) will indirectly affect forage quality by altering potential stands and weed infestation. However, these will only be transitory, as their results will rapidly be modified by other factors such as population dynamics (e.g. tillering).

Species Associations

Cool-season grasses are seldom grow in monocultures but are most often components of dynamic mixtures with other grasses, legumes and forbs. In a pasture or hayfield, mixture composition is transitory in response to prevailing biological and management factors. Important factors influencing mixture composition include animal and harvest management, insect predation, soil fertility and environmental conditions (e.g. temperature and rainfall) (Fales *et al.*, 1996). Because species differ in competitiveness and in adaptability to these factors, their persistence and abundance in mixtures can change dramatically as the level of stress factors change (Seguin, 1997). Consequently, it is typical to observe seasonal and year-to-year variation in composition (Yarrow and Penning, 1994; Kanneganti and Kaffka, 1995).

Although yield of legume–grass mixtures may not exceed those of monocultures of either species (Chamblee and Collins, 1988), in most cases legume–grass mixtures will tend to have a greater forage quality compared with grass monocultures. This is due to the generally superior forage quality of legumes and broadleaf weed species, relative to grasses. These differences are attributable to the lower cell wall concentration and higher CP concentration of legumes compared with cool-season grasses (Buxton, 1996). Thus, forage quality of mixtures will vary depending on the sward composition. There are several important agronomic advantages of planting mixtures containing a diversity of plant species.

Improved distribution of yield and quality during the growing season

Grass–legume mixtures and grass mixtures can improve the yield and quality distribution throughout the season due to different peaks of production and diverse adaptability to environmental conditions between grasses and legumes. For example, Sheaffer *et al.* (1981) reported that including reed canarygrass in a mixture with smooth bromegrass and orchardgrass improved total summer forage yields. Ta and Faris (1987a) reported that in a lucerne (*Medicago sativa* L.)–timothy mixture, timothy contribution to yield was greatest in the first harvest each year while lucerne yields were greatest at later cuttings. Although the introduction of legumes in grass swards might result in a better distribution of available forage quality, results are variable and depend on the legume and weed content (P. Seguin, P.R. Peterson, C.C. Sheaffer and D.L. Smith (1998) personal communication). Abdalla *et al.* (1988) reported that a 1 : 1 mixture of a cool-season grass with birdsfoot trefoil (*Lotus corniculatus* L.) had a better seasonal yield distribution and lower NDF concentration than grass monocultures; however there were no differences in ADF, CP and IVDDM concentrations.

Adaptation to diverse and adverse environmental conditions

Since species vary in their adaptation to soil and climatic conditions, mixtures with diverse species provide insurance against total stand loss, should one or more of the mixture components fail to persist. Diverse species may also be best able to utilize environmental resources. For example, compared with a white clover monoculture, a 1 : 1 white clover (*Trifolium repens* L.)–tall fescue mixture had a greater proportion of legume-N derived from the air (i.e. N_2 fixation), and generally greater productivity (Mallarino *et al.*, 1990). Tilman *et al.* (1996) reported that increased diversity in grasslands will result in better nutrient utilization, leading to greater productivity and sustainability. However, the complementarity of grass and legume species will vary. For example, Jones *et al.* (1988) observed that initially, lucerne and reed canarygrass competed for 'different space' and were complementary in their environmental resource exploitation. However, this relationship was transitory, lucerne having excluded reed canarygrass from the mixture by the end of the study. An opposite trend was observed with birdsfoot trefoil, with competition for the same resources at early harvests, but reaching an equilibrium by the end of the study. Additional aspects of the dynamic nature of grass–legume mixtures were reviewed by Haynes (1980).

Enhanced quality of pastures by dilution of antiquality factors

Cool-season grasses like reed canarygrass are frequently planted with lucerne or other bloat inducing legumes to reduce the incidence of frothy legume bloat (Sheaffer *et al.*, 1990a). Typically, it is recommended the total herbage biomass be 30% grass to dilute degradable protein concentrations. A contrasting example is provided by Hoveland *et al.* (1997) who reported that inclusion of as little as 6% of lucerne in mixture with an endophyte infected tall fescue increased animal performance by 30%. While using mixtures to reduce antiquality factor effects is a viable management practice, the animal performance outcome will also be influenced by opportunities for selectivity by grazing animals.

Enhanced quality and yield of grass pastures through the addition of legume N

Fixation of atmospheric N_2 provides a mechanism for self-sufficient protein production by legumes and an opportunity for transfer of N to grasses grown in mixtures. Legume-derived N supplied by transfer reduces grass dependence on N from soil mineralization and fertilizer. Heichel and Henjum (1991) reported that several perennial grasses derived an average of 36% (28 kg N ha^{-1}) of their N from legumes in the second year following seeding, with the greatest contribution by white clover of 53 kg N ha^{-1}. Other researchers have reported values as high as 30% (13 kg N ha^{-1}) for a lucerne–timothy mixture (Ta and Faris, 1987b) and 80% for a white clover–perennial ryegrass mixture (Boller and Nösberger, 1987). Ultimately,

N concentration in grass forage increases. Ta and Faris (1987b) reported a 50% increase in timothy N concentration due to growth in mixture with lucerne.

Challenges

The advantages of mixtures can only be achieved if a balance of components is achieved, however the composition of mixtures is unstable and greatly affected by environmental conditions and management. From an agronomic standpoint, knowledge of species characteristics is critical in solving problems with mixture instability. Potential problems can be averted or lessened if components are selected to fit known environmental and management regimes. Cool-season grasses vary in response to frequent defoliation because of differences in growth habit, presence of basal tillers, and root reserves. Marten and Hovin (1980) reported that reed canarygrass and orchardgrass persisted well when frequently cut (three or four times per year); whereas, smooth bromegrass had poor persistence. Consequently, in designing mixtures of grasses or grasses–legume mixtures, it is desirable to understand how they fit the potential grazing or haying system. Likewise, the relative rate of germination, growth and consequently competitiveness of species during establishment should be considered in developing mixtures. Blaser *et al.* (1986) identified annual and perennial ryegrass seedlings as very aggressive, tall fescue, orchardgrass and smooth bromegrass seedlings as aggressive, and timothy and Kentucky bluegrass seedlings as non-aggressive. It is likely that reed canarygrass seedlings would also be characterized as non-aggressive because of low seedling vigour (Sheaffer *et al.*, 1990a). These differences are most often practically accounted for by adjusting the proportion of species in mixtures, with aggressive species sown at lower seeding rates. Most perennial legumes are less persistent than grasses. Consequently, in many pastures, perennial legumes require periodic reintroduction. Strategies for legume establishment are described by Sheaffer (1989) and Seguin (1998).

Environmental Factors

Many environmental factors interact with management factors to significantly influence cool-season grass forage quality. Environmental effects can lessen or enhance forage quality differences among species. Within a harvest season, forage quality of grasses is frequently influenced by temperature and moisture. Other factors such as light, pathogens and soils are discussed by Buxton and Fales (1994) and in Chapter 6.

Temperature

Temperature has a greater effect on forage quality than other environmental factors because of its influence on plant metabolism. It determines the rate of plant development and maturity, and thus the relative proportion of leaves and stems. For most cool-season grasses, the optimum temperature for dry matter accumulation and growth is about 20–25°C (McWilliam, 1978) and the greater challenges to forage quality occur with temperatures above optimum. Temperatures above optimum will lower digestibility with the effect greater in stems than leaves (Deinum and Dirven, 1974). Ames *et al.* (1993) reported similarly that lignin content of both leaves and stems increased with a temperature rise from 10°C to 30°C and that this increase was greater for the stems. The difference in forage quality between timothy cultivars was lower at an optimal (20°C) than suboptimal temperature (10°C). Thorvaldsson (1992) established that the effect of temperature on digestibility of timothy was independent of the grass growth stage with a 0.06% decrease in digestibility for each degree increase in temperature. Buxton and Fales (1994) in reviewing the literature concluded that above optimum temperatures increase NDF and IVDDM but that the cause of reduced dry matter digestibility was uncertain. They suggested that it could be an effect on enzyme-mediated cell wall synthesis, or other related biochemical or physiological processes.

Soil moisture

Soil moisture deficits (drought) affect growth and development of cool-season perennial grasses. A common plant response is to reduce growth, specifically leaf area expansion to conserve moisture, since leaves are a primary site for transpiration. Drought can also influence carbohydrate partitioning in plants and increase cuticle thickness (Frank *et al.*, 1996). Leaf area reduction can decrease forage quality especially when severe moisture stress results in leaf senescence. However, moisture deficits often have either no effect or a beneficial effect on forage quality because maturity is delayed and the proportion of leaves is increased due to reduced stem development (Wilson, 1981). Sheaffer *et al.* (1992) reported that under periods of moderate moisture deficit stress, reed canarygrass, smooth bromegrass, timothy and orchardgrass had a greater leaf proportion than well watered controls. Leaf number and weight were not affected by moisture deficits. Leaf proportion of reed canarygrass, smooth bromegrass, and timothy was increased at each regrowth. In contrast, leaf percentage of orchardgrass was increased by moisture deficits only at the initial spring regrowth, but not at subsequent ones during the season, due to a culmless vegetative regrowth. The impact of moisture deficits on forage quality of the species was inconsistent in part because of differences in duration of stress.

When moisture deficits occurred throughout the regrowth cycle, CP concentration of leaves, stems and total forage of all grasses was increased; however, effects on fibre and lignin concentrations were less consistent among grasses. Moisture deficits decreased leaf, stem and total forage NDF and ADF concentrations of reed canarygrass and total forage NDF and ADF concentrations of smooth bromegrass, but did not affect fibre levels in other species.

Bittman *et al.* (1988) reported the effect of prolonged drought on forage quality of several grasses in the Canadian prairies. Smooth bromegrass, crested wheatgrass (*Agropyron cristatum* (L.) Beauv spp. Pectinatum (Bieb.) Tzvel.), and Altai wildrye (*Leymus angustus* (Trin.) Pilger) declined in quality due to a decrease in leaf : stem ratio, stem and leaf quality, and increased leaf senescence. Water stress increased the rate of decline in N and P concentration in herbage due to translocation from senesced leaves. However, drought increased the digestibility of all grasses by lowering ADF and lignin concentration in the forage.

Animal Disorders

Chemical and mineral constituents of cool-season grasses affect livestock performance by altering animal metabolism and neurological functions. They have been grouped in the category designated by Marten (1985) as antiquality components. Important antiquality chemical factors in cool-season grasses include intrinsic and extrinsic chemical toxicants and mineral imbalances (Mayland and Cheeke, 1995). Chemical toxicants are often found in plants as a defence mechanism against herbivory by mammals and invertebrates. Mineral imbalances are due to soil deficiencies or mineral interactions and environmental stress effects on plants (see Chapter 6). We discuss fungal and intrinsic toxins as examples of chemical toxins and grass tetany and nitrate poisoning as examples of mineral disorders.

Fungal toxins

Endophytic fungi are symbiotically associated with many cool-season grasses including tall fescue, timothy, ryegrass, orchardgrass, and tall fescue (Bacon, 1994). Bacon (1994) described the paradox of endophytes as a 'two edged sword' of mutualism and antiquality. Mutualism describes the mutually beneficial association of two organisms in which the fitness of each participant is increased by the action of the other. For example, the association of tall fescue with the endophytic fungus *Neotyphodium coenophialum* Glenn, Bacon et Hanlin enhances plant persistence, tolerance to drought stress, N use efficiency, and defence against insects and nematodes. However, the association has a negative effect on cattle performance when seed or forage is eaten. For example, Strahan *et al.* (1987) reported that cows fed endophyte infected tall fescue hay had lower intake and lower milk production than cows fed uninfected tall fescue hay. Other commonly cited symptoms of tall fescue toxicosis in cattle include fescue foot, fat necrosis, and summer slump (Mayland and Cheeke, 1995). Ergopeptide alkaloids (ergovaline) produced by the fungus reduce prolactin production by the pituitary gland and ultimately affect smooth muscle contraction in the body. Approaches being used to lessen the problem include eradication of infected fescue, managing grazing intensity and frequency to maintain forage in a vegetative state, dilution of the diet with other grasses, and development of endophyte-free cultivars (Sleper and West, 1996; Hoveland *et al.*, 1997).

Another endophytic fungus, *Neotyphodium lolii* Glenn, Bacon et Hanlin, in perennial ryegrass produces lolitrems, a chemical that inhibits neurotransmitters in the brain and causes 'ryegrass staggers' (Gallagher *et al.*, 1984). Animals consuming forage display symptoms of incoordination, staggering, head shaking, and collapse. This disorder has been frequently observed in New Zealand but has also been reported in California and Oregon (Mayland and Cheeke, 1995). Although ryegrass staggers is generally not a fatal malady, animal management and performance is affected. Another fungus, *Neotyphodium uncinanum* Glenn, Bacon, Price et Hanlin, has been reported to infect meadow fescue *(Festuca pratensis* Huds.). Unlike the other endophytes of this genus, it has not been reported to have toxic effects on livestock (Daccord *et al.*, 1995). However, it has been demonstrated to increase the competitive ability of the infected grass (Malinowski *et al.*, 1997).

Intrinsic plant toxins

In *Phalaris* spp. naturally occurring indole alkaloids have been associated with significant health problems of grazing animals. In reed canarygrass, indole alkaloids have been associated with lack of palatability and weight loss by grazing animals (Wittenberg *et al.*, 1992). Weight loss is often associated with diarrhoea induced by tryptamine and carboline alkaloids. Marten *et al.* (1981) reported that the threshold indole alkaloid level beyond which sheep will reject reed canarygrass is about 0.6% dry weight and that gain is reduced if indole alkaloid concentration is at or above 0.2% dry weight. There are several strategies to reduce the impact of alkaloids on animal performance. Cultivars have been developed with reduced total and tryptamine alkaloids and that are more palatable to livestock and result in improved animal performance compared with unimproved cultivars (Marten *et al.*, 1981). In addition, alkaloid levels in forage can be reduced by drying herbage and harvesting it as hay or silage (Sheaffer and Marten, 1995). Problems with alkaloids can also be avoided by not grazing forage of plants subject to drought or plants fertilized with high rates of N. Ingestion of *Phalaris aquatica* L. and *Phalaris caroliniana* Walt has produced 'phalaris staggers' and 'sudden death syndrome' in sheep and cattle. The symptoms are most often associated with grazing but have also been reported following feeding of *Phalaris* hay. Symptoms of phalaris staggers includes incoordination, confusions, and mortality caused by tryptamine alkaloid interference with brain function. Sudden death syndrome is associated with cardiac arrest and several chemical toxins are suspected agents.

Nitrate poisoning

Nitrate (NO_3) is the primary form of N normally taken up by plants. Within plants, it is converted to NH_4 that is used in the synthesis of amino acids and proteins. When plants take up excessive NO_3, and transformation to protein is disrupted by drought or low temperature stress, NO_3 can accumulate in plants. Concentrations are usually greatest in the stem bases, less in the leaves, and lacking in the caryopsis (Wright and Davison, 1964). Nitrate in forages is converted to NO_2 in the rumen. Under normal conditions, NO_2 is utilized by rumen microbes for microbial protein production but, at excessive concentrations NO_2 is absorbed through the rumen wall into the blood stream. Nitrite reacts with blood haemoglobin forming methaemoglobin which cannot carry oxygen. Consequently, cellular oxygen needs are not met and animals die of asphyxiation. The blood typically turns brown and membranes turn blue. Forage should be considered toxic when forage contains over 2000 µg N g^{-1} although animal health problems have been observed at much lower concentrations when animal condition has been poor or high NO_3 forage is a high proportion of the diet (Bush *et al.*, 1971). Cattle have also been known to adapt to high levels of NO_3 and tolerate high concentrations in the ration.

Grass tetany (hypomagnesaemia)

Grass tetany is a complex disorder of cattle associated with magnesium (Mg) deficiency in cattle grazing cool-season grasses. Progressive symptoms including staggering gait, nervousness, frequent urination, muscular tremors, and death occur as blood plasma levels decrease to less than 18 mg l^{-1} (Mayland and Cheeke, 1995). Herbage Mg concentration and availability are decreased by potassium (K) and N fertilization. Although, a number of animal and plant factors interact to produce grass tetany, agronomic strategies such as splitting applications of K and N fertilization, use of dolomitic limestone, and use of supplemental Mg fertilizers have been used to lessen the frequency of the disorder (Robinson *et al.*, 1989).

Summary

Quality of cool-season grass forage fed to dairy cattle is influenced by management, the environment, as well as species and cultivars. Because of the negative relationship between forage quality and grass maturity, harvest timing probably is the most significant management factor. However, other factors such as cutting height, stockpiling, and mixture stability can influence quality. Environmental conditions such as air temperature and soil moisture interact with management factors and grass species to influence quality.

Grass species and cultivars differ in quality due to inherent differences in chemical composition and presence of symbiotic relationships with fungi. Knowledge of these differences and their interaction with management will allow producers to develop systems to produce forage of consistent quality to meet the needs of dairy cattle.

References

Abdalla, H.O., Fox, D.G. and Seaney, R.R. (1988) Protein distribution in four cool-season grass varieties alone or in combination with trefoil. *Journal of Animal Science* 66, 2325–2329.

Akin, D.E. (1989) Histological and physical factors affecting digestibility of forages. *Agronomy Journal* 81, 17–25.

Åman, P. and Lindgren, E. (1983) Chemical composition and in vitro degradability of individual chemical constituents of six Swedish grasses harvested at different stages of maturity. *Swedish Journal of Agricultural Research* 13, 221–227.

Ames, N., McElroy, A.R. and Erfle, J. (1993) The effect of temperature on quality characteristics in timothy (*Phleum pratense* L.) genotypes. *Canadian Journal of Plant Science* 73, 1017–1026.

Archer, K.A. and Decker, A.M. (1977). Autumn-accumulated tall fescue and orchardgrass. I. Growth and quality as influenced by nitrogen and soil temperature. *Agronomy Journal* 69, 605–609.

Bacon, C.W. (1994) Fungal endophytes, other fungi, and their metabolites as extrinsic factors of grass quality. In: Fahey, G.C., Collins, M., Mertens, D.R. and Moser, L.E. (eds) *Forage Quality, Evaluation, and Utilization*. American Society of Agronomy, Madison, Wisconsin, pp. 318–366.

Balasko, J.A. (1977) Effects of N, P, and K fertilization on yield and quality of tall fescue forage in winter. *Agronomy Journal* 69, 425–428.

Bélanger, G. (1996) Morphogenetic and structural characteristics of field-grown timothy cultivars differing in maturity. *Canadian Journal of Plant Science* 76, 277–282.

Bélanger, G. and McQueen, R.E. (1996) Digestibility and cell wall concentrations of early- and late-maturing timothy (*Phleum pratense* L.) cultivars. *Canadian Journal of Plant Science* 76, 107–112.

Bidlack, J.E. and Buxton, D.R. (1992) Content and deposition rates of cellulose, hemicellulose, and lignin during regrowth of forage grasses and legumes. *Canadian Journal of Plant Science* 72, 809–818.

Bittman, S., Simpson, G.M. and Mir, Z. (1988). Leaf senescence and seasonal decline in nutritional quality of three temperate forage grasses as influenced by drought. *Crop Science* 28, 546–552.

Blaser, R.E., Hammes, R.C., Fontenot, J.P, Bryant, H.T., Polan, C.E., Wolf, D.D., McClaugherty, F.S., Kline, R.G. and Moore, J.S. (1986) Forage-animal management systems. *Virginia Agricultural Experiment Station Bulletin 86–7*, 90pp.

Boller, B.C. and Nösberger, J. (1987). Symbiotically fixed nitrogen from field grown white and red clover mixed with ryegrass at low levels of ^{15}N-fertilization. *Plant and Soil* 104, 219–226.

Bolsen, K.K., Berger, L.L., Conway, K.L. and Riley, J.G. (1976) Wheat, barley, and corn silages for growing steers and lambs. *Journal of Animal Science* 42, 185–191.

Bruckner, P.L. and Hanna, W.W. (1990) In vitro digestibility of fresh leaves and stems of small-grain species and genotypes. *Crop Science* 30, 196–202.

Bush, L., Boling, J. and Yates, S. (1971) Animal disorders. In: Buckner, R.C. and Bush, L.P. (eds) *Tall Fescue*. American Society of Agronomy, Madison, Wisconsin, pp. 247–249.

Buxton, D.R. (1990) Cell-wall components in divergent germplasms of four perennial forage grass species. *Crop Science* 30, 402–408.

Buxton, D.R. (1996) Quality-related characteristics of forages as influenced by plant environment and agronomic factors. *Animal Feed Science and Technology* 59, 37–49.

Buxton, D.R. and Fales, S.L. (1994) Plant environment and quality. In: Fahey, G.C., Collins, M., Mertens, D.R. and Moser, L.E. (eds) *Forage Quality, Evaluation, and Utilization*. American Society of Agronomy, Madison, Wisconsin, pp. 155–199.

Buxton, D.R. and Marten, G.C. (1989) Forage quality of plant parts of perennial grasses and relationship to phenology. *Crop Science* 29, 429–435.

Chamblee, D.S. and Collins, M. (1988) Relationships with other species in a mixture. In: Hanson, A.A., Barnes, D.K. and Hill, R.R. (eds) *Alfalfa and Alfalfa Improvement*. American Society of Agronomy, Madison, Wisconsin, 439–461.

Cherney, J.H. and Marten, G.C. (1982a) Small grain crop forage potential: I. Biological and chemical determinants of quality, and yield. *Crop Science* 22, 227–231.

Cherney, J.H. and Marten, G.C. (1982b) Small grain crop forage potential: II. Interrelationships among, biological, chemical, morphological, and anatomical determinants of quality. *Crop Science* 22, 240–245.

Cherney, D.J.R., Cherney, J.H. and Lucey, R.F. (1993) In vitro digestion kinetics and quality of perennial grasses as influenced by forage maturity. *Journal of Dairy Science* 76, 790–797.

Cuomo, G.J., Johnson, D.G. and Sheaffer, C.C. (1997). Smooth bromegrass management for stockpiled forage. In: *Agronomy Abstracts*. American Society of Agronomy, Madison, Wisconsin, pp. 103.

Daccord, D., Arrigo, Y., Gutzwiller, A. and Schmidt, D. (1995) Les endophytes: un facteur limitant les performances du ruminant? *Revue Suisse d'Agriculture* 27, 197–199.

Decker, A.M., Jung, G.A., Washko, J.B., Wolf, D.D. and Wright, M.J. (1967) Management and productivity of perennial grasses in the northeast. I. Reed

Canarygrass. *West Virginia University Agricultural Experiment Station Bulletin 550T,* 43pp.

Deinum, B. and Dirven, J.G.P. (1974) A model for the description of the effects of different environmental factors on the nutritive value of forages. In: Iglovikov, V.G. and Moscsyants, A.P. (eds) *Proceedings of the XIIth International Grassland Congress,* Moscow, USSR, Isd-vo MIR, Moscow, pp. 89–97.

Durand, J. and Surprenant, J. (1993) Relations entre les charactères morphologiques et la qualité chez la fléole des prés (*Phleum pratense* L*.*). *Canadian Journal of Plant Science* 73, 803–814.

Edwards, G.R., Parsons, A.J., Newman, J.A. and Wright, I.A. (1996) The spatial pattern of vegetation in cut and grazed grass/white clover pastures. *Grass and Forage Science* 51, 219–231.

Fairey, N.A. (1985) Productivity and quality of perennial and hybrid ryegrass, orchardgrass and reed canarygrass grown in the lower mainland of British Columbia. *Canadian Journal of Plant Science* 65, 117–124.

Fales, S.L. and Hoover, R.J. (1990) Manipulating seasonal growth distribution and nutritive quality of timothy and orchardgrass using the growth regulator Imazethapyr. *Canadian Journal of Plant Science* 70, 501–508.

Fales, S.L., Laidlaw, A.S., and Lambert, M.G. (1996) Cool-season grass ecosystems. In: Moser, L.E., Buxton, D.R. and Casler, M.D. (eds) *Cool-Season Forage Grasses.* American Society of Agronomy, Madison, Wisconsin, pp. 267–296.

Frank, A.B., Bitman, S. and Johnson, D.A. (1996) Water relations of cool-season grasses. In: Moser, L.E., Buxton, D.R. and Casler, M.D. (eds) *Cool-Season Forage Grasses.* American Society of Agronomy, Madison, Wisconsin, pp. 127–164.

Gallagher, R.T., Hawkes, A.D., Steyn, P.S. and Vlegaar, R. (1984) Tremorgenic neurotoxins from perennial ryegrass causing ryegrass staggers disorder of livestock: Structure elucidation of lolitrem B. *Journal of the Chemical Society, Chemical Communications* 1984(9), 614–616.

Grabber, J.H. and Allinson, D.W. (1992) Anatomical structure and digestibility of reed canarygrass cultivars and hybrid ryegrass. *Grass and Forage Science* 47, 400–404.

Hacker, J.B. and Minson, D.J. (1981) The digestibility of plant parts. *Herbage Abstracts* 51, 459–482.

Haynes, R.J. (1980) Competitive aspects of the grass–legume association. *Advances in Agronomy* 33, 227–261.

Heichel, G.H. and Henjum, K.I. (1991) Dinitrogen fixation, nitrogen transfer, and productivity of forage legume–grass communities. *Crop Science* 31, 202–208.

Hides, D.H., Lovatt, J.A. and Hayward, M.V. (1983) Influence of stage of maturity on the nutritive value of Italian ryegrass. *Grass and Forage Science* 38, 33–38.

Hockensmith, R.L., Sheaffer C.C., Marten G.C. and Halgerson, J.L. (1997). Maturation effects on forage quality of Kentucky Bluegrass. *Canadian Journal of Plant Science* 77, 75–80.

Hodgson, J. (1985) The significance of sward characteristics in management of temperate sown pastures. In: *Proceedings XV International Grassland Congress, 24–31 August, Kyoto, Japan.* Japanese Society of Grassland Science, Nishinasumo, Tochigi-ken, Japan, pp. 63–67.

Hoveland, C.S., McCann, M.A. and Bouton, J.H. (1997) Influence of endophyte, alfalfa, and grazing pressure on steer performance and plant persistence of Jesup tall fescue. *Journal of Production Agriculture* 10, 546–550.

Hume, D.E. (1991) Primary growth and quality characteristics of *Bromus willdenowii* and *Lolium multiflorum. Grass and Forage Science* 46, 313–324.

Hume, D.E. and Brock, J.L. (1997). Morphology of tall fescue (*Festuca arundinacea*) and perennial ryegrass (*Lolium perenne*) plants in pastures under sheep and cattle grazing. *Journal of Agricultural Science, Cambridge* 129, 19–31.

Jones, M.B., Collett, B. and Brown, S. (1982) Sward growth under cutting and continuous stocking management: sward canopy structure, tiller density and leaf turnover. *Grass and Forage Science* 37, 67–73.

Jones, T.A., Carlson, I.T. and Buxton, D.R. (1988) Reed canarygrass binary mixtures with alfalfa and birdsfoot trefoil in comparaison to monocultures. *Agronomy Journal* 80, 49–55.

Kanneganti, V.R. and Kaffka, S.R. (1995) Forage availability from a temperate pasture managed with intensive rotational grazing. *Grass and Forage Science* 50, 55–62.

Kunelius, H.T. (1990) Dry matter production, fibre composition and plant characteristics of cool-season grasses under two harvest systems. *Journal of Agricultural Science, Cambridge* 115, 321–326.

McWilliam, J.R. (1978) Response of pasture plants to temperature. In: Wilson, J.R. (ed.) *Plant Relations in Pastures.* Commonwealth Scientific and Industrial Research Organization, East Melbourne, Australia, pp. 17–34.

Malinowski, D., Leuchtmann, A., Schmidt, D. and Nösberger, J. (1997) Symbiosis with *Neotyphodium uncinatum* endophyte may increase the competitive ability of meadow fescue. *Agronomy Journal* 89, 833–839.

Mallarino, A.P., Wedin, W.F., Goyenola, R.S., Perdomo, C.H. and West, C.P. (1990) Legume species and proportion effects on symbiotic dinitrogen fixation in legume–grass mixtures. *Agronomy Journal* 82, 785–789.

Marten, G.C. (1985) Factors influencing feeding value and effective utilization of forages for animal production. In: *Proceedings of the XV International Grassland Congress, 24–31 August, 1985, Kyoto, Japan.* The Japanese Society of Grassland Science, Nishi-nasuno, Tochigi-ken, Japan, pp. 89–96.

Marten, G.C. and Hovin A.W. (1980) Harvest schedule, persistence, yield and quality interactions among four perennial grasses. *Agronomy Journal* 72, 378–387.

Marten, G.C., Jordan, R.M. and Hovin A.W. (1981) Improved lamb performance associated with breeding for alkaloid reduction in reed canarygrass. *Crop Science* 21, 295–298.

Matches, A.G. (1979) Management. In: Buckner, R.C. and Bush, L.P. (eds) *Tall Fescue.* American Society of Agronomy, Madison, Wisconsin, pp. 171–199.

Matches, A.G. (1992) Plant response to grazing: A review. *Journal of Production Agriculture* 5, 1–7.

Matches, A.G. and Tevis, J.B. (1973) Yield and quality of tall fescue stockpiled for winter grazing. In: *Research in Agronomy-1973.* University of Missouri Miscellaneous Publications 73-5, Columbia, Missouri, pp. 54–57.

Mayland, H.F. and Cheeke, P.R. (1995) Forage induced animal disorders. In: Barnes, R.F., Miller, D.A. and Nelson, C.J. (eds) *Forages Volume II. The Science of Grassland Agriculture,* 5th edn. Iowa State University Press, Ames, Iowa, pp.121–135.

Mays, D.A. and Washko, J.B. (1960) The feasibility of stockpiling legume–grass pasturage. *Agronomy Journal* 52, 190–192.

Míka, V. (1983) A comparaison of the nutritive values of early and late varieties of timothy. *Grass and Forage Science* 38, 67–71.

Miller, D.W. (1986) Alfalfa-perennial grass mixtures: yield, botanical composition, forage quality, and PAR competition between components. MSc thesis, University of Minnesota, St Paul, Minnesota.

Moore, K.J. and Moser, L.E. (1995) Quantifying developmental morphology of perennial grasses. *Crop Science* 35, 37–43.

Mowat, D.N., Fulkerson, R.S., Tossell, W.E and Winch, J.E. (1965) The *in vitro* digestibility and protein content of leaf and stem portions of forages. *Canadian Journal of Plant Science* 45, 321–331.

Mullahey, J.J., Waller, S.S., Moore, K.J., Moser, L.E. and Klopfenstein, T.J. (1992) In situ ruminal protein degradation of switchgrass and smooth bromegrass. *Agronomy Journal* 84, 183–188.

Ocumpaugh, W.R. and Matches, A.G. (1977) Autumn-winter yield and quality of tall fescue. *Agronomy Journal* 69, 639–643.

Robinson, D.L., Kappel, L.C. and Boling, J.A. (1989) Management practices to overcome the incidence of grass tetany. *Journal of Animal Science* 67, 3470–3484.

Sanderson, M.A. and Wedin, W.F. (1989) Phenological stage and herbage quality relationships in temperate grasses and legumes. *Agronomy Journal* 81, 864–869.

Seguin, P. (1997) Pasture renovation: introduction of legumes in a grass dominated pasture with physical suppression of the resident vegetation. MSc thesis, McGill University, Montréal, Québec, Canada.

Seguin, P. (1998) Review of factors determining legumes sod-seeding outcome during pasture renovation in North America. *Biotechnologie, Agronomie, Société et Environnement* 2, 120–127.

Sheaffer, C.C. (1989) Legume establishment and harvest management in the USA. In Marten, G.C., Matches, A.G., Barnes, R.F., Brougham, R.W., Clements, R.J. and Sheath, G.W. (eds) *Persistence of Forage Legumes*. American Society of Agronomy, Madison, Wisconsin, pp. 277–289.

Sheaffer, C.C. and Marten, G.C. (1986) Effect of mefluidide on cool-season perennial grass forage yield and quality. *Agronomy Journal* 78, 75–79.

Sheaffer, C.C. and Marten, G.C. (1995) Reed canarygrass. In: *Forages Volume I: An Introduction to Grassland Agriculture*, 5th edn. Iowa State University Press, Ames, Iowa, pp. 335–343.

Sheaffer, C.C., Hovin, A.W. and Rabas, D.L. (1981) Yield and composition of orchardgrass, tall fescue, and reed canarygrass mixtures. *Agronomy Journal* 73, 101–106.

Sheaffer, C.C., Marten, G.C., Rabas, D.L., Martin, N.P., and Miller, D.W. (1990a) Reed canarygrass. *Minnesota Agricultural Experiment Station, Station Bulletin 595-1990*, St Paul, Minnesota, 6pp.

Sheaffer, C.C., Wyse, D.L., Marten, G.C. and Westra, P.H. (1990b) The potential of quackgrass for forage production. *Journal of Production Agriculture* 3, 256–259.

Sheaffer, C.C., Peterson, P.R., Hall, M.H. and Stordahl, J.B. (1992). Drought effects on yield and quality of perennial grasses in the North Central United States. *Journal of Production Agriculture* 5, 556–561.

Sleper, D.A. and C.P. West (1996) Tall fescue. In: Moser, L.E., Buxton, D.R. and Casler, M.D. (eds) *Cool-Season Forage Grasses*. American Society of Agronomy, Madison, Wisconsin, pp. 471–502.

Smouter, H., Simpson, R.J. and Pearce, G.R. (1996) Water-soluble carbohydrates and in vitro digestibility of annual ryegrass (*Lolium rigidum* Gaudin) sown at varying densities. *Australian Journal of Agricultural Research* 46, 611–625.

Strahan, S.R., Hemken, R.W., Jackson, J.A., Jr, Buckner, R.C., Bush, L.P. and Siegel, M.R. (1987) Performance of lactating dairy cows fed tall fescue forage. *Journal of Dairy Science* 70, 1228–1234.

Surprenant, J., Drapeau, R. and Fernet, C. (1993) Cultivar-by-management interaction effects on timothy yield and quality evaluation. *Canadian Journal of Plant Science* 73, 445–460.

Ta, T.C. and Faris, M.A. (1987a) Effects of alfalfa proportions and clipping frequencies on timothy–alfalfa mixtures. I. Competition and yield advantages. *Agronomy Journal* 79, 817–820.

Ta, T.C. and Faris, M.A. (1987b) Effects of alfalfa proportions and clipping frequencies on timothy–alfalfa mixtures. II. Nitrogen fixation and transfer. *Agronomy Journal* 79, 820–824.

Taylor, T.H. and Templeton, W.C., Jr. (1976) Stockpiling Kentucky bluegrass and tall fescue forage for winter pasturage. *Agronomy Journal* 68, 235–239.

Terry, R.A. and Tilley, J.M.A. (1964) The digestibility of the leaves and stems of perennial ryegrass, cocksfoot, timothy, tall fescue, lucerne and sainfoin, as measured by an in vitro procedure. *Journal of the British Grassland Society* 19, 363–372.

Thorvaldsson, G. (1992) The effects of temperature on digestibility of timothy (*Phleum pratense* L.), tested in growth chambers. *Grass and Forage Science* 47, 306–308.

Tilman, D., Wedin, D. and Knops, J. (1996) Productivity and sustainability influenced by biodiversity in grassland ecosystems. *Nature* 379, 718–720.

Turner, K.E., Belesky, D.P., Fedders, J.M. and Rayburn, E.B. (1996) Canopy management influences on cool-season grass quality and simulated livestock performance. *Agronomy Journal* 88, 199–205.

Undersander, D.J. and Paine, L.K. (1997) Forage availability and nutritional quality of rotationally and continuously grazed pastures in the Upper Midwest. In: *Proceedings of the American Forage and Grassland Council, 13–15 April, Fort Worth, Texas*. American Forage and Grassland Council, Georgetown, Texas, 6, 136.

Van Keuren, R.W. (1972) All-season forage systems for beef cow herds In: *27th Soil Conservation Society of America Meeting, 6–9 August, 1972, Portland, Oregon*. Soil Conservation Society of America, Ankeny, Iowa, pp. 39–44.

Vogel, K.P. and Sleper, D.A. (1994) Alteration of plants via genetics and plant breeding. In: Fahey, G.C., Collins, M., Mertens, D.R. and Moser, L.E. (eds) *Forage Quality, Evaluation, and Utilization*. American Society of Agronomy, Madison, Wisconsin, pp. 891–921.

Wedin, W.F., Carlson, I.T. and Vetter, R.L. (1966). Studies on nutritive value of fall-saved forage using rumination and chemical analysis. In: Normand, N.J.T. (ed) *Xth International Grassland Congress Proceedings, 7–16 July, 1966, University of Helsinki, Finland*. Valtioneuvoston Kirjepaino, Helsinki, pp. 424–428.

Willms, W.D. and Beauchemin, K.A. (1991) Cutting frequency and cutting height effects on forage quality of rough fescue and Parry oat grass. *Canadian Journal of Animal Science* 71, 87–96.

Wilman, D., Acuna, P.G.H. and Michaud, P.J. (1994) Concentrations of N, P, K, Ca, Mg and Na in perennial ryegrass and white clover leaves of different ages. *Grass and Forage Science* 49, 422–428.

Wilman, D., Gao, Y. and Altimimi, M.A.K. (1996a) Differences between related grasses, times of year and plant parts in digestibility and chemical composition. *Journal of Agricultural Science, Cambridge* 127, 311–318.

Wilman, D., Mtengeti, E.J. and Moseley, G. (1996b) Physical structure of twelve forage species in relation to rate of intake by sheep. *Journal of Agricultural Science, Cambridge* 126, 277–285.

Wilson, J.R. (1981) Effects of water stress on herbage quality. In: Smith, J.A. and Hayes, V.W. (eds) *Proceedings of the XIVth International Grassland Congress, June 15–24, 1981, Lexington, KY.* Westview Press, Boulder, Colorado, pp. 470–472.

Wilson, J.R. (1994) Cell wall characteristics in relation to forage digestion by ruminants. *Journal of Agricultural Science, Cambridge* 122, 173–182.

Wittenberg, K.M., Duynisveld, G.W. and Tosi, H.R. (1992) Comparison of alkaloid content and nutritive value for tryptamine- and carboline-free cultivars of reed canarygrass (*Phalaris arundinaceae* L). *Canadian Journal of Animal Science* 72, 903–909.

Wright, M.J. and Davison, K.L. (1964) Nitrate accumulation in crops and nitrate poisoning in animals. *Advances in Agronomy* 16, 197–247.

Yarrow, N.H. and Penning, P.D. (1994) Managing grass/clover swards to produce differing clover proportions. *Grass and Forage Science* 49, 496–501.

Tropical and Subtropical Grass Management and Quality

5

R.T. Cowan[1] and K.F. Lowe[2]

[1]Australian Tropical Dairy Institute, University of Queensland, Gatton, Queensland 4345, Australia
[2]Australian Tropical Dairy Institute, Department of Primary Industries, M. S. 825, Peak Crossing, Queensland 4306, Australia

Introduction

Approximately 40% of the world milk production is in tropical and subtropical regions, and production is increasing at 3.9% annually compared with a 1.4% increase in temperate regions (Sere and Steinfeld, 1996). Much of this increase is occurring as a result of intensification in mixed farming systems, using cereal stubbles, cut and carry grasses and various by-products. Around 15% of production in the region comes from specialized grazing systems. The forage base in rations for milking ruminants is usually a grass product, whether cut or grazed fresh grass, or a by-product of a grass crop such as cereal or sugar cane (Preston and Leng, 1987).

This chapter is primarily concerned with the use of cut or grazed fresh tropical grass for dairy production. Though extremely important, the use of products such as cereal stubble and sugar cane for cattle feed has been the topic of a number of books (Sundstol and Owen, 1984; APO, 1990), and the information will be repeated here only to expand a result applying to all grass products in tropical regions. We concentrate on those management factors that impact on the quality of fresh grasses.

Adaptation

The tropical grasses have evolved mechanisms which enable them to survive, and at times thrive, in what are generally harsh environments with variable climates (Cameron, 1983). There are some examples of successful breeding of new tropical grasses (Burton *et al.*, 1967), but generally animal production is based on grasses which have evolved in that niche or been transferred there from a similar niche ('t Mannetje, 1982; Chapter 3). Consequently, the grasses often have similar characteristics which enhance their adaptation to tropical and subtropical latitudes (Snaydon, 1991), such as high fibre levels, drought tolerance and rapid growth and flowering responses to high soil moisture and nitrogen levels (Skerman and Riveros, 1990). These features are often associated with the relatively low nutritive value of tropical grasses (Minson, 1971). Attempts to breed for higher nutritive value have often resulted in a loss of one or more of these characteristics and limited commercial acceptance of the grass (Clements *et al.*, 1987; Lowe *et al.*, 1991a). Even with annual grasses, such as some sorghums (*Sorghum* spp.), and millets (*Pennisetum* and *Echinochloa* spp.), intensive plant breeding has resulted in plants of higher growth potential and in some cases higher leaf to stem ratios, or delayed flowering, but retaining characteristics such as high fibre levels, drought tolerance and rapid response to soil moisture and nitrogen (Stuart, 1993).

Growth rate of tropical grasses increases rapidly with mean environmental temperatures above 18°C (Bryan and Sharpe, 1965), with maximum growth in temperature bands from 28 to 33°C (Sweeney and Hopkinson, 1975). This contrasts with temperate grasses where maximum growth occurs at approximately 20°C (Whyte *et al.*, 1959; Belesky and Fedders, 1995). The efficiency of tropical grasses in extracting water and nitrogen from the soil is high (Lowe, 1976a; Cowan *et al.*, 1995a).

Tropical grasses are very effective competitors once established, but in the establishment phase weeds often have superior vigour. This is true for pastures established through seed or vegetative propagation, and is in part due to the relatively slow early growth of the root mass of tropical grasses (Silcock, 1980; Ayala *et al.*, 1987) limiting the ability to compete for water and nutrients (Hawton, 1980). Grass seedlings are responsive to phosphorus applied at planting on many tropical and subtropical soils (Lowe *et al.*, 1981; Cook and Stillman, 1981), although nitrogen applications at this time can be counter-productive, encouraging weeds more than grass growth (Cook and Stillman, 1981).

The grasses are utilized in a wide range of production systems. The most obvious difference in these systems is the choice of cut and carry or grazing as the method of harvesting the grass. Napier grass (*Pennisetum purpureum* Schumach.) and some of the *Panicum* cultivars are pre-eminent in cut and carry systems, as they have a rapid, erect growth habit and are perennial (Omaliko, 1980; Singh *et al.*, 1995), while sorghums (*Sorghum* spp.) and millets (*Pennisetum* and *Echinochloa* spp.) are predominately used as annuals (Fribourg, 1985). There is a wide range of grasses suited to grazing systems in the tropics and subtropics (Skerman and Riveros, 1990). Many of these grasses have stolons and rhizomes as an added mode of growth, e.g. signal grass (*Brachiaria decumbens* Stapf.), kikuyu (*P. clandestinum* Chiov.) and Rhodes grass (*Chloris gayana* Kunth). In seeding grasses, the ability to produce adequate quantities of seed is an important adaptation character for maintaining or regenerating stands, although during reproduction leaf production is reduced and overall levels of quality are lower (Whyte *et al.*, 1959).

Plant Structure

Grasses grow through a process of bud development in the apical growing point. The habit of growth varies greatly between species, being particularly influenced by the extent of internode elongation and the direction of the elongated stems (Whyte *et al.*, 1959). Tropical species generally have greater stem elongation during normal growth than do temperate species. In many tropical grasses flowering occurs during much of the growing season (Tothill and Hacker, 1983). This contrasts to the situation with improved temperate grasses, such as ryegrass (*Lolium* spp.), where stem elongation and flowering do not occur until late in the growing season (Anslow, 1966). Figure 5.1 gives the 'typical' structure of a tropical grass

Fig. 5.1. Morphology of an actively growing tropical grass, showing stem elongation, branching from stems, runners, and distribution of leaf along stems (for *Chloris gayana*, from Tothill and Hacker, 1983).

plant, showing the growing tips at the end of elongated stems and nodes for the initiation and expansion of leaf.

Grass Quality – Leafiness

It has generally become accepted that there is a substantial difference in the quality of grass leaf and stem (Cowan *et al.*, 1993a). Tropical grass leaf, although containing a relatively higher proportion of bundle sheath and vascular tissue and a lower proportion of thin walled mesophyll cells compared with temperate grass leaf (Wilson *et al.*, 1989), has substantially less bundle sheath and vascular tissue than stem (Wilson and Ng, 1975). Some differences in indices of nutritive value for leaf and stem are shown in Table 5.1. Leaf is consistently higher in those parameters associated with higher

Table 5.1. Some differences in quality parameters between leaf and stem of tropical grasses.

Parameter	Leaf	Stem	Selected reference
Chemical (% DM)			
crude protein	9.4–16.3	3.0–12.7	Grainger *et al.* (1996)
in vitro dry matter digestibility	59–69	43–58	Grainger *et al.* (1996)
acid detergent fibre (ADF)	38	44	Minson *et al.* (1993)
neutral detergent fibre (NDF)	63	72	Cowan *et al.* (1993a)
phosphorus	0.21	0.22	Cowan *et al.* (1995a)
potassium	1.63	2.08	Cowan *et al.* (1995a)
Physical			
density (g cm^{-2})	0.080	0.211	Laredo and Minson (1973)
surface area (cm^2 g^{-1})	132	47	Laredo and Minson (1973)
Rumen			
dry matter content (kg)	7.5	5.1	McLeod *et al.* (1990)
fluid passage (l h^{-1})	8.6	4.4	McLeod *et al.* (1990)
retention time (h)	24	32	McLeod *et al.* (1990)
rumen load (g kg $W^{-0.75}$)	383	367	Laredo and Minson (1973)
Energy expenditure			
grinding (J g^{-1})	201	337	Laredo and Minson (1973)
chewing (No. g^{-1} DMI)	5.5	10.1	Chacon and Stobbs (1976)
Selection			
selection index[1]	2–6	0.3–1.0	Cowan *et al.* (1986)
Intake			
fed separately (g kg $W^{-0.75}$)	72	53	Minson (1990b)
grazed – long term (kg $cow^{-1}day^{-1}$)	9.8	1.3	Ehrlich *et al.* (1996)

[1](% component in diet)(% component in the pasture) $^{-1}$.

nutritive value, such as crude protein and digestibility, undergoes more rapid and extensive degradation in the rumen, requires less energy for chewing, promotes a higher rumen load of dry matter and is eaten in greater quantities than stem. When offered only leaf or stem, cattle eat in the order of 20% more leaf than stem. When given a choice, such as in grazing or cut and carry of long herbage, animals will spend time selecting for leaf and reject a high proportion of stem (Stobbs, 1973a; Cowan *et al.*, 1986). Annual grass forage crops (such as *Sorghum* and *Pennisetum* spp.) had a similar nutritive value and gave similar levels of milk production to tropical grass pastures (Stobbs, 1975).

In both short (Stobbs, 1977) and longer term studies (Cowan and O'Grady, 1976; Cowan and Stobbs, 1976), there was a curvilinear relationship between total pasture yield on offer and milk yield, with the maximum pasture intake occurring at total pasture yields of 35 kg DM cow^{-1} day^{-1} or 2500 kg DM ha^{-1}. These pasture yields are considerably above those associated with maximum levels of milk production in temperate grass swards (Greenhalgh *et al.*, 1966) and largely reflect the relatively high proportion of stem in these pastures (Stobbs, 1977). Leaf yield has consistently been correlated with milk yield (Davison *et al.*, 1985b, 1993; Cowan *et al.*, 1986), although there is often a close relationship between leaf and stem yield (Cowan *et al.*, 1995c). In some situations where stem yields are very high (Davison *et al.*, 1993) or when tropical grass is rotationally grazed without removal of stem build up (Ehrlich *et al.*, 1996), there is a negative association between amount of stem on offer and milk yield.

The optimum yield of leaf in a pasture will influence quality and accessibility of leaf to cows. The profile of many tropical grass pastures, particularly in clump-forming types, is relatively tall, with a small decrease in proportion of leaf in the dry matter with height. However the top 20% of the profile often has a very low leaf to stem ratio as tropical grasses frequently send up flowering stems during active growth (Stobbs, 1973b). Leaf density declines progressively from 60–80 kg ha^{-1} cm^{-1} in the lower profiles to < 10 kg ha^{-1} cm^{-1} in the top 20 cm. Leaf density is low in the 0–10 cm profile and this reduces regrowth after severe defoliation which occurs in cut and carry systems. Cows are adept at selecting leaf from within the profile under a range of pasture conditions (Cowan *et al.*, 1986; Lowe *et al.*, 1991a; Ehrlich *et al.*, 1996), though there is empirical evidence that a large build up of stem can restrict access to leaf (Lowe *et al.*, 1991b; Ehrlich *et al.*, 1996). Stobbs (1973a) showed cows need to harvest approximately 0.3 g organic matter each bite to avoid a serious restriction to dry matter intake, and intakes from mature or very stemy pastures are often below this. There are also limits to the time cows will spend grazing each day, with a maximum time in the order of 12 h (Goldson, 1963; Cowan, 1975) or 36,000 bites (Stobbs, 1973a). Effort spent grazing is affected by pasture yield (Cowan, 1975; Cowan and O'Grady, 1976).

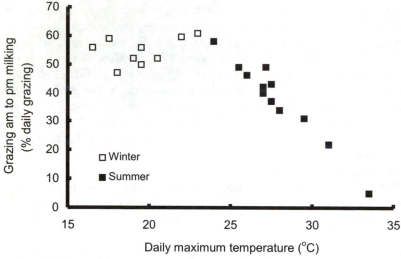

Fig. 5.2. Relation of the proportion of grazing done between morning and afternoon milking and the daily maximum temperature (from Cowan, 1995).

On days when maximum temperature exceeded 32°C very little effective grazing was done by Holstein Friesian cows between morning and evening milking (Fig. 5.2) and it is imperative that cows have access to high quality pasture during the night. Rees *et al.* (1972) showed that milk fat production increased with proportion of the farm grazed at night, reaching a maximum when 85% of the farm area was used for night grazing.

Grass Quality – Nutritive Value

Traditionally, crude protein (N × 6.25) has been the primary measure of quality in tropical grasses. Many tropical grasses have only modest levels of crude protein when green, and this falls to very low values (<1% N) following maturity (Hardison, 1966; Norton, 1982). Crude protein levels have consistently been correlated with measures of energy supply from grasses, such as dry matter digestibility and fibre content (Minson, 1971). Four-week-old regrowth of adequately fertilized tropical grasses contains about 20% crude protein, 60% neutral detergent fibre (NDF), 28% acid detergent fibre (ADF), 10% water soluble carbohydrates (Walker *et al.*, 1996) and 65% digestible organic matter in the dry matter (Reeves *et al.*, 1996). These values change rapidly in the 2 weeks after leaf initiation or the first 4 weeks of seedling growth (Fig. 5.3) but, after this, the changes in quality are much slower.

In all but the most intensive of rotational grazing systems, fresh new leaf forms only a small proportion of total leaf on offer. The mean chemical composition of grass leaf and the diet selected by grazing cows do not change substantially even when grasses are grazed at a range of regrowth intervals (Table 5.2).

Mature, or standover, tropical grass forage (Ostrowski, 1969) and annual grasses (Bishnoi *et al.*, 1993) usually contain very low levels of crude protein and high levels of fibre. However, this material is a valuable resource in many tropical and subtropical feeding systems where seasonal variation in pasture growth rate is extremely large (Sere and Steinfeld,

Table 5.2. Effect of regrowth interval of a Rhodes grass (*Chloris gayana*) pasture on nutritive value (% DM) of grass leaf and the diet of grazing cows. Pastures were nitrogen fertilized, irrigated, grazed at 3.5 Holstein Friesian cows ha^{-1} and mulched to 10 cm stubble height after each grazing (Ehrlich *et al.*, 1996).

| | Regrowth interval (weeks) | | | | LSD |
Parameter	2	4	6	Open[1]	($P < 0.05$)
Leaf					
pasture	44.0	48.0	46.0	30.0	5.0
diet	76.0	82.0	83.0	72.0	8.0
Crude protein					
pasture leaf	15.6	15.2	15.5	12.5	0.7
diet	14.3	13.8	14.9	10.7	0.9
NDF	64.1	63.7	63.1	65.3	2.5
pasture leaf	65.6	64.9	65.3	68.0	2.2
diet					
	56.4	55.9	55.9	53.3	1.2
IVDMD	55.6	55.5	55.3	51.5	2.1
pasture leaf					
diet					

[1]A 2-week rotation using a lower stocking rate (1.7 cows ha^{-1}) and no mulching.

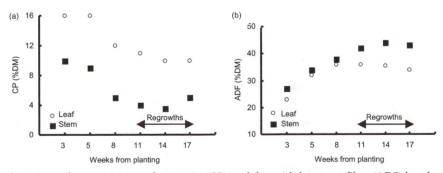

Fig. 5.3. Change in (a) crude protein (CP) and (b) acid detergent fibre (ADF) levels in leaf and stem of forage sorghum (Superdan) from planting to the fourth grazing (from Cowan *et al.*, 1993a).

Table 5.3. Calculated effect of differences in mean summer temperatures on digestibility and potential milk production from kikuyu grass (*Pennisetum clandestinum*) across its growth range in eastern Australia. Changes in digestibility calculated from the equation of Minson and McLeod (1970).

Centre	Latitude (°S)	Mean summer temperature (°C)	Change in digestibility (%DM)	Change in milk output[1] (kg cow^{-1} day^{-1})
Atherton	17	24.8	0	0
Brisbane	28	21.6	+3.0	+2.5
Berry	37	17.1	+7.5	+6.0

[1]Using estimated intake values in SCA (1990) and milk conversion rates in NRC (1989).

1996), yet supply of milk is demanded throughout the year. Mature grass is used in conjunction with supplements (Preston, 1995) or irrigated green pasture (Moss and Lowe, 1993) to maintain production during dry or cold periods.

Though some differences have been demonstrated between tropical grass species in quality parameters (Wilson *et al.*, 1983; Lowe *et al.*, 1991a), these do not appear to substantially influence milk production on an annual or lactation basis (Stobbs, 1975; Robbins and Bushell, 1986; Lowe *et al.*, 1991b). Across tropical grass species, temperature exerts a more substantial effect on quality, increasing the fibrous structure of tropical grasses and reducing digestibility (Minson and McLeod, 1970; Wilson and Ford, 1973; Dirven and Dienum, 1977). Temperature may lead to a 5 percentage unit difference in dry matter digestibility of a tropical grass across various altitudes and latitudes where it is grown (Table 5.3). In practice, there is little that can be done to moderate these effects compared with the influences of fertilization, stocking rate and management practices. These are dealt with in more detail in the following section.

Most of the tropical grasses have a low content of water soluble carbohydrates, in the range 3–10% DM (Noble and Lowe, 1974; Reeves *et al.*, 1996; Walker *et al.*, 1996). An exception is a high sugar variety of sorghum (*Sorghum bicolor* (L.) Moench s. lat.) which has a water soluble carbohydrate content in the top growth at flowering of more than 20% in the DM (Stuart, 1993). In perennial grasses, levels are normally at a maximum in new growth, then decline with age (Walker *et al.*, 1996). There is a diurnal variation, with higher values (6–10% DM) present in the afternoon than in the morning (2–5% DM) (Reeves *et al.*, 1996).

Tropical grasses are relatively low, compared with temperate grasses, in some of the minerals needed by milking cows, particularly phosphorus, sodium and calcium (McDowell *et al.*, 1977; Norton, 1982; Minson, 1990a). Application of nitrogen fertilizer consistently depresses the phosphorus content of grass (Davison *et al.*, 1985a; Cowan *et al.*, 1995a) but may increase the calcium content (Cowan *et al.*, 1995a). Reason and Chaseling (1993) showed these effects are more pronounced in some grasses but not in others. Certain grasses, for example kikuyu, have very low calcium and sodium content (Reeves *et al.*, 1996; Davison *et al.*, 1997), while others accumulate minerals such as sodium (Little, 1982; Minson, 1990b). Some tropical grasses, particularly *Setaria* spp., have a high oxalic acid content which can induce an acute calcium deficiency (Jones *et al.*, 1970). For cows grazing well managed tropical grasses, supplements of calcium, phosphorus and sodium would normally be required (Reason and Chaseling, 1993; Abate, 1994).

Use of nitrogen fertilizer on tropical grasses can result in increased frost damage to pasture in autumn or winter in those areas where frost occurs. Cowan *et al.* (1995a) demonstrated a linear increase in frost damage to Rhodes grass with level of nitrogen fertilizer applied over the preceding 5 years. The nutritional value of tropical grass deteriorates rapidly after frosting (Jones, 1967; Lowe, 1976b; Cowan *et al.*, 1995a). In situations where frost is expected in autumn, frost tolerant tropical grasses should be sown (Jones, 1967; Lowe, 1976b).

Pasture Productivity

Measured dry matter yields of tropical grasses under realistic farming practices have been well below the potential yields as reported by Cooper (1970). Nitrogen fertilized grasses have consistently yielded 11–24 t DM ha^{-1} annually; yield is strongly dependent on nitrogen fertilizer level and soil moisture status (Buchanan and Cowan, 1990). The incremental response from a wide range of tropical grasses to level of applied nitrogen averaged 23 ($\pm$ 0.2) kg DM kg^{-1} N, a response consistent over applied nitrogen levels from 0 to 400 kg N ha^{-1} year^{-1} (Buchanan and Cowan, 1990; Teitzel *et al.*, 1991). Grass response to nitrogen fertilizer was greater under grazing than cutting, and the effect increased with time (Stephens, 1967). Amount of rainfall received has been found to interact with level of applied nitrogen. Gilbert and Clarkson (1993) reviewed 31 studies and demonstrated a maximum response at 300 kg N ha^{-1} year^{-1} in a 500 mm summer rainfall zone, compared with 600 kg N ha^{-1} year^{-1} in a 1000 mm summer rainfall zone. These authors also demonstrated that in dry years the response averaged 23 kg DM kg^{-1} N, and in wet years the response increased to 31 kg DM kg^{-1} N. The difference between wet and dry years was 0.2 t DM ha^{-1} where no nitrogen was applied and 2.8 t DM ha^{-1} where 300 kg N ha^{-1} year^{-1} was applied.

Where tropical grasses are cut infrequently, the response to applied nitrogen can be high, in the order of 100 kg DM kg^{-1} N (Henzell, 1963; Grof, 1969). However, there is a penalty in terms of the quality of forage harvested (Fig. 5.4), and the optimum combination of nitrogen fertilizer

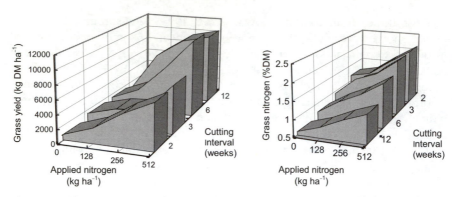

Fig. 5.4. The interaction of level of applied nitrogen and cutting interval on yield and nitrogen content of a tropical grass under a cut and removal system of feeding (from Blunt and Haydock, 1978). Note the reversal of the axis for cutting interval in the graph describing nitrogen content.

level and cutting interval will depend on the relative needs for quantity and quality. Many experiments have measured this interaction in terms of DM yield (Oji and Ughenighe, 1992; Cuomo *et al.*, 1996a, b), and it is generally considered that a 6-week interval between harvests is the maximum time compatible with high quality in cut tropical grasses (Singh, 1993). However there is a need to verify this assumption with animal production experiments that take into account losses of forage through stem rejection and differences in nutritional value of grass eaten.

Under rainfed conditions where rainfall is limiting or unreliable, the milk production responses to nitrogen fertilizer appear to be reduced at high stocking rates. Cowan *et al.* (1995c) measured a reduction in pasture yield response to applied nitrogen at a high stocking rate on Rhodes grass (*C. gayana*) pastures, and this effect was related to initial pasture yield. Where pasture yield was less than 1000 kg DM ha^{-1}, nitrogen had very little effect on dry matter yield. Murtagh *et al.* (1980) reported a diminution in carrying capacity of kikuyu grass pasture when both level of applied nitrogen and stocking rate were high. They attributed this to greater drought susceptibility in these pastures, although other research suggests that this may be due to excess volatilization of nitrogen as ammonia in overstocked pastures (Catchpoole *et al.*, 1983; Cowan *et al.*, 1995a). The maintenance of a substantial ground cover appears to enhance the response of grass to applied nitrogen when rainfall is limiting.

Where irrigation is applied or rainfall is high and consistent, these effects do not appear to be evident. Growth was maintained at high levels of nitrogen and stocking rates up to 7.9 cows ha^{-1} on irrigated pangola grass in a tropical lowland environment (Chopping *et al.*, 1976). In relatively high rainfall environments, the response of pasture to 400–600 kg N ha^{-1} was

Table 5.4. The effects of daily strip grazing on a 28-day rotation and mulching to 10 cm stubble following grazing on aspects of leaf quality of a Rhodes grass pasture (*Chloris gayana*). Stocking rate was 3.3 Holstein Friesian cows ha^{-1} (R.T. Cowan, 1997, unpublished).

| | Grazing rotation | | | | |
| | Open[1] | | Rotational | | LSD |
Mulching post grazing	−	+	−	+	(*P* < 0.05)
Leaf					
pasture content (%DM)	44^a	57^c	47^b	59^d	1.8
crude protein (%DM)	14.7^a	16.9^b	13.8^a	17.4^b	1.9
NDF (%DM)	65.6^c	63.9^b	66.3^c	61.8^a	1.5
Milk yield (kg cow^{-1} day^{-1})	15.0^b	14.9^b	13.6^a	16.7^c	1.2

[1]Two paddock system, one week grazing, one week rest.
[a-d] Means with different letters are statistically different (*P* < 0.05).

consistent at stocking rates from 2 to 3.5 cows ha^{-1} (Vicente-Chandler and Caro-Costas, 1974; Davison *et al.*, 1985a).

There have been a number of studies of grazing management on tropical grass pastures (Cowan *et al.*, 1993a), but detailed information on growth rates of grass under grazing is not available. Forage is strongly affected by management. For example, forage availability of grass stem was substantially reduced when slashing or mulching was practised (Davison *et al.*, 1981; Ehrlich *et al.*, 1996), while quality parameters of grass leaf were improved under the combination of intensive rotational grazing and mulching (Davison *et al.*, 1981; Cowan, R.T., Queensland, 1994, unpublished). The combination of intensive rotational grazing and removal of standing residue after grazing has increased leaf digestibility by 4 percentage units (Davison *et al.*, 1981), increased leaf percentage in the pasture and reduced fibre levels in leaf (Table 5.4). Management of grass–legume combinations can also interact and affect forage availability. Ottosen *et al.* (1975) found that forage availability of legumes declined in mixed tropical grass–legume pastures. On the other hand, Lowe *et al.* (1978) found that the presentation yield of Tinaroo glycine (*Neonotonia wightii* (Arn.) Lackey) was substantially increased under rotational, compared with continuous, grazing at a relatively high stocking rate (4.2 steers ha^{-1}).

Milk Production

Associated with the rapid development of tropical pasture technology between the 1960s and the 1980s, research attempted to define the potential milk production per cow and the optimum stocking rates these pastures could support. Hardison (1966) concluded that tropical pasture grasses contained sufficient digestible energy for 5 kg milk daily, though later experiments have shown levels of 9–13 kg are feasible (Winks *et al.*, 1970; Stobbs, 1971; Swain, 1971; Cowan *et al.*, 1974). Kikuyu grown at lower latitudes may support slightly higher levels of production (Reeves *et al.*, 1996). About 1.5 kg additional milk is produced on grass–legume mixed pastures, compared with pure grass swards (Davison and Cowan, 1978).

Quality limitations of tropical grasses are reflected in the shape of lactation curves. Cows calving onto tropical grasses in northern Australia show an immediate fall in milk production in the month following calving, followed by a linear decline through the rest of lactation (Papajcsik and Bodero, 1988). By contrast cows calving during winter, when irrigated temperate pastures are actively growing (Moss and Lowe, 1993), show the classical peak and fall (Papajcsik and Bodero, 1988). This difference is associated with differences in the ability of cows to consume metabolizable energy (ME) on the two pasture types, estimated by Grainger *et al.* (1996) to be 110 and 200 MJ ME daily for cows in mid summer and mid winter, respectively.

Stocking Rate

Payne (1963) calculated it would be possible on 'good humid tropical pastures' to maintain five dairy cows per hectare throughout the year, producing 13,600 kg milk ha^{-1} year^{-1}. In most instances this has proved to be well above achievements to date, largely due to instability of pastures (Cowan *et al.*, 1975; Cowan *et al.*, 1995c) or extreme stress on animals during the dry season at high stocking rates (Cowan and Stobbs, 1976; Davison *et al.*, 1985b). Growth rate of twining tropical legumes, such as *Neotonia wightii*, decreased linearly with stocking rate, and at stocking rates above 1.6 cows ha^{-1}, they made very little contribution to the animals' diet or to the nitrogen cycle of pastures (Cowan and Stobbs, 1976). Consequently, the development of tropical pasture systems has been based on nitrogen fertilized grasses (Table 5.5).

In a tropical upland area with 1250 mm annual rainfall, Davison *et al.* (1985b) considered a stocking rate of 3.0 cows ha^{-1} to be sustainable at an applied nitrogen rate of 400 kg N ha^{-1} year^{-1}, with an average output over 3 years of 8550 kg milk ha^{-1} year^{-1}. At higher stocking rates, there was rapid weight loss by pregnant cows during the dry season and weed invasion of

Table 5.5. Examples of levels of milk production achieved using feeding systems based on tropical grasses.

Method of grass use	Nitrogen levels (kg ha⁻¹ year⁻¹)	Stocking rate (cows ha⁻¹)	Milk production		References
			(kg cow⁻¹ year⁻¹)	(kg ha⁻¹ year⁻¹)	
Grass–legume pastures, grazed	–	1.1–1.6	2,500–5,000	2,700–5,600	Cowan et al. (1974), Cowan et al. (1975), Cowan and Stobbs (1976)
Dryland grass pastures, grazed	0–400	2.0–4.9	2,300–5,300	4,000–10,400	Davison et al. (1985b), Cowan et al. (1995a), Cowan et al. (1995c), Davison et al. (1989), Colman and Kaiser (1974), Ruiz et al. (1995)
Irrigated grass pastures					
Grazed	300–670	3.5–7.9	2,000–3,800	12,000–20,000	Chopping et al. (1976), Lowe et al. (1991a, b), Anindo and Potter (1986), Muinga et al. (1992), Randel (1993), Mogaka (1995), Odhuba (1989)
Cut and carry	–	–	1,500–4,500[1]	–	

[1]Extrapolated to a 300 day lactation.

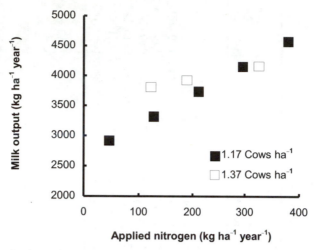

Fig. 5.5. A high stocking rate reduced the efficiency of use of nitrogen fertilizer applied to a tropical grass pasture (from Cowan *et al.*, 1995c).

the pasture. In a lower rainfall (800 mm), subtropical environment, Cowan *et al.* (1995c) considered a stocking rate of 2 cows ha^{-1} optimum for a Rhodes grass pasture receiving 300 kg N ha^{-1} year^{-1}. In this experiment a higher stocking rate caused stress in cows during the dry season, and also reduced the response to applied nitrogen fertilizer (Fig. 5.5).

The shape of the response curve of milk production to level of applied nitrogen is therefore altered by stocking rate and, probably, other factors such as rainfall, as is the case with the grass dry matter response (Gilbert and Clarkson, 1993). There have been very few studies which included more than two levels of applied nitrogen. Reason and Chaseling (1993) studied the effects of three levels of applied nitrogen (300, 400 or 500 kg N ha^{-1} annually) on a portion of each of 12 commercial farms; the incremental response from 400 to 500 N was greater than that from 300 to 400 N. Rees *et al.* (1972) and Kerr and Chaseling (1992) measured linear responses to level of applied nitrogen in analyses of survey data from commercial farms, where levels ranged from 0 to 150 kg N ha^{-1} year^{-1} (Fig. 5.6).

Where grass was irrigated, Chopping *et al.* (1976) reported milk yields up to 19,000 kg ha^{-1} year^{-1} at a stocking rate of 7.9 Holstein Friesian cows ha^{-1} on pangola grass pasture receiving 672 kg N ha^{-1} year^{-1}. Lowe *et al.* (1991b) demonstrated milk yields of 14,000 kg ha^{-1} over summer and autumn in a subtropical environment on Rhodes grass and paspalum (*Paspalum dilatatum* Poir.) pastures receiving 400 kg N ha^{-1} year^{-1} and stocked at 7.5 Holstein Friesian cows ha^{-1}.

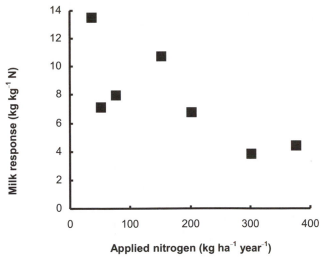

Fig. 5.6. Change in the incremental milk response to nitrogen applied to a tropical grass pasture in relation to the mean level of applied nitrogen (data from seven studies reviewed by Cowan *et al.*, 1993b).

Efficiency of Pasture Use

The literature on tropical pastures contains many practical experiments where animal production was measured, but almost no studies have looked in detail at pasture utilization. One approximation of the level of pasture utilization can be obtained by comparing measurements of pasture dry matter yield with leaf intakes calculated from estimated nutrient intakes and ME content of grass leaf (Table 5.6). These estimates suggest that 30–50% of pasture is consumed by cows, but if the premise is accepted that only leaf is eaten by the cow then much or all of the leaf is eaten. At high stocking rates, as used by Chopping *et al.* (1976), stem would appear to make up a substantial proportion of the diet, an observation consistent with the low milk yields per cow in these experiments. These calculations suggest that an increased leaf to stem ratio in pasture, or greater pasture growth rates, would be necessary to further increase milk output per hectare. Murtagh *et al.* (1980) demonstrated that, at least for kikuyu pasture on red volcanic soils, the capacity for leaf growth is far greater than is being produced under the present levels of stocking.

Grazing Management

A common system of grazing tropical grass pastures in Australia is to have the pasture area divided into approximately 10 paddocks, on a farm of

Table 5.6. Estimation of the efficiency of utilization of pasture by cows grazing tropical pastures (from Cowan et al., 1993a).

Pasture type	Pasture yield (kg DM ha^{-1} year^{-1})	Legume yield (kg DM ha^{-1} year^{-1})	Grass leaf yield (kg DM ha^{-1} year^{-1})	Utilization		Reference
				Pasture (%)	Grass leaf (%)	
Grass–legume	10,000	3500	1630	29	57	Cowan et al. (1975)
Grass–nitrogen	9,400	–	4700	28	56	Cowan et al. (1995b)
Grass–nitrogen	15,360	–	5100	36	107	Colman and Kaiser (1974)
Irrigated grass–nitrogen	23,000	–	7525	49	148	Chopping et al. (1976)

80 ha, and graze these paddocks in a rapid rotation of 2–4 weeks. This provides light defoliation at each grazing and the animal is able to select a diet high in leaf content (Moss and Murray, 1984). There is a build up of pasture during the growing period, and after growth ceases in autumn, animals consume a diet of progressively higher stem and fibre content.

In short-term studies, cows selected a diet of higher nutritive value from short, leafy swards than from tall, mature pastures (Stobbs, 1971; Stobbs, 1973a; Davison *et al.*, 1981). However, in milk production studies over a longer period of time there have been very few benefits from either pasture manipulation such as mulching (Davison *et al.*, 1981; Hernandez and Rosete, 1985) or from grazing methods different from that described above (Thurbon *et al.*, 1971; Ottosen *et al.*, 1975; Chopping *et al.*, 1978; Buchanan *et al.*, 1985; Humphreys, 1991). No difference was measured between 2-, 4- and 6-week rotations on Rhodes grass (Ehrlich *et al.*, 1996). There were significant effects of a 28-day rotation and mulching after grazing on aspects of pasture quality (see Table 5.4), and these were reflected in milk output. Rotational grazing without mulching reduced milk yield, probably as a consequence of stem build up restricting the cow's access to leaf.

Integration with Legumes

The search for a tropical legume that will form the basis of a productive perennial pasture has been intense over the last 30 years (Hutton, 1970). However, progress has not been rapid and the search continues.

Twining tropical legumes are capable of producing 2–4 t DM ha^{-1} in mixtures with tropical grasses (Jones, 1974). This contribution is dependent on maintenance of a low stocking rate or long rest periods in spring and summer (Minson *et al.*, 1993), and optimum stocking rates have been in the order of one-third of those on nitrogen fertilized grasses (Cowan *et al.*, 1993a). The most likely situation where twining legumes can contribute to dairy production is in relatively large areas of inaccessible, broken and hilly land which can be kept aside for grazing during the dry season.

The potential of shrub legumes, such as leucaena (*Leucaena leucocephala* (Lam.) Wit), to combine with grasses in productive systems would appear to be high (Bray *et al.*, 1988; Wilden, 1989), though examples ·of successful development of dairy production systems are rare. A number of tropical countries have developed the concept of a 'protein bank', where approximately 10% of the land area is set aside for a pure stand of shrub legumes. These are not used intensively during the growing season but are available for cutting or grazing during the dry season (Preston and Leng, 1987). An admixture of shrub legumes and grasses is more difficult to manage. The legume requires a rest period of 4–6 weeks following defoliation to maintain vigour (Jones, 1979), a management procedure which is not optimum for utilization of the grass. Optimal row spacing, density and height of shrubs are uncertain in grazing systems (Jones, 1979; Wilden, 1989). The shrub legumes are also often difficult or slow to establish, resulting in land being out of production for 1–3 years. These difficulties mean there is still much developmental work to be done with these legumes before they will form a substantial part of grazing systems.

Recently, emphasis has shifted to selecting tropical legumes which withstand heavy grazing, instead of selecting for high forage production. The genus *Arachis*, a tropical legume which spreads by rhizomes or stolons, is showing promise. As a group they are well accepted by cattle (Lascano and Thomas, 1988), are persistent under heavy grazing (Jones and Carter, 1989) and produce high quality forage for milk production (Lowe *et al.*, 1993). *Aeschynomene* is another genus which could produce potentially useful cultivars for dairying.

Conservation

As with grazing management, many concepts of fodder conservation have been brought to the tropics from temperate areas. Consequently there have been repeated attempts, invariably failed, to develop grass silage conservation systems from tropical grasses. Silage from tropical grass contains a high proportion of stem, is difficult to consolidate (Catchpoole and Henzell, 1971), has an acetic acid fermentation pattern (Kaiser, 1984), and is of low digestibility when fed back to animals (Moss *et al.*, 1984). Consequently, when the practice was examined as part of a grazing system, the response to silage feeding was nil or low, and the practice could not be economically justified (Hamilton *et al.*, 1978; Davison *et al.*, 1984).

Economic gains are more likely where the conserved product is at least of similar quality to grazed grass. Consequently, as originally suggested by Hardison (1966), maize silage has increased the productivity of subtropical dairy farms based on pasture grazing systems (Cowan *et al.*, 1991) and has been widely adopted (Kaiser *et al.*, 1993). Alternatively, legume crops such as soybeans and cowpeas have been successfully used for conservation as whole crop silage (Desborough and Ayres, 1988).

Tropical grass hay is widely used in dry cow feeding in subtropical areas, and in recent years as the basis for intensive cut and carry systems of milk production in the middle east.

Supplementary Feeding

Milk yield of cows grazing tropical pastures is limited by a restriction in digestible energy intake (Hamilton *et al.*, 1970) and is consistently increased when highly digestible energy supplements such as grains are fed (Table 5.7). Response is variable in short-term feeding trials but more consistent

Table 5.7. A comparison of milk responses to modest levels of concentrate in short- and longer-term experiments.

Study	Duration of feeding (weeks)	Amount fed (kg day⁻¹)	Milk response (kg kg⁻¹ DM)
Short term			
Stobbs (1971)	1.5	4.0	0.5
Royal and Jeffrey (1972)	2	2.7	0.5
Cowan and Davison (1978)	4	3.0	0.5 FCM[1]
McLachlan *et al.* (1991)	8	3.9	0.4
Longer term			
Colman and Kasier (1974)	24	2.5	0.9
Chopping *et al.* (1976)	40	3.6	0.9
Cowan *et al.* (1977)	36	6.0	1.03 FCM[1]
Chopping *et al.* (1980)	36	3.6	0.8
McLachlan *et al.* (1994)	36	4.0	0.8
Davison *et al.* (1986)	40	3.0	1.1

[1]FCM, fat corrected milk.

over a longer period (i.e. one year or more, Davison and Elliot, 1993). Molasses has given a response equal to 70% that of grain-based concentrate (Cowan and Davison, 1978; Chopping *et al.*, 1980). In a three-lactation, farm system study, Walker *et al.* (1992) found this response to level of concentrate feeding established in the second year, while Davison *et al.* (1982) showed that for cows grazing abundant pasture the incremental response to level of concentrate slowly increased throughout a 16-week experimental period. Since the energy density of tropical grass and grain are so different, the delayed onset of the response to supplementation is due to rumen adaptation, which can take considerable time (Preston and Leng, 1987). Mayer (1982) demonstrated that, for cows moving from a tropical to a temperate pasture in autumn, it took approximately 100 days before cows were milking to the potential of the temperate pasture.

The long-term nature of the response to supplementary feed has reinforced the need to evaluate the practice as part of a farming system, rather than as an isolated input. The maintenance of milk markets and nutritional issues such as associative effects of feeds, rumen adaptation, cow partitioning of nutrients between body reserves and milk, and long-term pasture utilization are relevant to an assessment of feeding systems. Supplementary feeding also may enable an increase in stocking rate, thus leading to more efficient use of pasture on the farm (Davison and Elliott, 1993).

Many by-products are used in dairy cow rations in the tropics. By-products of sugar cane, oil palm, coconut, cotton seeds, banana, cassava and other crops have been incorporated in rations of cows on grass pastures (Preston and Leng, 1987; Ehrlich *et al.*, 1990; Sanda and Methu, 1990;

Ehrlich *et al.*, 1992; Davison *et al.*, 1994; Preston, 1995). Each by-product
has a unique nutritional profile, making it impossible to generalize on the
milk responses. However, their use has encouraged more attention to ration
formulation, matching the nutrients in the by-product with those supplied
from pasture (Preston, 1995).

The economics of including protein meals in concentrates are much
less clear than those for concentrates as a whole. Moss *et al.* (1992, 1994)
found a substantial milk and milk protein response to cottonseed meal
supplementation of grain-based concentrates while cows grazed tropical
pastures but no difference between protein meals of different rumen
degradability (Moss *et al.*, 1996). Davison *et al.* (1991) measured a small
response to meat meal, a meal with a relatively high level of bypass protein,
given to cows grazing tropical pastures during early lactation. Milk responses have been measured when cows on predominantly grass diets were
given protein supplements resistant to degradation in the rumen (Hamilton
et al., 1992). The level and type of protein is critical to the cost of milk production, and present evidence would suggest that there is little or no economic benefit to the specific addition of protein of low degradability.

Sustainability

The Standing Committee on Agriculture (1991) defines sustainable agriculture as 'the use of farming practices and systems which maintain or enhance
the economic viability of agricultural production, the natural resource base,
and other ecosystems which are influenced by agricultural activities'. Grass
pastures can be very effective in maintaining the economic viability of
animal production and the natural resource base. One concern expressed
about the sustainability of grass pastures is the potential for applied nutrient
fertilizers to move and affect other ecosystems. The other is the longevity of
the pasture. Experience in tropical and subtropical regions suggests that
grass pastures will not persist for long periods unless they are grown in
association with effectively nodulated legumes or supplied with regular
applications of nitrogen fertilizer. The level of nitrogen required is estimated
to be in the order of 100 kg N ha^{-1} year^{-1} (Robbins and Bushell, 1986;
Cowan *et al.*, 1995a).

Economic analyses of regional feeding systems invariably show that
pastures are a cost-effective feed source for cows and often considerably
cheaper than feeds purchased from off the farm (Busby and Lake, 1996).
Cowan *et al.* (1995b) showed that margin over all feed input costs increased
from $400 ha^{-1} year^{-1}, where no nitrogen was applied, to $800 ha^{-1} year^{-1}
where 200 kg N ha^{-1} was applied annually to a Rhodes grass pasture grazed
by dairy cows. The economic benefits of pasture improvement have long
been recognized in temperate areas (Beever, 1993). Studies of pasture productivity on dairy farms in northern Australia have shown present levels of

Table 5.8.   Average levels of milk production from pastures (kg ha^{-1} year^{-1}) on northern Australian dairy farms, and their relation to levels achieved in research (R.T. Cowan, unpublished; estimates derived from the knowledge-based decision support system DAIRYPRO (Kerr and Chaseling, 1995)).

	Production level		
	Farm average	Reasonable[1]	Research trials
Summer pastures	2,000	3,000	5,600
Winter pastures	5,600	8,500	18,000

[1]Suggested by experts as a conservative estimate attainable on all farms. It is close to 50% of research results.

production well below potential, thus offering opportunities for rapid gains in economic viability (Table 5.8).

Grass-based pastures are effective in rebuilding soil carbon content and improving water infiltration rates in soils degraded through the processes of continuous cropping and erosion (Dalal *et al.*, 1991). Grasses are more effective in this than are legumes (Clarke *et al.*, 1967). Barnes (1981) measured increases in soil organic carbon, total nitrogen, exchangeable bases, and crop yields following a ley pasture of *Cynadon nlemfuensis* v. *robustus* Vandcryst. Increased water infiltration rates have also been measured in response to a grass pasture ley (Wilkinson, 1975). There are a number of economic and social factors to consider in the choice of ley pastures, but it is likely their use will increase as part of a sustainable use of tropical and subtropical soils (Humphreys, 1994).

The use of zero till methods to introduce cool season growing species such as ryegrass, oats and annual clovers (*Trifolium* spp.) into tropical grass swards in the autumn has gained widespread acceptance in northern Australia and southern Africa (Lowe and Hamilton, 1985). Advantages of this combination include an improved continuity of forage supply compared with conventional tillage, less exposure of soil to erosion during the high rainfall months of the year, and more flexibility in grazing animals on the firm base associated with minimum tillage compared with the soft base associated with cultivation.

Higher temperatures in the subtropics and tropics change the rate of nutrient cycling (Lavelle and Swift, 1993) and promote a more active process of nitrogen fixation associated with grasses (Weier, 1980). The interaction of soil type with nutrient has not been extensively studied, but present results suggest that, at modest levels of nitrogen application (i.e. up to 300 kg N ha^{-1}) the rate of nitrogen leaching through the profile is very slow (Cowan *et al.*, 1995a), even in extremely porous soils (B. Prove, 1996, Queensland, personal communication). At present the proportions of pasture areas in the tropics and subtropics that are receiving high levels of applied fertilizer are in general much lower than in intensively farmed temperate areas.

Losses of surface applied urea to the air can be substantial, in the order of 30–70% of that applied (Catchpoole *et al.*, 1983). The loss is closely related to reliability of rainfall following application (Murtagh, 1975) and amount of ground cover of grass or trash (Fenn and Hossner, 1985). Present indications are that gaseous losses are likely to be a more important loss than leaching through soil profiles. Methane production is higher with grass-fed animals than with those fed grains (Blaxter and Wainman, 1964), though with the increased levels of production associated with well fertilized pastures, the amount of gas produced per unit of livestock product is decreased (Sibbald and Hutchings, 1994). The flux of methane in tropical agricultural systems is not well understood. For example, Mosier *et al.* (1991) demonstrated a capacity of tropical soils to take up and oxidize methane.

Integration

It is true to say there are very few areas of the tropics where tropical grasses alone form the sole feed input to a dairy production system. Rather, tropical grasses are well suited to providing a bulk of feed which both enhances the number of animals which can be maintained in the system and reduces the risks associated with maintaining a feed supply (Rachmat *et al.*, 1992). In sophisticated feeding systems using irrigated temperate pastures and maize silage, tropical grasses are usually needed to maintain continuity of feed supply during times of changeover from summer to winter feeds (Fig. 5.7). The system is characterized by high capital cost, high output and substantial control over production and quality in each month (Moss and Lowe, 1993).

There is continuing debate as to the merits or otherwise of intensive and extensive systems of dairy production in tropical countries (Simpson and Conrad, 1993; Nicholson *et al.*, 1995). The debate includes aspects of social development and sustainability, in addition to those of economic production (Griffith and Zepeda, 1994), though present trends in Australia, South Africa, Brazil and other countries suggest an increase in the use of intensive systems.

Conclusion

There are a large number of tropical grasses used for dairy production, often associated with particular feeding systems or soil types. The grasses share many quality characteristics, being tall, of low leaf density, and with leaf growth generally being initiated at nodes along elongated stems. There

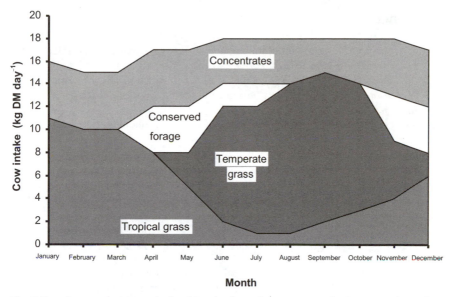

Fig. 5.7. Seasonal change in feed intake for a dairy cow producing 5200 kg milk per lactation in a typical grazing system in subtropical Australia.

are large differences in nutritional quality between leaf and stem, and leaf yield is more closely associated with milk output than total pasture yield.

Grass leaf growth is strongly stimulated by nitrogenous fertilizer, and stocking rate is critical to both the response by grass to applied nitrogen and the efficient use of grass leaf. Grazing management is best kept simple, allowing cows to select a diet of high leaf content, though infrequent slashing may be beneficial in removing excess stem and providing cows improved access to leaf.

Energy supplements, such as cereal grains or molasses, invariably provide milk responses in cows. At times, responses are also likely to protein and mineral supplements. The conservation of tropical grasses has generally not been economically successful, though in some geographic areas, grasses form a valuable source of forage if stood over from the growing to the dry season, or if made into hay for dry cows.

The main contribution of these grasses to feeding systems has been in their ability to produce a high yield of forage and their availability during periods when alternative feeds are unavailable. In smallholder units, the grasses are grown on small areas of the farm unsuited to cropping and are fed with crop residues to cattle. Alternatively, as in Australia, the grasses are integrated with maize silage, irrigated temperate pasture species and concentrates. The price paid to the farmer for milk, and the resources of money, land and water available, have a strong effect on the type of feeding system developed.

References

Abate, A. (1994) An assessment of the mineral supply to dairy cows fed mixed diets. *East African Agriculture and Forestry Journal* 59, 235–240.

Anindo, D.O. and Potter, H.L. (1986) Milk production from Napier grass (*Pennisetum purpureum*) in a zero grazing feeding system. *East Africa Agriculture and Forestry Journal* 52, 106–111.

Anslow, R.C. (1966) The rate of appearance of leaves on tillers of the *Gramineacae*. *Herbage Abstracts* 36, 149–155.

[APO] Asian Productivity Organization (1990) *Animal Feed Resources in Asia and the Pacific.* Asian Productivity Organization, Tokyo, 430pp.

Ayala, J.R., Sistacho, M. and Tuero, R. (1987) Effect of sowing distance on King grass establishment (*Pennisetum purpureum* × *P. typhoida*). *Cuban Journal of Agricultural Science* 21, 309–314.

Barnes, D.L. (1981) Residual effect of grass leys on the productivity of sandy granite-derived soils. 2. Grazed leys. *Zimbabwe Journal of Agricultural Research* 19, 51–67.

Beever, D.E. (1993) Ruminant animal production from forages: present position and future opportunities. In: Baker, M.J. (ed.) *Grasslands for our World.* SIR Publishing, Wellington, New Zealand, pp. 158–164.

Belesky, D.P. and Fedders, J.M. (1995) Comparative growth analysis of cool and warm season grasses in a cool-temperate environment. *Agronomy Journal* 87, 974–980.

Bishnoi, U.R., Oka, G.M. and Fearon, A.L. (1993) Quantity and quality of forage and silage of pearl millet in comparison to sudax, grain and forage sorghums harvested at different growth stages. *Tropical Agriculture, Trinidad* 70, 98–105.

Blaxter, K.L. and Wainman, F.W. (1964) The utilization of the energy of different rations by sheep and cattle for maintenance and for fattening. *Journal of Agricultural Science, Cambridge* 63, 113–128.

Blunt, C.G. and Haydock, K.P. (1978) Effect of irrigation, nitrogen and defoliation on pangola grass in the dry season at the Ord Valley, north-western Australia. *Australian Journal of Experimental Agriculture and Animal Husbandry* 18, 825–833.

Bray, R.A., Cooksley, D.G., Hall, T.J. and Ratcliff, R.J. (1988) Performance of 14 *Leucaena* lines at five sites in Queensland. *Australian Journal of Experimental Agriculture* 28, 69–76.

Bryan, W.W. and Sharpe, J.P. (1965) The effect of urea and cutting treatments on the production of Pangola grass in south-eastern Queensland. *Australian Journal of Experimental Agriculture and Animal Husbandry* 5, 433–441.

Buchanan, I.K., Arnold, G., Brown, G.W. and Maroske, M. (1985) Effect of pasture slashing on milk production from summer grasses. *Queensland Agricultural Journal* 111, 41–44.

Buchanan, I.K. and Cowan, R.T. (1990) Nitrogen level and environmental effects on the annual dry matter yield of tropical grasses. *Tropical Grasslands* 24, 299–304.

Burton, G.W., Hart, R.H. and Lowrey, R.S. (1967) Improving forage quality in Bermuda grass by breeding. *Crop Science* 7, 329–332.

Busby, G. and Lake, M. (1996) *Queensland Dairy Accounting Scheme Handbook.* Information Series QI96005, Queensland Department of Primary Industries, Brisbane, Australia, 32pp.

Cameron, D.F. (1983) To breed or not to breed. In: McIvor, J.G. and Bray, R.A. (eds) *Genetic Resources of Forage Plants*, CSIRO, Melbourne, pp. 237–250.

Catchpoole, V.R. and Henzell, E.F. (1971) Silage and silage making from tropical herbage species. *Herbage Abstracts* 41, 213–221.

Catchpoole, V.R., Oxenham, D.J. and Harper, L.A. (1983) Transformation and recovery of urea applied to a grass pasture in south-eastern Queensland. *Australian Journal of Experimental Agriculture and Animal Husbandry* 23, 80–86.

Chacon, E. and Stobbs, T.H. (1976) Influences of progressive defoliation of a grass sward on the eating behaviour of cattle. *Australian Journal of Agricultural Research* 27, 709–727.

Chopping, G.D., Deans, H.D., Sibbick, R., Thurbon, P.N. and Stokoe, J. (1976) Milk production from irrigated nitrogen fertilised pangola grass. *Proceedings of the Australian Society of Animal Production* 11, 481–484.

Chopping, G.D., Moss, R.J. Goodchild, I.K. and O'Rourke, P.K. (1978) The effect of grazing systems and nitrogen fertiliser regimes on milk production from irrigated pangola-couch pastures. *Proceedings of the Australian Society of Animal Production* 12, 229.

Chopping, G.D., Smith, L.J., Buchanan, I.K. and O'Rourke, P.K. (1980) Molasses supplementation of Friesian cows grazing irrigated couch-pangola pastures. *Proceedings of the Australian Society of Animal Production* 13, 401–405.

Clarke, A.L., Greenland, D.J. and Quirk, J.P. (1967) Changes in some physical properties of the surface of an impoverished red-brown earth under pasture. *Australian Journal of Soil Research* 5, 59–68.

Clements, R.J, Walker, B. and Cowan, R.T. (1987) Nutritive value improvement in tropical forages. In: Hutchinson, K.J. (ed.) *Improving the Nutritive Value of Forages.* SCA Technical Report Series, No. 20, CSIRO, Melbourne, pp. 74–79.

Colman, R.L. and Kaiser, A.G. (1974) The effect of stocking rate on milk production from kikuyu grass pastures fertilized with nitrogen. *Australian Journal of Experimental Agriculture and Animal Husbandry* 14, 155–160.

Cook, B.G. and Stillman, S.L. (1981) Effects of nitrogen, planting material and sowing rates on establishment and early yield of kikuyu grass and Kenya white clover on phyllite soils in south-eastern Queensland. *Tropical Grasslands* 15, 157–162.

Cooper, J.P. (1970) Potential production and energy conversion in temperate and tropical grasses. *Herbage Abstracts* 40, 1–15.

Cowan, R.T. (1975) Grazing time and pattern of grazing of Friesian cows on a tropical grass-legume pasture. *Australian Journal of Experimental Agriculture and Animal Husbandry* 15, 32–37.

Cowan, R.T. (1995) Milk production from grazing systems in northern Australia. In: de Assis, A.G. (ed.) *International Symposium on the Future Systems on Milk Production in Brasil.* Juis de Fora, Brazil, pp. 41–54.

Cowan, R.T. and Davison, T.M. (1978) Milk yields of cows fed maize and molasses supplements to tropical pastures at two stocking rates. *Australian Journal of Experimental Agriculture and Animal Husbandry* 18, 12–15.

Cowan, R.T. and O'Grady, P. (1976) Effect of presentation yield of a tropical grass–legume pasture on grazing time and milk yield of Friesian cows. *Tropical Grasslands* 10, 213–218.

Cowan, R.T. and Stobbs, T.H. (1976) Effects of nitrogen fertiliser applied in autumn and winter on milk production from a tropical grass–legume pasture grazed at four stocking rates. *Australian Journal of Experimental Agriculture and Animal Husbandry* 16, 829–837.

Cowan, R.T., O'Grady, P., Moss, R.J. and Byford, I.J.R. (1974) Milk and fat yields of Jersey and Friesian cows grazing tropical grass–legume pastures. *Tropical Grasslands* 8, 367–370.

Cowan, R.T., Byford, I.J.R. and Stobbs, T.H. (1975) Effects of stocking rate and energy supplementation on milk production from tropical grass–legume pasture. *Australian Journal of Experimental Agriculture and Animal Husbandry* 15, 740–746.

Cowan, R.T., Davidson, T.M. and O'Grady, P. (1977) Influence of level of concentrate feeding on milk production and pasture utilization by Friesian cows grazing tropical grass–legume pasture. *Australian Journal of Experimental Agriculture and Animal Husbandry* 17, 373–379.

Cowan, R.T., Davison, T.M. and Shepherd, R.K. (1986) Observations on the diet selected by Friesian cows grazing tropical grass and grass legume pastures. *Tropical Grasslands* 20, 183–192.

Cowan, R.T., Kerr, D.V. and Davison, T.M. (1991) Maize silage for dairy systems in northern Australia. In: Moran, J. (ed.) *Proceedings of the First Australian Maize Conference*. Moama, pp. 228–235.

Cowan, R.T., Moss, R.J. and Kerr, D.V. (1993a) Northern Dairy Feedbase 2001. 2. Summer feeding systems. *Tropical Grasslands* 27, 150–161.

Cowan, R.T., Davison, T.M., Lowe, K.F., Reason, G.K. and Chopping, G.D. (1993b) Integrating pasture technology research with farm management. In: Baker, M.J. (ed.) *Proceedings of the XVII International Grassland Congress, 8–21 February, 1993, Hamilton, NZ and Rockhampton, Australia*. New Zealand Grassland Association, Palmerston North, pp. 1286–1288.

Cowan, R.T., Lowe, K.F., Ehrlich, W., Upton, P.C. and Bowdler, T.M. (1995a) Nitrogen-fertilised grass in a subtropical dairy system. 1. Effect of level of nitrogen fertiliser on pasture yield and soil chemical characteristics. *Australian Journal of Experimental Agriculture* 35, 125–135.

Cowan, R.T., Lowe, K.F., Ehrlich, W., Upton, P.C. and Bowdler, T.M. (1995b) Nitrogen-fertilised grass in a subtropical dairy system. 2. Effect of level of nitrogen fertiliser on animal production. *Australian Journal of Experimental Agriculture* 35, 137–143.

Cowan, R.T., Lowe, K.F., Ehrlich, W., Upton, P.C. And Bowdler, T.M. (1995c) Nitrogen-fertilised grass in a subtropical dairy system. 3. Effect of stocking rate on the response to nitrogen fertiliser. *Australian Journal of Experimental Agriculture* 35, 145–151.

Cuomo, G.J., Blouin, D.C. and Beatty, J.F. (1996a) Forage potential of dwarf napiergrass and a pearl millet*napiergrass hybrid. *Agronomy Journal* 88, 434–438.

Cuomo, G.J., Blouin, D.C., Corkern, D.L., McCoy, J.E. and Walz, R. (1996b) Plant morphology and forage nutritive value of 3 bahiagrasses as affected by harvest frequency. *Agronomy Journal* 88, 85–89.

Dalal, R.C., Strong, W.M., Weston, E.J. and Gaffney, J. (1991) Sustaining multiple production systems. 2. Soil fertility decline and restoration of cropping lands in sub-tropical Queensland. *Tropical Grasslands* 25, 173–180.

Davison, T.M. and Cowan, R.T. (1978) Comparitive summer milk production from tropical grass–legume and nitrogen fertilized pastures. *Proceedings of the Australian Society of Animal Production* 12, 230.

Davison, T.M. and Elliot, R. (1993) Response of lactating cows to grain based concentrates in northern Australia. *Tropical Grasslands* 27, 229–237.

Davison, T.M., Cowan, R.T. and O'Rourke, P.K. (1981) Management practices for tropical grasses and their effects on pasture and milk production. *Australian Journal of Experimental Agriculture and Animal Husbandry* 21, 196–202.

Davison, T.M., Cowan, R.T. and Chopping, G.D. (1982) Milk responses to supplementation under tropical pasture feeding systems. *Proceedings of the Australian Society of Animal Production* 14, 110–115.

Davison, T.M., Orr, W.N. and Clark, R. (1984) Changes in silage use on the Atherton Tablelands, North Queensland. In: Kempton, T.J., Kaiser, A.G. and Trigg, T.E. (eds) *Proceedings National Workshop 'Silage in the Eighties'*. P.G. Print, Armidale, Australia, pp. 392–396.

Davison, T.M., Cowan, R.T., Shepherd, R.K. and Martin, P. (1985a) Milk production from cows grazing on tropical grass pastures. 1. Effects of stocking rate and level of nitrogen fertiliser on the pasture and diet. *Australian Journal of Experimental Agriculture and Animal Husbandry* 25, 505–514.

Davison, T.M., Cowan, T.R. and Shepherd, R.K. (1985b) Milk production from cows grazing on tropical grass pastures. 2. Effects of stocking rate and level of nitrogen fertiliser on milk yield and pasture–milk yield relationships. *Australian Journal of Experimental Agriculture and Animal Husbandry* 25, 515–523.

Davison, T.M., Isles, D.H. and McGuigan, K.R. (1986) The effect of dietary supplementation with molasses and Christmas Is. Phosphate on milk yield of cows grazing tropical pastures. *Proceedings of the Australian Society of Animal Production* 16, 179–182.

Davison, T., Orr, W.N., Silver, B.A. and Duncalfe, F. (1989) Phosphorus fertilizer and the long term productivity of nitrogen fertilized dairy pastures In: Desroches, R. (ed.) *Proceedings of the XVIth International Grassland Congress, 4–11 October, 1989, Nice, France*. French Grassland Society, Versailles Cedex, France, pp. 1133–1134.

Davison, T.M., Williams, D., Orr, W.N. and Lisle, A.T. (1991) Responses in milk yield from feeding grain and meat-and-bone meal to cows grazing tropical pastures. *Australian Journal of Experimental Agriculture* 31, 159–163.

Davison, T.M., Orr, W.N., Doogan, V.J. and Lowe, K.F. (1993) Relationships between milk yield, pasture on offer and diet selection in tropical grass pastures. In: Baker, M.J. (ed.) *Proceedings of the XVII International Grassland Congress. 8–21 February, 1993, Hamilton, NZ and Rockhampton, Australia*. New Zealand Grassland Association, Palmerston North, pp. 706–708.

Davison, T.M., Ehrlich, W.K., Orr, W.N. and Ansell, J. (1994) Palm kernel expeller as a substitute for grain in dairy cow rations. *Proceedings of the Australian Society of Animal Production* 20, 372.

Davison, T.M., Frampton, P.J., Orr, W.N., Martin, P. and McLachlan, B. (1997) An evaluation of kikuyu–clover pastures as a dairy production system. 2. Milk production and system comparisons. *Tropical Grasslands* 30, 15–23.

Desborough, P.J. and Ayres, J.F. (1988) Cultivar and growth stage effects on the nutritive value of soybean hay. *Proceedings of the Australian Society of Animal Production* 17, 388.

Dirven, J.G.P. and Deinum, B. (1977) The effect of temperature on the digestibility of grasses. An analysis. *Forage Research* 3, 1–17.

Ehrlich, W.K., Upton, P.C., Cowan, R.T. and Moss, R.J. (1990) Copra meal as a supplement for grazing dairy cows. *Proceedings of the Australian Society of Animal Production* 18, 196–199.

Ehrlich, W.K., Moss, R.J., Ansell, J. and Martin, P. (1992) Hominy based concentrate as a supplement for grazing dairy cows. *Proceedings of the Australian Society of Animal Production* 19, 103–105.

Ehrlich, W., Cowan, T. and Lowe, K. (1996) *Improving the Quality of Irrigated Tropical Grass Forage during Autumn*. Final report, DAQ 093, Dairy Research and Development Corporation, Melbourne.

Fenn, L.B. and Hossner, L.R. (1985) Ammonia volatilisation from ammonium or ammonium-forming nitrogen fertilisers. *Advances in Soil Science* 1, 123–169.

Fribourg, H.A. (1985) Summer annual grasses. In: Heath, M.E., Metcalfe, D.S. and Barnes, R.F. (eds) *Forages: The Science of Grassland Agriculture*, 4th edn, Iowa State University Press, Ames, Iowa, pp. 278–286.

Gilbert, M.A. and Clarkson, N.M. (1993) Efficient nitrogen fertilizer strategies for tropical beef and dairy production using summer rainfall and soil analyses. In: Baker, M.J. (ed.) *Proceedings of the XVII International Grassland Congress. 8–21 February, 1993, Hamilton, NZ and Rockhampton, Australia*. New Zealand Grassland Association, Palmerston North, pp. 1552–1554.

Goldson, J.R. (1963) Observations on the grazing behaviour of grade dairy cattle in a tropical climate. *East African Agricultural and Forestry Journal* 29, 72–77.

Grainger, C., Dellow, D. and Cowan, T. (1996) Dairy cow nutrition. In: Schelling, S. (ed.) *The Comprehensive Reference to the Australian Dairy Industry*. Morescope Publishing, East Hawthorn, Victoria, Australia, pp. 161–190.

Greenhalgh, J.F.D., Reid, G.W., Aitken, J.N. and Florence, E. (1966) The effects of grazing intensity on herbage consumption and animal production. 1. Short-term effects in strip grazed dairy cows. *Journal of Agricultural Science, Cambridge* 67, 13–24.

Griffith, K. and Zepeda, L. (1994) Farm level tradeoffs of intensifying tropical milk production. *Ecological Economist* 9, 121–126.

Grof, B. (1969) Elephant grass for warmer and wetter lands. *Queensland Agricultural Journal* 95, 227–234.

Hamilton, R.I., Lambourne, L.J., Roe, R. and Minson, D.J. (1970) Quality of tropical grasses for milk production. In: Norman, M.J.T. (ed.) *Proceedings of the XIth International Grassland Congress*. Surfers Paradise, Australia, pp. 860–863.

Hamilton, R.I., Catchpoole, V.R., Lambourne, L.J. and Kerr, J.D. (1978) The preservation of a Nandi Setaria silage and its feeding value for dairy cows. *Australian Journal of Experimental Agriculture and Animal Husbandry* 18, 16–24.

Hamilton, B.A., Ashes, J.R. and Carmichael, A.W. (1992) Effect of formaldehyde treated sunflower meal on the milk production of grazing dairy cows. *Australian Journal of Agricultural Research* 43, 279–287.

Hardison, W.A. (1966) *Chemical Composition, Nutrient Content and Potential Milk Producing Capacity of Fresh Tropical Herbage*. Technical Bulletin No. 1, Dairy Training and Research Institute, University of the Philippines, 51pp.

Hawton, D. (1980) The effectiveness of some herbicides for weed control in *Panicum maximum* and *Brachiaria decumbens* and some factors affecting the atrazine tolerance of these species. *Tropical Grasslands* 14, 34–39.

Henzell, E.F. (1963) Nitrogen fertiliser responses of pasture grasses in south-east Queensland. *Australian Journal of Experimental Agriculture and Animal Husbandry* 3, 290–299.

Hernandez, D. and Rosete, A. (1985) Milk production with *Cynadon dactylon*. Integral analysis of the rotational cycle and the rest period. *Pastos y Forrajes* 8, 423–434.

Humphreys, L.R. (1991) *Tropical Pasture Utilization*. Cambridge University Press, Cambridge, 227pp.

Humphreys, L.R. (1994) *Tropical Forages: Their Role in Sustainable Agriculture*. Longman Scientific and Technical, Harlow, Essex, 400pp.

Hutton, E.M. (1970) Australian research in pasture plant introduction and breeding. In: Norman, M.J.T (ed.) *Proceedings of the XIth International Grassland Congress*. Surfers Paradise, Australia, pp. A1–A12.

Jones, R.J. (1967) Losses of dry matter and nitrogen from autumn-saved improved pasture during winter at Samford, south-eastern Queensland. *Australian Journal of Experimental Agriculture and Animal Husbandry* 22, 88–94.

Jones, R.J. (1974) Effect of previous cutting interval and leaf area remaining after cutting on regrowth of *Macroptilium atropurpureum* cv. Siratro. *Australian Journal of Experimental Agriculture and Animal Husbandry* 14, 343–348.

Jones, R.J. (1979) The value of *Leucaena leucocephala* as a feed for ruminants in the tropics. *World Animal Review* 31, 13–23.

Jones, R.M. and Carter, E.D. (1989) Demography of pasture legumes. In: Martin, G.C., Matches, A.G., Barnes, R.F., Brougham, R.W., Clements, R.J. and Sheath, G.W. (eds) *Persistence of Forage Legumes*. American Society of Agronomy, Crop Science Society of America, Soil Science Society of America, Madison, Wisconsin, pp. 139–158.

Jones, R.J., Seawright, A.A. and Little, D.A. (1970) Oxalate poisoning in animals grazing the tropical grass *Setaria sphacelata*. *Journal of the Australian Institute of Agricultural Science* 36, 41–43.

Kaiser, A.G. (1984) The influence of silage fermentation on animal production. In: Kempton, T.J., Kaiser, A.G. and Trigg, T.E. (eds) *Silage in the 80's. Proceedings of a National Workshop*. Armidale, New South Wales, pp. 106–135.

Kaiser, A.G., Havilah, E.J., Chopping, G.D. and Walker, R.W. (1993) Northern Dairy feedbase 2001. 4. Feeding systems during winter and spring. *Tropical Grasslands* 27, 180–211.

Kerr, D.V. and Chaseling, J. (1992) A study of the level and efficiency of production in relation to inputs for dairy farms in Queensland. Final report DAQ 077, Dairy Research and Development Corporation, Melbourne, 36pp.

Kerr, D.V. and Chaseling, J. (1995) DAIRYPRO – A knowledge based decision support system for strategic planning on dairy farms in Queensland Australia. In: *Workshop on Artificial Intelligence and the Environment, Eighth Australian Joint Conference on Artificial Intelligence (AI95) November 1995*. Canberra, Australia, pp. 81–87.

Laredo, M.A. and Minson, D.J. (1973) The voluntary intake, digestibility, and retention time by sheep of leaf and stem fractions of five grasses. *Australian Journal of Agricultural Research* 24, 875–888.

Lascano, C.E. and Thomas, D. (1988) Forage quality and animal selection of *Arachis pintoi* in association with tropical grasses in the eastern plains of Colombia. *Grass and Forage Science* 43, 433–439.

Lavelle, P. and Swift, M.J. (1993) Origin and regulation of nutrient supply to plants in humid tropical grasslands. In: Baker, M.J. (ed.) *Grasslands for Our World*. SIR Publishing, New Zealand, pp. 524–530.

Little, D.A. (1982) Utilization of minerals. In: Hacker, J.B. (ed.) *Nutritional Limits to Animal Production from Pastures*. CAB, Farnham Royal, UK, pp. 259–283.

Lowe, K.F. (1976a) The growth, nutritive value and water use of irrigated pasture mixtures at Gatton, South-eastern Queensland (Thesis summary). *Journal of the Australian Institute of Agricultural Science* 42, 105.

Lowe, K.F. (1976b) The value of a frost tolerant setaria component in mixed pastures for autumn saved feed in south-eastern Queensland. *Tropical Grasslands* 10, 334–336.

Lowe, K.F. and Hamilton, B.A. (1985) Dairy pastures in the Australian Tropics and Sub-tropics. In: Murtagh, M.G.J. and Jones, R.M. (eds) *Proceedings of the Third Australian Conference on Tropical Pastures, Rockhampton*. Tropical Grassland Society of Australia, Occasional Publication No. 3, pp. 68–79.

Lowe, K.F., Burns, M.A. and Bowdler, T.M. (1978) *Productivity of a Tropical Mixture under Continuous and Rotational Grazing at Gatton, South East Queensland*. Queensland Department of Primary Industries, Agriculture Branch Report P-22-78.

Lowe, K.F., Bruce, R.C. and Bowdler, T.M. (1981) Effect on phosphorus on the establishment of Siratro and green or Gatton panic pastures on duplex soils of sub-coastal Queensland. *Tropical Grasslands* 15, 101–107.

Lowe, K.F., Moss, R.J., Cowan, R.T., Minson, D.J. and Hacker, J.B. (1991a) Selecting for nutritive value in *Digitaria milanjiana*. 4. Milk production from an elite genotype compared with *Digitaria eriantha* ssp. *pentzii* (pangola grass). *Australian Journal of Experimental Agriculture* 31, 603–608.

Lowe, K.F., Cowan, R.T., Bowdler, T.M., Buchanan, I. and Moss, R.J. (1991b) Tropical grasses for an irrigated forage system in a subtropical environment. *Proceedings of the Third International Symposium on the Nutrition of Herbivores, August 1991*. Penang, Malaysia, pp. 59.

Lowe, K.F., Bowdler, T.M., Chapman, C., Moss, R.J. and Hovey, R. C. (1993) *Performance of grazing peanut (*Arachis pintoi *cv. Amarillo) for subtropical dairy production*. Final report, Project DAQ 067, Dairy Research and Development Corporation, Melbourne, 39pp.

McDowell, L.R., Conrad, J.H., Thomas, J.E., Harris, L.E. and Fick, K.R. (1977) Nutritional composition of latin American forages. *Tropical Animal Production* 2, 273–279.

McLachlan, B.P., Ehrlich, W.K., Cowan, R.T., Davison, T.M., Silver, B.A. and Orr, W.N. (1994) Effect of level of concentrate fed once or twice daily on the milk production of cows grazing tropical pasture. *Australian Journal of Experimental Agriculture* 34, 301–306.

McLauchlin, B., Orr, W. and Silver, B. (1991) Short term responses in solids not fat to grain fed in addition to molasses for dairy cows grazing tropical pastures. *Australian Journal of Experimental Agriculture* 31, 191–194.

McLeod, N.M, Kennedy, P.M. and Minson, D.J. (1990) Resistance of leaf and stem fractions of tropical forage to chewing and passage in cattle. *British Journal of Nutrition* 63, 105–119.

't Mannetje, L. (1982) Problems of animal production from tropical pastures. In: Hacker J.B. (ed.) *Nutritional Limits to Animal Production from Pastures*. CAB, Farnham Royal, UK, pp. 67–85.

Mayer, D.G. (1982) Simulating milk production of a dairy farm in a subtropical environment. In: *Proceedings of the 5th Biennial Conference of the Simulation Society of Australia*. Armidale, 5, 83–90.

Minson, D.J. (1971) The nutritive value of tropical pasture. *Journal of the Australian Institute of Agricultural Science* 37, 255–263.

Minson, D.J. (1990a) The chemical composition and nutritive value of tropical grasses. In: Skerman, P.J. and Riveros, F. (eds) *Tropical Grasses*. FAO, Rome, pp. 163–180.

Minson, D.J. (1990b) *Forage in Ruminant Nutrition*. Academic Press, London, 483pp.

Minson, D.J. and McLeod, M.N. (1970) The digestibility of temperate and tropical grasses. In: Norman, M.J.T (ed.) *Proceedings of the XIth International Grassland Congress*. Surfers Paradise, Australia, pp. 719–722.

Minson, D.J., Cowan, T. and Havilah, E. (1993) Northern dairy feedbase 2001. 1. Summer pasture and crops. *Tropical Grasslands* 27, 131–149.

Mogaka, L.M. (1995) An evaluation of zero grazing under a Napier grass-based system in high yielding cows at small-holder level. *East Africa Agriculture and Forestry Journal* 60, 201–206.

Mosier, A., Schimel, D., Valentine, D., Bronson, K. and Parton, W. (1991) Methane and nitrous oxide fluxes in native, fertilised and cultivated grasslands. *Nature* 350, 330–332.

Moss, R.J. and Lowe, K.F. (1993) Development of forage systems for dairying in subtropical Australia. In: *Proceedings of the 17th International Grasslands Congress*, Hamilton, New Zealand and Rockhampton, Australia, pp. 1991–1992.

Moss, R.J. and Murray, R.M. (1984) Diet selection by Friesian weaners grazing an irrigated tropical grass pasture. *Proceedings of the Australian Society of Animal Production* 15, 476–479.

Moss, R.J., Byford, I.J.R., Winks, L., Chambers, G. and O'Grady, P. (1984) *In vivo* digestibility of tropical grass–legume pastures conserved as silage or hay. In: Kempton, T.J., Kaiser A.G. and Trigg, T.E. (eds) Silage in the 80's. *Proceedings of a National Workshop*. Armidale, New South Wales, pp. 137–140.

Moss, R.J., Ehrlich, W.E., Martin, P.R. and McLachlan, B.P. (1992) Responses to protein supplementation by dairy cows grazing nitrogen fertilized forages. *Proceedings of the Australian Society of Animal Production* 19, 100–102.

Moss, R.J., Martin, P.R. and Chapman, N.D. (1994) Protein level in concentrates for dairy cows in subtropical pasture systems supplemented with maize silage. *Proceedings of the Australian Society of Animal Production* 20, 124–127.

Moss, R.J., Martin, P.R. and Ansell, J. (1996) Amount and source of protein in concentrates fed to high producing dairy cows in a subtropical feeding system. *Proceedings of the Australian Society of Animal Production* 21, 95–98.

Muinga, R.W., Thorpe, W. and Topps, J.H. (1992) Voluntary feed intake, liveweight change, and lactation performance of crossbred dairy cows given *ad libitum*

Penissetum purpureum (Napier grass, var. Bana) supplemented with leucaena forage in the lowland semi humid tropics. *Animal Production* 55, 331–337.

Murtagh, G.J. (1975) Environmental effects on the short term response of tropical grasses to nitrogen fertiliser. *Australian Journal of Experimental Agriculture and Animal Husbandry* 15, 679–688.

Murtagh, G.J., Kaiser, A.G., Huett, D.O. and Hughes, R.M. (1980) Summer growing components of a pasture system in a subtropical environment. 1. Pasture growth, carrying capacity and milk production. *Journal of Agricultural Science, Cambridge* 94, 645–663.

Nicholson, C.F., Blake, R.W. and Lee, D.R. (1995) Livestock, deforestation and policy making: Intensification of cattle production systems in Central America revisited. *Journal of Dairy Science* 78, 719–734.

Noble, A. and Lowe, K.F. (1974) Alcohol-soluble carbohydrates in various tropical and temperate pasture species. *Tropical Grasslands* 8, 179–187.

Norton, B.W. (1982) Differences between species in forage quality. In: Hacker, J.B. (ed.) *Nutritional Limits to Animal Production from Pastures.* CAB, Farnham Royal, UK, pp. 89–110.

[NRC] National Research Council (1989) *Nutrient Requirements of Dairy Cattle.* Revised, 6th edn. National Academy Press, Washington, DC, 157pp.

Odhuba, E.K. (1989) Napier grass ensiled with broiler poultry waste as a feed for dairy cows. *East Africa Agriculture and Forestry Journal* 55, 1–4.

Oji, C.K. and Ughenighe, P.O. (1992) Effect of nitrogen fertiliser and cutting height on forage yield and quality of Maiiwa pearl millet. *Tropical Agriculture, Trinidad* 69, 11–14.

Omaliko, C.P.E. (1980) Influence of initial cutting date and cutting frequency on yield and quality of star, elephant and guinea grass. *Grass and Forage Science* 35, 139–145.

Ostrowski, H. (1969) Standover tropical grasses. *Queensland Agricultural Journal* 95, 508–516.

Ottosen, E.M., Brown, G.W. and Maroske, M.R. (1975) Strip grazing – advantage or disadvantage. *Queensland Agricultural Journal* 101, 569–570.

Papajcsik, I.A. and Bodero, J. (1988) Modelling lactation curves of Friesian cows in a subtropical climate. *Animal Production* 47, 201–207.

Payne, W.J.A. (1963) The potential for milk production in tropical areas. *Proceedings of the World Congress of Animal Production*, Rome 3, 204.

Preston, T.R. (1995) *Tropical Animal Feeding: a Manual for Research Workers.* FAO Animal Production and Health paper 126. FAO, Rome, 157pp.

Preston, T.R. and Leng, R.A. (1987) *Matching Ruminant Production Systems with Available Resources in the Tropics and Sub-tropics.* Penambul Books, Armidale, Australia, 245pp.

Rachmat, R., Stur, W.W. and Blair, G.J. (1992) Cattle feeding systems and limitations to feed supply in South Sulawesi, Indonesia. *Agricultural Systems* 39, 409–419.

Randel, P.F. (1993) Mixed chopped and ground grass hay for dairy cows vs. grazing plus concentrates. *Journal of Agriculture of the University of Puerto Rico* 77, 193–200.

Reason, G.K. and Chaseling, J. (1993) Effect of nitrogen fertilizer on dairy farm productivity and viability in the subtropics. In: Baker, M.J. (ed.) *Proceedings of the XVII International Grassland Congress, 8–21 February, 1993, Hamilton, NZ and Rockhampton, Australia.* New Zealand Grassland Association, Palmerston North, pp. 520–521.

Rees, M.C., Minson, D.J and Kerr, J.D. (1972) Relation of dairy productivity to feed supply in the Gympie district of south-eastern Queensland. *Australian Journal of Experimental Agriculture and Animal Husbandry* 12, 553–560.

Reeves, M., Fulkerson, W.J. and Kellaway, R.C. (1996) Forage quality of kikuyu (*Pennisetum clandestinum*): the effect of time of defoliation and nitrogen fertiliser application and in comparison with perennial ryegrass (*Lolium perenne*). *Australian Journal of Agricultural Research* 47, 1349–1359.

Robbins, G.B. and Bushell, J.J. (1986) Productivity of morphologically different sown tropical grass pastures grown under similar conditions. In: *Proceedings of the XV International Grassland Congress, 24–31 August, 1985, Kyoto, Japan.* The Japanese Society of Grassland Science, Nishi-nasuno, Tochigi-ken, Japan, pp. 1000–1002.

Royal, A.J.E. and Jeffery, H. (1972) Energy and protein supplements for dairy cows grazing tropical pasture. *Proceedings of the Australian Society of Animal Production* 9, 292–295.

Ruiz, T.M., Rivera, B. and Corchado, N. (1995) Two feeding systems for lactating dairy cows grazing fertilised grasses. *Journal of Agriculture of the University of Puerto Rico* 79, 99–110.

Sanda, A.I. and Methu, J.N. (1990) Evaluation of cassava as an energy source in concentrate feed for dairy cows in Kenya. *East Africa Agriculture and Forestry Journal* 55, 135–140.

[SCA] Standing Committee on Agriculture (1990) *Feeding Standards for Australian Livestock, Ruminants.* Standing Committee on Agriculture and Resource Management, Ruminants subcommittee. CSIRO, Australia, 266pp.

[SCA] Standing Committee on Agriculture (1991) *Sustainable Agriculture.* Australian Agricultural Council, SCA Technical report 36, CSIRO, Melbourne, 118pp.

Sere, C. and Steinfeld, H. (1996) *World Livestock Production Systems: current status, issues and trends.* FAO Animal Health and Production Paper 127. FAO, Rome.

Sibbald, A.R. and Hutchings, N.J. (1994) The integration of environmental requirements into livestock systems based on grazed pastures in the European community. In: Gibon, A. and Flamant, J.C. (eds) *The Study of Livestock Farming Systems in a Research and Development Framework.* EAAP Publication No. 63, pp. 86–100.

Silcock, R.G. (1980) Seedling characteristics of tropical pasture species and their implications for ease of establishment. *Tropical Grasslands* 14, 174–180.

Simpson, J.R. and Conrad, J.H. (1993) Intensification of cattle production systems in Central America: why and when. *Journal of Dairy Science* 26, 1744–1752.

Singh, D.K., Singh, V. and Sale, P.W.G. (1995) Effect of cutting management on yield and quality of different selections of Guinea grass (*Panicum maximum* (Jacq.) L.) in a humid subtropical environment. *Tropical Agriculture, Trinidad* 72, 181–187.

Singh, V. (1993) Role of perennial forage crops for year round forage supply in India. *Asian Livestock* 18, 141–142.

Skerman, P.J. and Riveros, F. (1990) *Tropical Grasses.* FAO, Rome, 832pp.

Snaydon, R.W. (1991) The productivity of C3 and C4 plants: A reassessment. *Functional Ecology* 5, 321–330.

Stephens, D. (1967) Effects of fertilisers on grazed and cut elephant grass leys at Kawanda Research Station, Uganda. *East Africa Agriculture and Forestry Journal* 32, 383–392.

Stobbs, T.H. (1971) Quality of pasture and forage crops for dairy production in the tropical regions of Australia. 1. Review of the literature. *Tropical Grasslands* 5, 159–170.

Stobbs, T.H. (1973a) The effect of plant structure on the intake of tropical pastures 1. Variation in the bite size of grazing cattle. *Australian Journal of Agricultural Research* 24, 809–819.

Stobbs, T.H. (1973b) The effect of plant structure on the intake of tropical pastures. II Differences in sward structure, nutritive value, and bite size of animals grazing *Setaria anceps* and *Chloris gayana* at various stages of growth. *Australian Journal of Agricultural Research* 24, 821–829.

Stobbs, T.H. (1975) A comparison of Zulu sorghum, bulrush millet and white panicum in terms of yield, forage quality and milk production. *Australian Journal of Experimental Agriculture and Animal Husbandry* 15, 211–218.

Stobbs, T.H. (1977) Short term effects of herbage allowance on milk production, milk composition and grazing time of cows grazing nitrogen-fertilized tropical grass pasture. *Australian Journal of Experimental Agriculture and Animal Husbandry* 17, 892–898.

Stuart, P.N. (1993) *The Forage Book*. Pacific Seeds, Toowoomba, Australia, 60pp.

Sundstol, F. and Owen, E. (1984) *Straw and Other Fibrous By-products as Feed*. Elsevier, Amsterdam, 604pp.

Swain, F.G. (1971) Dairy production systems relevant to the tropical regions of Australia. 1. Review of the literature. *Tropical Grasslands* 5, 269–280.

Sweeney, F.C. and Hopkinson, J.M. (1975) Vegetative growth of nineteen tropical and subtropical pasture grasses and legumes in relation to temperature. *Tropical Grasslands* 9, 209–217.

Thurbon, P.N., Cowan, R.T. and Chopping, G.D. (1971) Animal utilization aspects of pasture and forage crop management in tropical Australia. *Tropical Grasslands* 5, 255–264.

Teitzel, J.K., Gilbert, M.A. and Cowan, R.T. (1991) Nitrogen fertilised grass pastures. *Tropical Grasslands* 25, 111–118.

Tothill, J.C. and Hacker, J.B. (1983) *The Grasses of Southern Queensland*. Tropical Grassland Society of Australia: University of Queensland Press, Brisbane, 300pp.

Vicente-Chandler, J. and Caro-Costas, R. (1974) Puerto Rican dairy research: high milk yields with tropical grass alone. *World Farming* 16, 30–31.

Walker, R.G., Knight, R.I., Lehmann, B. and Cowan, R.T. (1992) Pasture based systems for milk production in the tropics. *Proceedings of the Australian Society of Animal Production* 19, 118.

Walker, R.G., Cowan, R.T., Orr, W.N. and Silver, B.A. (1996) Effect of access to irrigated ryegrass on the milk yield and composition of autumn calving cows in a tropical upland environment. *Tropical Grasslands* 30, 249–256.

Weier, K.L. (1980) Nitrogen fixation associated with grasses. *Tropical Grasslands* 14, 194–201.

Whyte, R.O., Moir, T.R.G. and Cooper, J.P. (1959) *Grasses in Agriculture*. FAO, Italy, 417pp.

Wilden, J.H. (1989) Trees in forage systems. In: Kang, B.T. and Reynolds, L. (eds) *Alley Farming in the Humid and Subhumid Tropics*. IDRC, Ottawa, Canada, pp. 71–81.

Wilkinson, G.E. (1975) Effect of grass fallow rotations on the infiltration of water into a savanna zone soil of northern Nigeria. *Tropical Agriculture* 52, 97–103.

Wilson, J.R. and Ford, C.W. (1973) Temperature influences on the *in vitro* digestibility and soluble carbohydrate accumulation of tropical and temperate grasses. *Australian Journal of Agricultural Research* 24, 187–198.

Wilson, J.R. and Ng, T.T. (1975) Influence of water stress on parameters associated with herbage quality of *Panicum maximum* var. *trichoglume*. *Australian Journal of Agricultural Research* 26, 127–136.

Wilson, J.R., Brown, R.H. and Windham, W.R. (1983) Influence of leaf anatomy on the dry matter digestibility of C3, C4 and C3/C4 intermediate types of *Panicum* species. *Crop Science* 23, 141–146.

Wilson, J.R., Akin, D.E., McLeod, M.N. and Minson, D.J. (1989) Particle size reduction of the leaves of a tropical and a temperate grass by cattle. 2. Relation of anatomical structure to the process of leaf breakdown through chewing and digestion. *Grass and Forage Science* 44, 65–75.

Winks, L., O'Grady, P., Chambers, G.A. and Byford, I.J.R. (1970) Milk production from a tropical legume grass pasture in north Queensland. In: *Proceedings of the 18th International Dairy Congress*. Sydney, Australia, pp. 567.

Potassium Management

6

J.H. Cherney[1], D.J.R. Cherney[2] and T.W. Bruulsema[3]

[1]*Department of Soil, Crop and Atmospheric Sciences, 153 Emerson Hall, Cornell University, Ithaca, New York 14853-1901, USA*
[2]*Department of Animal Science, 327 Morrison Hall, Cornell University, Ithaca, New York 14853-4801, USA*
[3]*Potash and Phosphate Institute, Guelph, Ontario, Canada*

Introduction

Potassium is essential for the normal functioning of both grass plants and ruminant animals. Soil K availability has a dramatic influence on K status of both grass forage and the dairy cows that ingest grass. Excess K in dry cow diets recently has been identified as a primary factor in milk fever in dairy cattle (Goff and Horst, 1997a). Beede (1996) noted that the incidence of clinical milk fever was between 5% and 11%, while subclinical effects could be as high as 50% in USA dairy herds. On the other hand, in the southern USA, diets of lactating dairy and beef cattle can be deficient in K, particularly under heat stress (McDowell, 1985; Preston *et al.*, 1994).

Research has identified plant requirements for K to achieve economically optimum yields and plant persistence, but management of grass to produce high yields with a desirable K content has received much less attention. While the significance of K in relation to maintaining dry matter yield and grass persistence is widely recognized, the importance of K to ruminant nutrition has been less obvious.

Soil K

Most soils have an abundant reserve supply of K, but only a very small portion of the total K is readily available for plant uptake at any given time. Soil K content varies between < 0.01 and > 30 g K kg^{-1}, and up to

980 g kg^{-1} of the soil K is considered unavailable. The form of K in soil depends on soil mineral composition, leaching, and the degree of soil weathering. Availability of K to growing forage crops depends on the form and the amount of K in any given soil, as well as environmental factors such as temperature, soil water and soil redox potential. Although leaching losses from sandy soil can be substantial, there is no indication that losses of K have any adverse environmental impact.

K-supplying capacity of soils

Soil K consists of (i) K in soil solution, (ii) readily exchangeable K, (iii) K exchangeable with difficulty, and (iv) mineral K (Barber, 1995). Available soil K (Fig. 6.1), as defined by Jeffrey (1988), is the soil K that is predictably accessible to the grass crop during the current growing season, and includes soil solution K and readily exchangeable K. In alluvial soils with high levels of K, it may be possible to get significant release of difficultly exchangeable K during the growing season (Robinson, 1985; Barber, 1995). Physical, chemical and biological processes result in the release of mineral K over time. Soil minerals are more weathered in older soils, and consequently these soils have lower K reserves. In North America, this includes most of the eastern USA, especially the southeast, and parts of southern Canada (Haby *et al.*, 1990). Even a relatively small geographic area can have soils with a wide range of K-supplying power (Plate 1). Organic soils, as well as sandy soils, are typically very low in K. There is generally less subsoil K available compared with topsoil K, but subsoil may be either depleted or enriched in K, following long-term addition or removal of K from the surface horizon. Subsoil K availability is used by some soil analysis laboratories as one of the components determining fertilizer K recommendations, but availability of subsoil K is highly dependent on climatic factors (Haby *et al.*, 1990).

Available soil K can come from release of difficultly exchangeable K, or by the addition of K from commercial fertilizer, animal excreta, or unharvested residue following decomposition (Fig. 6.1). A relatively small amount of K is lost from the farm system through removal of animal products, compared with potential leaching or erosion losses. Leaching losses on heavy-textured soils were very small, compared with coarse-textured soils, and increased precipitation increased K leaching (Ylaranta *et al.*, 1996). Use of a grass crop reduced K leaching, compared with fallow land (Ylaranta *et al.*, 1996). Three years after fertilizing with 117 kg K ha^{-1} over a 2-year period, an equivalent amount of K had leached from the soil in this high rainfall area (Cuttle and James, 1995). Since available soil K is replenished to some extent during the growing season, calculated recovery of fertilizer K often approaches or exceeds 100% (Robinson *et al.*, 1962; Robinson, 1985). Depending on the terrain, however, there is potential for considerable loss

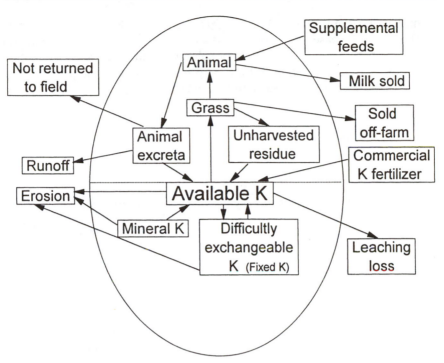

Fig. 6.1. Potassium cycle on a dairy farm.

of K due to runoff or erosion. Obviously, grass sold off-farm would be a significant loss of K. Potassium in excreta non-uniformly returned to the land by grazing dairy cattle can be considered a temporary loss of K to the system.

Soil water affects K availability by affecting the expansion and contraction of clay minerals. Increased soil water results in a significant linear increase in the supply of K to plant roots by diffusion and mass flow related to plant water use and transpiration rate (Barber, 1995). However, a saturated soil with insufficient soil aeration would usually result in decreased K uptake, owing to both reduced root activity and, in montmorillonitic clays, to increased fixation of K resulting from the chemical reduction of iron (Roth, 1996; Stucki, 1996).

Soil pH has little effect on K uptake by plants (Fleming, 1973; Barber, 1995), unless the pH is below approximately 5.2. Soluble Al at lower pH values causes more displacement of K into solution (Barber, 1995), but inhibits plant absorption of K (Follett and Wilkinson, 1995).

Increased soil temperature increases the supply of K to plants by increasing diffusion of K through the soil and by increasing K uptake (Barber, 1995). In the range between 7°C and 25°C, K concentration and the K : (Ca+Mg) equivalent ratio increased in grasses in response to increased temperature (Reinbott and Blevins, 1994). However, K concentration may decline with increased temperature over a longer period of time (Reid and Horvath,1980). Although the effect of soil temperature on K concentration in grasses is relatively short-lived, it may have a dramatic impact on incidence of grass tetany (Miyasaka and Grunes, 1990).

K fertilizer sources

Approximately 95% of the K fertilizer used in North America is potassium chloride (KCl), with potassium sulphate (K_2SO_4) or potassium-magnesium sulphate ($K_2SO_4 \cdot 2MgSO_4$) used occasionally (Griffith and Murphy, 1996).

Potassium recommendations for grass

Most commercial K fertilizer is applied to grain and oilseed crops. The primary source of K for grasses on most dairy farms is animal manure. Much of the K ingested by dairy cattle is excreted in faeces or urine, with most of the K contained in the urine (Hodgson, 1990). The proportion of a grazed sward covered by animal manure during the grazing season has been estimated at 5–50%, depending on the stocking rate. With consecutive grazing seasons this situation should improve somewhat, unless animals gather at the same location (e.g. a water source) year after year. Fertilizer effect of K in a urine patch is considered to last for 2 years (Brockman, 1971). Under this assumption, efficiency of urine K is between 50% and 90% of that of commercial fertilizer K, depending on stocking rate. Need for supplemental commercial K fertilizer on pastures primarily depends on K-supplying power of the soil and K removal rate by grazing. Winter, dry-season range or tropical pastures may be deficient enough in K to require K supplementation of animal diets (McDowell, 1985).

Grass swards that are mechanically harvested and grass swards under grazing often are considered different crops, from the standpoint of K fertilization. Intensively-managed perennial grass in the USA under a 3–5× per season cutting management could yield up to 16 t ha^{-1} (Allinson *et al.*, 1992). Quantity of K taken up by grass in a single growing season can approach or exceed 500 kg K ha^{-1} (Allinson *et al.*, 1992; Follett and Wilkinson, 1995), although 300 kg K ha^{-1} would be a more common removal rate.

Most soils will require some fertilizer K to maintain the available soil K at a level high enough to prevent K deficiency in succeeding grass crops. A common suggestion in temperate climates is to split apply K fertilizer in relation to the yield expected from each cutting. Unless soils are very deficient in K, however, a better suggestion is to split apply K fertilizer later in the season, with no K applied to spring growth. Spring growth of grass forage typically has a higher concentration of K than regrowth (Brown *et al.*, 1969), implying that spring growth has more K available to it per unit of forage yield. Also, fertilizer K applied in the mid or later portion of the growing season would have a more beneficial effect on the over-wintering capability of the grass.

The need for K fertilization of grass is often assessed in relation to the N fertilization regime (Allinson *et al.*, 1992). Based on grass studies on soils with low K-supplying power, Brown *et al.* (1969) concluded that 3 kg K_2O should be applied for every 4 kg N. Brockman (1971) found a response to 1 kg K for every 2 kg N fertilizer applied, but suggested that repeated cutting over years would eventually lead to 1 kg K being required for every 1 kg N applied. Bélanger *et al.* (1989) concluded that annual rates of 160 kg N ha^{-1} and 110 kg K ha^{-1} were needed to optimize yield and persistence of a long-term timothy sward for 25 years. Robson *et al.* (1989) suggested fertilizing cut swards with 1.5–2 kg K, depending on K-supplying power of the soil, for each 3 kg of fertilizer N. The recommended K : N fertilizer ratio may decrease as N application rate is increased. On the other hand, grass fertilized with little or no N requires little or no K fertilizer, regardless of the soil K-supplying power. Potassium fertilizer regimes should be controlled by K-supplying power of the soil and by total K removed per season.

Plant K

Potassium is an essential element in grass nutrition, with a vital role in many plant functions. While Jeffrey (1988) lists K as a Category 2 element; essential but usually non-limiting for plant growth, North American texts generally consider K a primary macronutrient, limiting crop growth in many soils (Follet and Wilkinson, 1995). The minimum concentration of K in grasses needed to maintain vital functions is fairly well-defined, but K concentration required for optimum stand longevity and drought resistance of perennial grasses is more open to debate, and varies by species. The goal of a best K management programme for grasses is to provide optimum K for plant functions without accumulating excess plant K.

K and plant metabolism

Potassium is a very mobile element that does not become a part of grass forage fibre. It has a metabolic role in protein synthesis, stomatal movement, turgor, water potential, electrical charge balance and permeability of cells. Transport of nutrients, particularly sugars, within the plant is dependent on K. Potassium also helps plants to resist disease, drought and cold temperatures.

Potassium deficiency symptoms are difficult to verify, and will occur only if grass K content drops to very low levels. Deficiency symptoms are not greatly different from those due to drought, fungal damage, or normal senescence. Brown *et al.* (1969) described K deficiency in timothy (*Phleum pratense* L.) as a yellowing and browning of the dying leaf tips and margins, which were observed when K concentration dropped below 8 g K kg^{-1} of dry matter. Visible K deficiency of grasses in New Zealand (McNaught, 1958), including chlorosis or tip burn of older leaves, was associated with a K concentration of 2.5 g kg^{-1}. Potassium deficiency in orchardgrass (*Dactylis glomerata* L.) was noted when K content of harvested forage was at 8.4 g kg^{-1} (Robinson *et al.*, 1962). Potassium deficiency symptoms were not readily apparent in reed canarygrass (*Phalaris arundinaceae* L.) with a K concentration of 8 g kg^{-1} (Cherney *et al.*, 1997).

Critical K values for grasses

Critical leaf K concentration for 90% of maximum yield in perennial ryegrass (*Lolium perenne* L.) was 13–20 g K kg^{-1} (McNaught, 1958) to as high as 28 g K kg^{-1} (Smith *et al.*, 1985). Optimum yield of reed canarygrass under a three cut management fertilized with 224 or 336 kg N ha^{-1} was obtained with a herbage K concentration of 25–27 g kg^{-1} (Allinson *et al.*, 1992). Cherney *et al.* (1997), however, found K concentration as low as 14 g kg^{-1} in reed canarygrass fertilized with 224 kg N ha^{-1} and sampled near anthesis, and as low as 8 g kg^{-1} in regrowth, with no significant yield reduction compared to K-fertilized treatments. A significant reduction in yield of timothy cut twice per year occurred whenever forage K content dropped below 12 g kg^{-1} (Brown *et al.*, 1969). Grant and MacLean (1966) concluded that timothy would persist over time with a concentration of 15–18 g K kg^{-1} DM in fully-headed first growth and 12–16 g K kg^{-1} DM in regrowth. Kelling and Matocha (1990) summarized critical K concentrations for 80–90% of max-imum yield in several cool-season annual and perennial grasses and found a range of 16–38 g K kg^{-1} DM. Mayland and Wilkinson (1996) listed critical, adequate and high K concentrations in whole-plant tissue of cool-season annual and perennial grasses as 12–15 g kg^{-1}, 16–25 g kg^{-1} and > 25 g kg^{-1}, respectively. Critical K concentrations of several warm-season annual and perennial grasses have been reported at between 12 and 20 g K kg^{-1} DM, depending on species and cultivar (Kelling and Matocha, 1990). Minimum K content for optimum growth of several warm-season grasses ranged from 3 g K kg^{-1} DM for weeping lovegrass (*Eragrostis curvula* (Schrad.) Nees var. *curvula* Nees) to 20 g K kg^{-1} DM for pangola grass (*Digitaria eriantha* Steud.) (Gammon, 1952).

Clearly, different critical K values are needed for the same grass species, depending on harvest system (grazing vs. cutting) and harvest intensity. Most grasses harvested twice per season should maintain high yields with approximately 15 g K kg^{-1} DM in spring growth forage (anthesis stage or later) and 12 g K kg^{-1} DM in regrowth forage. Grass harvested more frequently or grazed at vegetative stages may require higher concentrations (20–25 g kg^{-1}) of K at harvest to maintain high yields. New grass seedings following tillage will require much less K in relation to established swards, as some soil K is released by tillage, and yields of establishment year swards will be relatively low.

K concentration in grasses

Potassium concentration has been reported as low as 0.6 g K kg^{-1} DM in buffalograss (*Buchloe dactyloides* (Nutt.) Engelm.) as a winter range pasture (Preston and Linsner, 1985) to as high as 70 g K kg^{-1} DM in orchardgrass (MacLeod and Carson, 1965). Concentration of K is influenced by available soil K, grass species, forage age, time of season, and also interacts strongly with N fertilization.

Luxury consumption

Fleming (1973) noted that monocotyledonous species have a preference for monovalent cations. Grasses will absorb K in excess of plant requirements, depending on the quantity of available soil K and the availability of other elements, particularly N. Plant parts accumulate K to different degrees. Roots of timothy and perennial ryegrass were found to contain 2.7 g K kg^{-1} DM when shoot K averaged 25.8 g kg^{-1} (ap Griffith and Walters, 1966). Highest concentrations of K in mature plant parts were generally found in leaves, although stem K content may exceed that of leaves in K-deficient plants (McNaught, 1958).

Forage age

Grass forage sampled early in the growing season generally increases in K concentration initially and then declines with increased plant age (Fig. 6.2). Plant uptake of K early in the season is very high, but is then diluted by the rapid increase of structural and non-structural carbohydrates. This dilution effect will occur with adequate N fertilization regardless of soil K status, as K uptake cannot keep up with dry matter accumulation. Decline in K concentration with plant age appears to be less pronounced in tropical grasses (Gomide *et al.*, 1969; Reid *et al.*, 1979) than in temperate species (Fig. 6.2), but severity of decline will be primarily influenced by level of fertilizer N and available soil K status.

Time of season

Although it can be difficult to separate plant age effects from time of season effects on grass K content in field studies, K concentration generally is lower in regrowth forage later in the growing season compared with spring

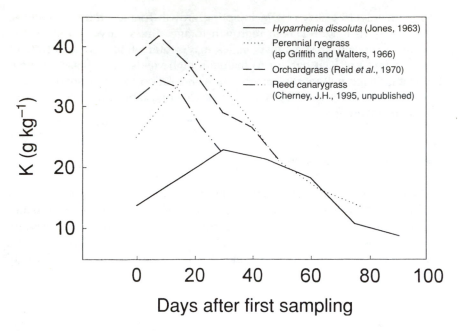

Fig. 6.2. Concentration of potassium in primary spring growth of grasses following application of potassium fertilizer.

growth for cool-season grasses (McNaught, 1958; Allinson *et al.*, 1992; Cherney *et al.*, 1997). Potassium accumulation in the autumn was lower than in spring for eight cool-season perennial grasses under controlled temperature conditions (Gross and Jung, 1978). Grass in regions of the world with wet and dry seasons will be higher in K concentration during the wet season, due to soil moisture effects on K availability, although such seasonal differences are not always consistent (Morillo *et al.*, 1989; Mtimuni *et al.*, 1990).

Species and variety
In general, reported K concentrations of warm-season grasses tend to be low, compared with cool-season grasses (Table 6.1; Gomide *et al.*, 1969; Serra *et al.*, 1996). This is due partly to low soil K in many tropical regions (McDowell, 1985) and partly to lower K concentrations in plant tissue of C_4 versus C_3 species (Blevins, 1985). Species differences in K concentration have been found by numerous researchers, but species ranking for K uptake has been somewhat inconsistent across studies (Table 6.1). Orchardgrass tends to be higher in K concentration than other cool-season grass species over a wide range of soil types and N fertilizer rates, while other species were more variable. Gomide *et al.*(1969) found no significant differences among six warm-season grasses in K concentration averaged over

five sampling dates in Brazil, even though there was a twofold difference in K content among species from 4.4 to 9.3 g K kg^{-1} DM. On the other hand, Reid *et al.* (1979) observed large differences among warm-season grass species harvested 13 times over 16 weeks of primary growth in Uganda. The same six species studied by Gomide *et al.* (1969) ranged from 16 to 38 g K kg^{-1} DM (Reid *et al.*, 1979), averaged over 13 sampling dates (Table 6.1). It should be possible to control K content in grass through species selection, although Reid and Horvath (1980) contend that species selection for mineral uptake may be more effective for overcoming mineral deficiencies than mineral excesses.

Differences have been found in K content among cultivars within grass species, with consistency of cultivar rank over a range of temperature and soil conditions (Vose, 1963; Gross and Jung, 1978). There was a significant range in K concentration among 45 progeny lines of Russian wildrye (*Psathyrostachys juncea* (Fisch.) Nevski) (Asay and Mayland, 1990). Significant differences in K content of 10 Rhodes grass (*Chloris gayana* Kunth) cultivars were mainly associated with ploidy level (Jones *et al.*, 1995). Varietal differences in K were relatively small and inconsistent for timothy, orchardgrass and perennial ryegrass in another study (ap Griffith and Walters, 1966). Saiga *et al.* (1992) found significant differences among four orchardgrass cultivars in concentration of N, P, Ca, and Mg, but not K. The possibility of developing grass cultivars with enriched or reduced K concentration exists, but rarely has been attempted. Selection for K content in grasses may be difficult, due to the significant effect of environment on K uptake, compared with environmental effects on uptake of other elements, such as Mg, Ca and Na (Mika, 1982; Sleper *et al.*, 1989).

Interactions of K with other nutrients

General trends in grass elemental concentrations as influenced by N and K fertilization are depicted in Fig. 6.3. Exceptions to most of these trends, however, can be found in the literature. Although researchers have concluded that N fertilization either increased K content (Reid *et al.*, 1970) or decreased K content of grasses (Grant and MacLean, 1966), a more complete literature review reveals an interaction between N fertilization and K content (Wolton, 1963; Allinson *et al.*, 1992; Table 6.2). If the soil contains sufficient available K, N fertilization increases K content of grass forage. High yields due to N fertilization quickly deplete available soil K and continued N fertilization results in decreased K content of forage (Cherney *et al.*, 1997). Nitrogen fertilization increased Mg, but had no consistent effect on Ca in orchardgrass forage (Reid *et al.*, 1966). Sodium content of forage (Fig. 6.3) has been consistently increased by N fertilization (Rahman *et al.*, 1960; Reith *et al.*, 1964).

Table 6.1.   Grass species rank for K concentration.

Rank (highest to lowest within each reference)	Number of sampling dates	Range in K concentration (g kg⁻¹ DM)	Country	Reference
Cool-season grasses				
Orchardgrass				
Tall fescue (*Festuca arundinacea* Schreb.)				
Tall oatgrass (*Arrhenatherum elatius* (L.) J.S. & C. Pres)				
Smooth bromegrass (*Bromus inermis* Leyss.)				
Timothy	5	14–20	USA	MacDonald, 1946
Orchardgrass				
Timothy				
Tall fescue				
Perennial ryegrass (*Lolium perenne* L.)				
Crested dogstail (*Cynosurus cristatus* L.)				
Chewings fescue (*Festuca rubra* subsp. *commutata* Gaud.)	5	20–28	UK	Thomas *et al.*, 1952
Bulbous canarygrass (*Phalaris aquatica* L.)				
Orchardgrass				
Tall fescue	1	40–52	USA	Odom *et al.*, 1980
Orchardgrass				
Tall fescue				
Reed canarygrass				
Smooth bromegrass				
Timothy	10	30–40	USA	Cherney *et al.*, 1997
Orchardgrass				
Timothy				
Smooth bromegrass				
Tall fescue	5	23–35	USA	Reid *et al.*, 1970
Warm-season grasses				
Pangola grass				
Carpetgrass (*Axonopus affinis* Chase)				
Bermudagrass (*Cynodon dactylon* (L.) Pers)				
Bahiagrass (*Paspalum notatum* Flugge)				
Weeping lovegrass	2	12–17	USA	Gammon, 1952

Table 6.1. *(contd.)*

Rank (highest to lowest within each reference)	Number of sampling dates	Range in K concentration (g kg^{-1} DM)	Country	Reference
Napier grass (*Pennisetum purpureum* Schumach.) Kikuyu grass (*Pennisetum clandestinum* Hochst. ex Chiov.) Semperverde grass (*Panicum maximum* Jacq) Molasses grass (*Melinis minutiflora* Beauv.) Pangola grass Bermudagrass	13	16–38	Uganda	Reid *et al.*, 1979
Urochloa pullulans Stapf *Sporobolus pyramidalis* Beauv. Bermudagrass *Hyparrhenia dissoluta*	8	10–17	Rhodesia	Jones, 1963

Table 6.2. Response of reed canarygrass to annually applied N and K in Connecticut (4-year averages recalculated from Allinson *et al.*, 1992).

Fertilizer K applied (kg ha^{-1})	Fertilizer N applied (kg ha^{-1})					
	112	224	336	112	224	336
	Dry matter yield (t ha^{-1})			Plant K concentration (g kg^{-1} DM)		
0	7.1	10.0	10.8	17.3	10.2	8.6
186	7.2	11.6	13.6	28.6	27.3	24.6
372	6.8	11.2	13.2	30.7	32.2	32.0

High rates of K fertilization significantly decreased herbage concentration of Mg and Na, and Ca to a lesser extent (Reid and Jung, 1974; Barber, 1995). Potassium fertilization increased plant N uptake (Barber, 1995), but high rates of K fertilization decreased herbage N content (MacLeod, 1965; MacLeod and Carson, 1965). Simultaneous fertilization with N and K results in offsetting effects on Na and Mg content in grasses (Fig. 6.3). Nitrogen and K fertilization also interact to affect the form of accumulated nitrogen in the

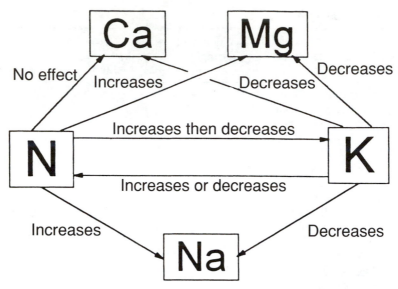

Fig. 6.3. Elemental concentration shifts in grass herbage when either nitrogen or potassium fertilizer is applied.

plant. High rate of N fertilization combined with low available K will result in an accumulation of nitrate nitrogen in grass herbage (MacLeod and Carson, 1965). Herbage P has not been consistently affected by N or K fertilization (Rahman *et al.*, 1960; Grant and MacLean, 1966; Brown *et al.*, 1969).

Fertilizer P, however, has been demonstrated to enhance uptake of Mg and Ca and to reduce K : (Ca+Mg) equivalent ratios in winter wheat seedlings (Reinbott and Blevins, 1991), in tall fescue (Sweeney *et al.*, 1996) and in ryegrass (Blevins and Sanders, 1994). Soil temperature also plays a role in the response of forage K content to P nutrition. Reinbott and Blevins (1997) reported that under greenhouse conditions, leaf K content decreased as treatment P levels increased when roots were warm (25°C), but not when they were cold (15°C). The same authors indicate that in the field, tall fescue forage K declined as soil test P increased in the October harvest (warm soil) but not in the May harvest (cool soil).

Fertilizer K source

Although information on grasses is scarce, there is evidence that the source of K may have an impact on the nutrients of importance to the dietary cation–anion difference (DCAD) and subsequently to the health of the cow. Rominger *et al.* (1976) compared sulphate versus chloride sources of K at very high rates on lucerne (*Medicago sativa* L.), and reported concentrations of all the major nutrients (Na, K, Cl, S) involved in the calculation of the DCAD (see section below on milk fever). Chloride concentration increased much more strongly in response to Cl addition than did S concentration to addition of sulphate. As a result, the chloride source of K produced forage with a lower DCAD than did the sulphate source. In corn (*Zea mays* L.) also, large increases in tissue chloride concentrations were produced by increasing rates of fertilization with KCl (Heckman, 1995).

Grass–legume mixtures

The K requirements for grasses and legumes are similar (Follett and Wilkinson, 1995), but the fibrous root system of grasses makes them more efficient at extracting available soil K. If available soil K is high, legumes will be similar in K content to grasses in a mixture. When available soil K is low, however, K content of grasses will be much higher than legumes in the mixture, and legume content of the mixture will decline over time. The primary effect of K fertilization of mixed pastures may be the shift in botanical composition (Reid and Jung, 1974). Potassium fertilization increased the legume content of a mixture, but had no effect on species content of the mixture in the presence of N fertilizer (Reith *et al.*, 1964). Thus, fertilization programmes for grass–legume mixtures require enough K to maintain the legume in the mixture, and will generally require more K than for either grass or legumes grown separately (McNaught, 1958; Follett and Wilkinson, 1995).

Animal K

Animal nutritionists have historically regarded K as a non-critical nutrient (Preston and Linsner, 1985), but its importance to dairy cattle nutrition has been greatly increasing over the past two decades. Excessive potassium in the dry cow dairy ration is a contributing factor to milk fever, subclinical hypocalcaemia and udder oedema. It also is a contributing factor to grass tetany (hypomagnesaemia or grass staggers) and wheat pasture poisoning, although these metabolic disorders are much more common in beef cattle than dairy cattle. Recent research has concentrated on the DCAD in dairy cattle rations as a primary indicator of milk fever predisposition. A number of strategies have been developed to influence the DCAD in rations.

K and animal metabolism

K requirements

McDowell (1985) notes that K is the third most abundant mineral in ruminants, but has been generally ignored when considering ruminant nutrition. Potassium is present primarily inside animal cells, and is involved in maintaining osmotic, water and acid–base balances (Miller, 1995). It also is involved in muscle contraction, nerve impulse conduction and several enzymatic reactions. Potassium content of ruminant animals is about 1.3–1.7 g kg^{-1} (Wilkinson and Mays, 1979) while the K content of milk is approximately 1.5 g kg^{-1} (NRC, 1989). The National Research Council (1989) lists K requirement for early lactation dairy cows as 10 g kg^{-1}, an increase from their former requirement of 8 g K kg^{-1} DM. The K requirement for dry pregnant cows is 6.5 g kg^{-1} of the total dry ration. Potassium in secreted milk is a major reason for the increased K requirement of lactating cows. Need for K under heat stress is considerably higher, however, with significant milk yield responses reported in rations up to 18 g K kg^{-1} DM (Schneider *et al.*, 1986). Similarly, Sanchez *et al.* (1994b) indicated that optimal dry matter intake and milk yield occurred at approximately 15 g K kg^{-1} DM in winter and at over 19 g K kg^{-1} DM during summer. Fisher *et al.* (1994) found no reduction in milk yield when dietary K for lactating cows was increased from 16 to 31 g kg^{-1} by addition of K_2CO_3, but increasing K to 46 g kg^{-1} of dry matter intake reduced milk yield and caused other metabolic problems.

The bioavailability of K is very high, with release of up to 990 g kg^{-1} of grass herbage K after 48 hours of rumen incubation (Emanuele *et al.*, 1991). Bioavailability of K in most forage grasses is high and generally exceeds 850 g kg^{-1} (Miller, 1995), compared with much lower availability of other minerals such as Ca, which may be only 250 g kg^{-1}. Potassium is not stored in the body to any significant extent and must be replenished daily. Most of the K is lost through excretion in the urine, although about 130 g kg^{-1} of K loss may be in the faeces, with another 120 g kg^{-1} of total K loss in milk (Preston and Linsner, 1985). Cows under heat stress may lose a significant amount of K in sweat and saliva (Beede *et al.*, 1992).

K and forage utilization

Few studies have investigated the effect of K or K fertilization on grass herbage quality *per se*. Different rates of K fertilization had no effect on digestibility of tall fescue, smooth bromegrass, orchardgrass or timothy (Reid and Jung, 1965; Calder and MacLeod, 1968). There does not appear to be any palatability response specifically due to K fertilization, as there is for Na fertilization (Chiy and Phillips, 1996). Wilson (1981) concluded that K fertilization generally has little influence on herbage digestibility.

Metabolic disorders involving K

As recently as 1985 it was concluded that K toxicity was not a practical problem for ruminants, because ingested K beyond that required is quickly excreted (McDowell, 1985). Although the role of K in metabolic disorders is not completely clarified, it is now obvious that excessive dietary K can have a strong negative effect on health of dairy cattle.

Hypomagnesaemia

Hypomagnesaemia, or grass tetany, results from a deficiency of available dietary Mg and occurs when animals are grazing cool-season grasses, typically primary spring growth. The incidence of grass tetany in dairy cattle is low in the USA, due to the relatively small percentage of animals grazing grass pastures, compared with some regions of Europe. Interest in grazing dairy cattle, as well as interest in grass for dairy cattle, is increasing in the USA which will undoubtedly increase the grass tetany risk. Both high K and a high milliequivalent ratio of K : (Ca + Mg) are associated with the incidence of grass tetany in dairy cattle (Grunes and Welch, 1989). Excessive available soil K has a twofold effect on magnesium status in dairy cows. High levels of available soil K consistently decrease Mg concentration in grass herbage (Robinson *et al.*, 1989), and high levels of K in herbage also decrease availability of Mg to ruminants by reducing absorption of Mg from the digestive tract (Martens and Kasebieter, 1983; Dua and Care, 1995; Dalley *et al.*, 1997).

Hypocalcaemia

Hypocalcaemia results from a deficiency in plasma Ca (hypocalcaemia) at the onset of lactation in dairy cows, and is a primary cause of a number of metabolic disorders (Goff and Horst, 1997a). Severe hypocalcaemia disrupts many metabolic and physiological functions, leading to coma and death if not treated (Horst *et al.*, 1997). Less severe, subclinical hypocalcaemia also results in metabolic disorders leading to reductions in milk production and/or reproductive performance (Fig. 6.4). Recent research has determined that reducing dietary K in prepartum diets is the most effective step that can be taken to prevent milk fever in dairy cattle, owing to the effect of K on the DCAD (Goff and Horst, 1997b). An increase in dietary K from 11 g kg^{-1} to 21 g kg^{-1} in prepartum diets increased the incidence of milk fever in older Jersey test cows from 10% to 50%. Interestingly, when dietary Ca was increased from 5 g kg^{-1} to 15 g kg^{-1}, there was no effect on milk fever except that when both K and Ca were high (31 g kg^{-1} and 15 g kg^{-1}, respectively), milk fever incidence declined slightly.

The DCAD, expressed as milliequivalents of [(Na + K) − (Cl + S)] per 100 g of dietary dry matter (Beede, 1996), has significant effects on dairy cow physiology and productivity. Prepartum cows in the last 3 weeks of pregnancy require a low or negative DCAD to reduce incidence of hypocalcaemia. Positive DCAD rations more than doubled both milk fever incidence and subclinical hypocalcaemia incidence, compared with negative DCAD rations (Beede *et al.*, 1992). To prevent milk fever one can reduce Na or K or increase Cl and S. As Na, Cl and S concentrations are quite low in grass herbage, it is K concentration that has the most significant effect on DCAD. Lactating dairy cows require a positive DCAD between +20 to +50 mEq per 100 g of dietary dry matter (Sanchez *et al.*, 1994a). The

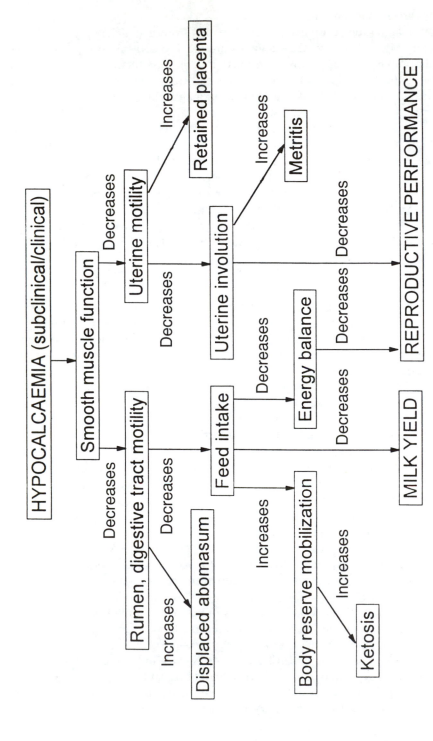

Fig. 6.4.   Early postpartum hypocalcaemic cascade of dairy cows (Beede, 1996).

higher K diets during lactation appear to reduce the need for mineral buffers to maintain milk fat and avoid rumen acidosis.

Feeding strategies

Feed composition tables are useless for estimation of K concentration in grasses. Herbage analysis is required for ration formulation. Caution is required when analysing K content using near infrared reflectance spectroscopy, as it will only estimate mineral composition of the organic portion of a mixed feed.

Potassium deficient diets

Perennial grass and legume forages are better sources of K than corn silage or many feed supplements, urea in particular. Some feed supplements, such as cane molasses, soybean meal and cottonseed meal, are relatively high K sources (Preston and Linsner, 1985). For cattle on pasture, a significant amount of soil ingestion can drastically affect mineral content of the diet (Cherney *et al.*, 1983). Soil ingestion will probably reduce the amount of available K in the total diet, as the average K content of soils is approximately 10 g kg^{-1}, and most of this K is in mineral forms unavailable to the animal. Grass herbage fed to lactating dairy cows is likely to contain sufficient K to avoid K deficiency in the diet, except for warm-season grasses grown on low K soils. Also, if a substantial portion of the forage in a diet is replaced by concentrates, it may be possible to have K deficiency.

Magnesium deficient diets

Dairy cattle grazing grass in the spring should be monitored for Mg consumption. If Mg intake is inadequate, or incidence of grass tetany occurs, free-choice Mg as a dietary supplement is the most cost-effective method of eliminating the deficiency (Robinson *et al.*, 1989). Short-term control of grass tetany can be achieved by foliar applications of Mg. Supplying supplemental Mg in drinking water also is a short-term solution to grass tetany. This may be somewhat ineffective as spring forage is high in water content, minimizing fluid water intake by grazing animals (Robinson *et al.*, 1989). It may be possible to control grass tetany through P fertilization to increase uptake of Mg and Ca (Reinbott and Blevins, 1994).

In Europe, where pasture is a main source of feed for dairy cattle, hypomagnesaemia is more common than in the USA. Sodium fertilization of grass pastures may reduce the herbage K : (Ca + Mg) ratio, depending on soil K and soil moisture availability (Cushnahan *et al.*, 1995; Chiy and Phillips, 1996). Sodium replaces some of the K in plants, and Na enhances Mg absorption across the rumen wall, while K decreases it (Martens and Kasebieter, 1983). Diluting the effects of high K in pastures by providing a stored forage source, such as corn silage, as a ration component also should be effective.

Hypocalcaemic diets

Excess K in lactating dairy cow diets probably will not result in metabolic disorders. Excess K in prepartum diets during the transition period 3–4 weeks before calving can be overcome by supplementation with anionic salts such as calcium sulphate, calcium chloride, magnesium sulphate, ammonium sulphate or ammonium chloride (Ender *et al.*, 1971; Beede, 1996). If K in the total diet is in excess of 18 g kg^{-1} it will be difficult to overcome using anionic salts, as large amounts of these salts can cause palatability problems.

K Management and the Soil–Grass–Animal System

Tremendous variation exists in soils for available K. Some soils may be able to support optimum grass yields without any K fertilization. No yield increase was produced with K fertilization of smooth bromegrass on prairie-derived soil, although available soil K concentration was as low as that produced by K release from the difficultly available soil K pool (George *et al.*, 1979). A strategic plan for minimizing animal metabolic disorders due to K in grasses is to provide as precisely as possible the amount of K required for optimum grass yield. Achieving such precision would require the use of both soil testing and forage analysis. At such levels, grass persistence should not be adversely affected. Fertilization of grass with K should not be done before primary grass growth in spring. Summer application of K evens out the K concentration in spring grass growth and regrowth (Robinson *et al.*, 1962). Potassium fertilization of other crops in a rotation also must be conservative.

Many dairy farms in the northern region of the USA are currently in a state of K surplus. Farms in a lucerne–corn silage system often use a portion of their land for grass production and for application of stored manure during the growing season. Grass is typically harvested at a mature stage after lucerne harvest, and is then stored as dry cow forage, resulting in risk of DCAD problems for prepartum cows. A tactical solution for these farms is to manage 10–15% of their cropland specifically for dry cow forage.

Dry cow forage management can be accomplished by first setting aside fields lowest in available soil K for low K grass production. Application of appropriate N fertilizer to maximize yield will quickly deplete excess available soil K in most soil types. Depending on available soil K levels, K content of grass forage may be very high through the first growing season. Initially, all commercial K fertilizer or manure applications should be avoided, with future K applications dependent on plant analysis, not soil analysis. When grass forage K content of harvested spring growth drops below approximately 17 g K kg^{-1} DM, commercial K fertilizer or manure should be applied to regrowth to provide sufficient K for winter survival of grass.

Species selection of timothy over orchardgrass, for example, also may reduce herbage K content, but species selection must also include persistence and yield considerations. Other species, such as reed canarygrass, may be a more appropriate choice than timothy under some environments and soil types. Forage produced that is lowest in K content should then be fed to dry cows in the transition phase just before calving. Consideration should also be given to increasing the energy content of the ration during the late transition phase, using corn silage or grain. The increased energy from these sources could help improve the cow's physical condition as well as dilute the higher K levels of other forages.

On the other hand, dairy operations in parts of the southern USA and in many tropical regions produce forages on K-deficient soils. Under heat stress and during lactation these low K forages may not meet the K requirements of dairy cattle. Adequate K fertilization is necessary to raise the K content of such forages, to help cattle tolerate heat and ensure optimal milk production.

Acknowledgements

The authors greatly appreciate the helpful comments of J.P. Goff and E.D. Thomas.

References

Allinson, D.W., Guillard, K., Rafey, M.M., Grabber, J.H. and Dest, W.M. (1992) Response of reed canarygrass to nitrogen and potassium fertilization. *Journal of Production Agriculture* 5, 595–601.

ap Griffith, G. and Walters, R.J.K. (1966) The sodium and potassium content of some grass genera, species and varieties. *Journal of Agricultural Science, Cambridge* 67, 81–89.

Asay, K.H. and Mayland, H.F. (1990) Genetic variability for elements associated with grass tetany in Russian wildrye. *Journal of Range Management* 43, 407–411.

Barber, S.A. (1995) *Soil Nutrient Bioavailability, a Mechanistic Approach*, 2nd edn. John Wiley, New York, USA, 414pp.

Beede, D.K. (1996) Cation–anion difference in dairy rations: dealing with high potassium content in alfalfa. In: *Proceedings of the 26th National Alfalfa symposium*. Certified Alfalfa Seed Council, Inc., Davis, California, USA, pp. 30–37.

Beede, D.K., Sanchez, W.K. and Wang, C. (1992) Macrominerals. In: Van Horn, H.H. and Wilcox, C.J. (eds) *Large Dairy Herd Management*. American Dairy Science Association, Champaign, Illinois, USA, pp. 272–286.

Bélanger, G., Richards, J.E. and Walton, R.B. (1989) Effects of 25 years of N, P and K fertilization on yield, persistence and nutritive value of a timothy sward. *Canadian Journal of Plant Science* 69, 501–512.

Blevins, D.G. 1985. Role of potassium in protein metabolism in plants. In: Munson, R.D. (ed.) *Potassium in Agriculture*. American Society of Agronomy, Crop Science Society of America and Soil Science Society of America Inc., Madison, Wisconsin, USA, pp. 413–424.

Blevins, D.G. and Sanders, J.L. (1994) Phosphorus and magnesium reduce grass tetany potential. *Better Crops* 78(1), 14–15.

Brockman, J.S. (1971) Potassium requirements of intensive grass as related to system of utilization. In: *Potassium and Systems of Grassland Farming*, 1st Colloquium of the Potassium Institute Ltd, Oxfordshire, pp. 63–71.

Brown, C.S., Carpenter, P.N. and Belyea, P.R. (1969) *Fertilization Responses of Hayland Grasses in Maine*. Bulletin 676. Maine Agricultural Experiment Station, Orono, Maine, USA, 28pp.

Calder, F.W. and MacLeod, L.B. (1968) In vitro digestibility of forage species as affected by fertilizer application, stage of development and harvest dates. *Canadian Journal of Plant Science* 48, 17–24.

Cherney, J.H., Robinson, D.L., Kappel, L.C., Hembry, F.G. and Ingraham, R.H. (1983) Soil contamination and elemental concentrations of forages in relation to grass tetany. *Agronomy Journal* 75, 447–451.

Cherney, J.H., Cherney, D.J.R. and Reid, W.S. (1997) Grass management for dry dairy cows. In: Williams, M.J. (ed.) *Proceedings of the American Forage and Grassland Council*. American Forage and Grassland Council, Georgetown, Texas, USA, 612–16.

Chiy, P.C. and Phillips, C.J.C. (1996) Effects of sodium fertiliser on the chemical composition of grass and clover leaves, stems and inflorescences. *Journal of the Science of Food and Agriculture* 72, 501–510.

Cushnahan, A., Bailey, J.S. and Gordon, F.J. (1995) Some effects of sodium application on the yield and chemical composition of pasture grown under differing conditions of potassium and moisture supply. *Plant and Soil* 176, 117–127.

Cuttle, S.P. and James, A.R. (1995) Leaching of lime and fertilisers from a reseeded upland pasture on a stagnogley soil in mid-Wales. *Agricultural Water Management* 28, 95–112.

Dalley, D.E. Isherwood, P., Sykes, A.R. and Robson, A.B. (1997) Effect of intraruminal infusion of potassium on the site of magnesium absorption within the digestive tract in sheep. *Journal of Agricultural Science, Cambridge* 129, 99–105.

Dua, K. and Care, A.D. (1995) Impaired absorption of magnesium in the aetiology of grass tetany. *British Veterinary Journal* 151, 413–426.

Emanuele, S.M., Staples, C.R. and Wilcox, C.J. (1991) Extent and site of mineral release from six forage species incubated in mobile dacron bags. *Journal of Animal Science* 69, 801–810.

Ender, F., Dishington, I.W. and Helgebostad, A. (1971) Calcium balance studies in dairy cows under experimental induction and prevention of hypocalcemic paresis puerperalis. *Zeitschrift für Tierphysiologie, Tierernaehrung, und Futtermittelkunde* 28, 233–256.

Fisher, L.J., Dinn, N., Tait, R.M. and Shelford, J.A. (1994) Effect of level of dietary potassium on the absorption and excretion of calcium and magnesium by lactating cows. *Canadian Journal of Animal Science* 74, 503–509.

Fleming, G.A. (1973) Mineral composition of herbage. In: Butler, G.W. and Bailey, R.W. (eds) *Chemistry and Biochemistry of Herbage*. Academic Press, London, UK, pp. 529–566.

Follett, R.F. and Wilkinson, S.R. (1995) Nutrient management of forages. In: Barnes, R.F., Miller, D.A. and Nelson, C.J. (eds) *Forages Vol. II. The Science of Grassland Agriculture,* 5th edn. Iowa State University Press, Ames, Iowa, USA, pp. 55–82.

Gammon, N., Jr (1952) Sodium and potassium requirements of pangola and other pasture grasses. *Soil Science* 76, 81–90.

George, J.R., Pinheiro, M.E. and Bailey, T.B., Jr (1979) Long-term potassium requirements of nitrogen-fertilized smooth bromegrass. *Agronomy Journal* 71, 586–591.

Goff, J.P. and Horst, R.L. (1997a) Physiological changes at parturition and their relationship to metabolic disorders. *Journal of Dairy Science* 80, 1260–1268.

Goff, J.P. and Horst, R.L. (1997b) Effects of the addition of potassium or sodium, but not calcium, to prepartum rations on milk fever in dairy cows. *Journal of Dairy Science* 80, 176–186.

Gomide, J.A., Noller, C.H., Mott, G.O., Conrad, J.H. and Hill, D.L. (1969) Mineral composition of six tropical grasses as influenced by plant age and nitrogen fertilization. *Agronomy Journal* 61, 120–123.

Grant, E.A. and MacLean, A.A. (1966) Effect of nitrogen, phosphorus, and potassium on yield, persistence, and nutrient content of timothy. *Canadian Journal of Plant Science* 46, 577–582.

Griffith, W.K. and Murphy, L.S. (1996) Macronutrients in forage production. In: Joost, R.E. and Roberts, C.A. (eds) *Proceedings of Nutrient Cycling in Forage Systems Symposium*. The Potash and Phosphate Institute and The Foundation for Agronomic Research, Manhattan, Kansas, USA, pp. 13–28.

Gross, C.F. and Jung, G.A. (1978) Magnesium, Ca and K concentration in temperate-origin forage species as affected by temperature and Mg fertilization. *Agronomy Journal* 70, 397–403.

Grunes, D.L. and Welch, R.M. (1989) Plant contents of magnesium, calcium and potassium in relation to ruminant nutrition. *Journal of Animal Science* 67, 3485–3494.

Haby, V.A., Russelle, M.P. and Skogley, E.O. (1990) Testing soils for potassium, calcium, and magnesium. In: Westerman, R.L. (ed.) *Soil Testing and Plant Analysis*. Soil Science Society of America, Madison, Wisconsin, USA, pp. 181–227.

Heckman, J.R. (1995) Corn responses to chloride in maximum yield research. *Agronomy Journal* 87, 415–419.

Hodgson, J.G. (1990) *Grazing Management: Science into Practice*. Longman Harlow, Essex, UK, 203pp.

Horst, R.L., Goff, J.P., Reinhardt, T.A. and Buxton, D.R. (1997) Strategies for preventing milk fever in dairy cattle. *Journal of Dairy Science* 80, 1269–1280.

Jeffrey, D.W. (1988) Mineral nutrients and the soil environment. In Jones, M.B. and Lazenby A. (eds) *The Grass Crop*. Chapman and Hall, London, UK, pp. 179–204.

Jones, D.I.H. (1963) The mineral content of six grasses from a *Hyparrhenia*-dominant grassland in northern Rhodesia. *Rhodesian Journal of Agricultural Research* 1, 35–38.

Jones, R.J., Loch, D.S. and LeFeuvre, R.P. (1995) Differences in mineral concentration among diploid and tetraploid cultivars of rhodesgrass (*Chloris gayana*). *Australian Journal of Experimental Agriculture* 35, 1123–1129.

Kelling, K.A. and Matocha, J.E. (1990) Plant analysis as an aid in fertilizing forage crops. In: Westerman, R.L. (ed.) *Soil Testing and Plant Analysis*. Soil Science Society of America, Madison, Wisconsin, USA, pp. 603–643.

MacDonald, H.A. (1946) *The Relationship of the Stage of Growth to the Yield and Chemical Composition of Forage Plants*. Cornell Agronomy Department Mimeo No. 17. Cornell University Agricultural Experiment Station, Ithaca, New York, USA, 38pp.

McDowell, L.R. (1985) Common salt (sodium and chlorine), magnesium, and potassium. In: McDowell, L.R. (ed.) *Nutrition of Grazing Ruminants in Warm Climates*. Academic Press, Inc., New York, USA, pp. 213–235.

MacLeod, L.B. (1965) Effect of nitrogen and potassium on the yield and chemical composition of alfalfa, bromegrass, orchardgrass, and timothy grown as pure species. *Agronomy Journal* 57, 261–266.

MacLeod, L.B. and Carson, R.B. (1965) Effect of source and rate of N and rate of K on the yield and chemical composition of alfalfa and orchardgrass. *Canadian Journal of Plant Science* 45, 557–569.

McNaught, K.J. (1958) Potassium deficiency in pastures. I. Potassium content of legumes and grasses. *New Zealand Journal of Agricultural Research* 1, 148–181.

Martens, H. and Kasebieter, H. (1983) *In vito* studies of the effects of sodium and potassium ions on magnesium transport across the isolated rumen mucosa of sheep. *Zentralblatt für Veterinaermedizin [A]* 30, 1–14.

Mayland, H.F. and Wilkinson, S.R. (1996) Mineral nutrition. In: Moser, L.E., Buxton, D.R. and Casler, M.D. (eds) *Cool-Season Forage Grasses*. American Society of Agronomy, Madison, Wisconsin, USA, pp. 165–191.

Mika, V. (1982) Sodium, potassium and calcium concentrations in varieties of *Dactylis glomerata*. *Grass and Forage Science* 37, 321–325.

Miller, E.R. (1995) Potassium bioavailability. In: Ammerman, C.B., Lewis, A.J. and Baker, D.H. (eds) *Bioavailability of Nutrients for Animals: Amino Acids, Minerals, Vitamins*. Academic Press, New York, USA, pp. 295–301.

Miyasaka, S.C. and Grunes, D.L. (1990) Root temperature and calcium level effects on winter wheat forage: II. Nutrient composition and tetany potential. *Agronomy Journal* 82, 242–249.

Morillo, D., McDowell, L.R., Chicco, C.F., Perdomo, J.T., Conrad, J.H. and Martin, F.G. (1989) Nutritional status of beef cattle in specific regions of Venezuela. I. Macrominerals and forage organic constituents. *Nutrition Reports International* 39, 1247–1262.

Mtimuni, J.P., Conrad, J.H., McDowell, L.R. and Martin, F.G. (1990) Effect of season on mineral concentration of soil, plant and animal tissues. *International Journal of Animal Science* 5, 181–189.

[NRC] National Research Council. (1989) *Nutrient Requirements of Dairy Cattle*, 6th edn. National Academy of Sciences, Washington, DC, USA, 157 pp.

Odom, J.W., Haaland, R.L., Hoveland, C.S. and Anthony, W.B. (1980) Forage quality response of tall fescue, orchardgrass, and phalaris to soil fertility level. *Agronomy Journal* 72, 401–402.

Preston, R.L. and Linsner, J.R. (1985) Potassium in animal nutrition. In: Munson, R.D. (ed.) *Potassium in Agriculture*. American Society of Agronomy, Crop Science

Society of America and Soil Science Society of America Inc., Madison, Wisconsin, USA, pp. 595–617.

Preston, R.L., Pratt, J.N., Cates, J. and Sanders, J.L. (1994) Hot days…cool cows: potassium helps make the difference. *Better Crops* 78(1), 12–13.

Rahman, H., McDonald, P. and Simpson, K. (1960) Effects of nitrogen and potassium fertilisers on the mineral status of perennial ryegrass (*Lolium perenne*). I. Mineral content. *Journal of the Science of Food and Agriculture* 11, 422–428.

Reid, R.L. and Horvath, D.J. (1980) Soil chemistry and mineral problems in farm livestock: A review. *Animal Feed Science and Technology* 5, 95–167.

Reid, R.L. and Jung, G.A. (1965) Influence of fertilizer treatment on the intake, digestibility and palatability of tall fescue hay. *Journal of Animal Science* 24, 615–625.

Reid, R.L. and Jung, G.A. (1974) Effects of elements other than nitrogen on the nutritive value of forage. In: Mays, D.A. (ed.) *Forage Fertilization*. American Society of Agronomy, Crop Science Society of America and Soil Science Society of America, Inc., Madison, Wisconsin, USA, pp. 395–435.

Reid, R.L., Jung, G.A. and Murray, S.J. (1966) Nitrogen fertilization in relation to the palatability and nutritive value of orchardgrass. *Journal of Animal Science* 25, 636–645.

Reid, R.L., Post, A.J. and Jung, G.A. (1970) *Mineral Composition of Forages*. West Virginia University Agricultural Experiment Station Bulletin 589T, Morgantown, West Virginia, USA, 35pp.

Reid, R.L., Post, A.J. and Olsen, F.J. (1979) *Chemical Composition and Quality of Tropical Forages*. West Virginia Agricultural Experiment Station Bulletin 669T, Morgantown, West Virginia, USA, 43pp.

Reinbott, T.M. and Blevins, D.G. (1991) Phosphate interaction with uptake and leaf concentration of magnesium, calcium, and potassium in winter wheat seedlings. *Agronomy Journal* 83, 1043–1046.

Reinbott, T.M. and Blevins, D.G. (1994) Phosphorus and temperature effects on magnesium, calcium, and potassium in wheat and tall fescue leaves. *Agronomy Journal* 86, 523–529.

Reinbott, T.M. and Blevins, D.G. (1997) Phosphate and magnesium fertilization interaction with soil phosphorus level: tall fescue yield and mineral element content. *Journal of Production Agriculture* 10, 260–265.

Reith, J.W.S., Inkson, R.H.E., Holmes, W., Maclusky, D.S., Reid, D., Heddle, R.G. and Copeman, G.J.F. (1964) The effects of fertilizers on herbage production. *Journal of Agricultural Science* 63, 209–219.

Robinson, D.L. (1985) Potassium nutrition of forage grasses. In: Munson, R.D. (ed.) *Potassium in Agriculture*. American Society of Agronomy, Crop Science Society of America and Soil Science Society of America Inc., Madison, Wisconsin, USA, pp. 895–914.

Robinson, D.L., Kappel, L.C. and Boling, J.A. (1989) Management practices to overcome the incidence of grass tetany. *Journal of Animal Science* 67, 3470–3484.

Robinson, R.R., Rhykerd, C.L. and Gross, C.F. (1962) Potassium uptake by orchardgrass as affected by time, frequency and rate of potassium fertilization. *Agronomy Journal* 54, 351–353.

Robson, M.J., Parsons, A.J. and Williams, T.E. (1989) Herbage production: grasses and legumes. In: Holmes, W. (ed.) *Grass: Its Production and Utilization*. Blackwell Scientific Publications, Oxford, UK, pp. 7–88.

Rominger, R.S., Smith, D., and Peterson, L.A. (1976) Yield and chemical composition of alfalfa as influenced by high rates of K topdressed as KCl and K_2SO_4. *Agronomy Journal* 68, 573–577.

Roth, C.B. (1996). Mineralogy/redox effects on yield maps. In: *Proceedings, Information Agriculture Conference, July 30–August 1, 1996.* Urbana, Illinois.

Saiga, S., Kikuchi, M., Kanesaka, M. and Watanage, K. (1992) Variation in the mineral elements concentrations among cultivars of orchardgrass (*Dactylis glomerata* L.). *Journal of Japanese Grassland Science* 38, 175–182.

Sanchez, W.K., Beede, D.K. and DeLorenzo, M.A. (1994a) Macromineral element interrelationships and lactational performance: empirical models from a large data set. *Journal of Dairy Science* 77, 3096–3110.

Sanchez, W.K., McGuire, M.A. and Beede, D.K. (1994b) Macromineral nutrition by heat stress interactions in dairy cattle: Review and original research. *Journal of Dairy Science* 77, 2051–2079.

Schneider, P.L., Beede, D.K. and Wilcox, C.J. (1986) Responses of lactating cows to dietary sodium source and quantity and potassium quantity during heat stress. *Journal of Dairy Science* 69, 99–110.

Serra, A.B., Serra, S.D., Fujihara, M., Orden, E.A., Cruz, L.C. Ichinohe, T. and Fujihara, T. (1996) Monthly nutrient variation of paragrass (*Brachiaria mutica*) and stargrass (*Cynodon plectostachyum*) collected from pastures grazed by goats. *Asian–Australasian Journal of Animal Sciences* 9, 203–210.

Sleper, D.A., Vogel, K.P., Asay, K.H. and Mayland, H.F. (1989) Using plant breeding and genetics to overcome the incidence of grass tetany. *Journal of Animal Science* 67, 3456–3462.

Smith, G.S., Cornforth, I.S. and Henderson, H.V. (1985) Critical leaf concentrations for deficiencies of nitrogen, potassium, phosphorus, sulphur, and magnesium in perennial ryegrass. *The New Phytologist* 101, 393–409.

Stucki, J.W. (1996) Potassium in the soil: Is it there or isn't it? Some new findings that may explain its behaviour. *Better Crops* 80(4), 16–19.

Sweeney, D.W., Moyer, J.L. and Havlin, J.L. (1996) Multinutrient fertilization and placement to improve yield and nutrient concentration of tall fescue. *Agronomy Journal* 88, 982–986.

Thomas, B., Thompson, A., Oyenuga, V.A. and Armstrong, R.H. (1952) The ash constituents of some herbage plants at different stages of maturity. *Empire Journal of Experimental Agriculture* 20, 10–22.

Vose, P.B. (1963) Varietal differences in plant nutrition. *Herbage Abstracts* 33, 1–13.

Wilkinson, S.R. and Mays, D.A. (1979) Mineral nutrition. In: Bucker, R.C. and Bush, L.P. (eds) *Tall Fescue*. Agronomy Series No. 20. American Society of Agronomy, Crop Science Society of America, Soil Science Society of America, Inc., Madison, Wisconsin, USA, pp. 41–73.

Wilson, J.R. (1981) Environmental and nutritional factors affecting herbage quality. In: Hacker, J.B. (ed.) *Nutritional Limits to Animal Production from Pastures.* Commonwealth Agricultural Bureaux, Farnham Royal, UK, pp. 111–131.

Wolton, K.M. (1963) Fertilizers and hypomagnesaemia. *National Agricultural Advisory Service Quarterly Review* 16, 122–130.

Ylaranta, T., Uusi-Kamppa, J. and Jaakkola, A. (1996) Leaching of phosphorus, calcium, magnesium and potassium in barley, grass and fallow lysimeters. *Acta Agriculturae Scandinavica, Section B, Soil and Plant Science* 46, 9–17.

Nitrogen Management and Sustainability

7

S.C. Jarvis

Institute of Grassland and Environmental Research, North Wyke Research Station, Okehampton, Devon EX20 2SB, UK

Introduction

One of the key management tools for dairy cow production systems over the past few decades which has influenced outputs and economic performance has been the use of chemically fixed nitrogen (N) fertilizer. When the other major controlling influences over plant growth, i.e. other nutrients, water supply and other climatic variables, are non-limiting, then grasses and many other forage species respond markedly to N fertilizer application in a classical response curve pattern. Dairy farmers have capitalized on this response to increase their output and productivity per unit area and/or to reduce production costs. As well as the increase in dry matter yield obtained, the directness of the response means that there has been much to be gained in terms of overall management flexibility of the system. Because of these advantages, rates of application increased substantially: over the period 1950–1980 average rates applied to cut swards increased from 50 to 253 kg ha^{-1} in the UK (van Burg and Prins, 1981) and although recently there has been a slight decline, rates of application are still high. The driving force for the high rates of N used has therefore been directed primarily at increased dry matter production (= increased carbon (C) fixation) in order to sustain carbohydrate supply for high yielding dairy cows, and not because there has been a direct need for extra N to enhance animal performance *per se* (see Chapter 13). This, as shown later, means that there are often large excesses of N within the totality of a dairy farm system which invite opportunities for large scale inefficiency and losses and, as a consequence, opportunities to improve efficiency and reduce inputs. The objective of the present chapter is to examine current principles determining the use of N within intensive dairy production management in the light of current knowledge and to explore opportunities for increasing the efficiency and the sustainability of the system. It is not an objective to prescribe practical protocols for on-farm management of N supply, but to describe the basis on which changes in management, where these are required, should be based.

As with all nutrient management requirements, the three major management decisions about N inputs are (i) an assessment of current nutrient status (e.g. soil/plant N status), (ii) the establishment of economic input thresholds and probable financial returns and (iii) the development of strategies for ensuring adequate nutrition for plant and ruminant while protecting soil, water and air resources and quality. While (ii) has been the over-riding determinant for ensuring the economic viability of enterprises, increasingly, more attention has been given to (iii) because of widespread concern and legislative pressures over leakage of various N materials to the wider environment. It is likely that this concern will continue to grow, because of direct effects, perceived or otherwise, on water and atmospheric quality and also because of a general interest in lower input production.

The amounts of N involved in any annual production system are large, and especially so in intensive dairying. Table 7.1 is a modelled estimate of the animal total of N entering, circulating or recycled and contained within an 'average' UK dairy farm as defined by Jarvis (1993). The amounts shown are for a moderately intensively managed system (with 250 kg N ha^{-1}): even in low input managements the amounts of N involved would be large, and for many of the categories shown in Table 7.1, would be greater than would be present in an equivalent area devoted to arable production. The key to increasing N-use efficiency within the dairy system is to maximize the utilization of those amounts which are circulating within the system; because of the chemical nature of N and its interactions with the biological components of the cycle, the recycling phases often present opportunities for inefficiency (Table 7.2).

The grassland N cycle has received much attention over recent years and the detail of our current understanding has been thoroughly reviewed either as discrete components of the whole system, namely mineralization (Jarvis *et al.*, 1996a), grazing impact (Jarvis *et al.*, 1995), gaseous losses from farm slurries (Stevens and Laughlin, 1997) or *in toto* (Whitehead, 1995). Whitehead provides a detailed description of each of the separate component parts contributing to the behaviour of N in grassland. There are also models that can be used to describe the cycle in its entirety or in part and at various levels of complexity ranging from the highly mechanistically based and integrating C and N flows (Thornley and Verberne, 1990), through to mass balance/budgetary approaches (Aarts *et al.*, 1992; Jarvis, 1993; Weissbach and Ernst, 1994; Jarvis *et al.*, 1996b), based on empirical data or

Table 7.1. Nitrogen in components of a dairy farming system (calculated for a typical 76 ha UK dairy farm: data abstracted from Jarvis, 1993).

	Nitrogen flows and pools	
	kg ha^{-1}	Farm total (t)
Inputs		
Fertilizers	250	19.00
Feeds and bedding	52	3.93
Others (atmospheric deposition + fixation)	35	2.66
Internal transfers		
Forage		
grazed	171	13.03
ensiled	117	8.93
Excreta		
dung	99	7.56
urine	172	13.10
Manures: stored	120	9.14
Present in animals[1]	28	2.16
Soil		
Total content[2]	5239	398.13
Released from mineralization	157	11.96

[1] Assumes an average of 81.7 kg crude protein per mature cow (from Gibb *et al.*, 1992).
[2] 0–10 cm only: calculated from a range of values for soil under different managements (from Gill *et al.*, 1995).

models such as that described by Scholefield *et al.* (1991). Nitrogen cycling is under a complex set of controls: Fig. 7.1 provides a simple representation of the flows within the dairy farm. It would be possible to produce much more elaborate diagrams for each component of this system to demonstrate in a mechanistic way all the controlling and interacting variables and factors involved, but for present purposes, Fig. 7.1 provides a useful, simple conceptual description of the major constituents for dairying.

Components of the System

The manner in which the management of the farm system interacts with N flows (Fig. 7.1) is illustrated in Fig. 7.2. This demonstrates the need not only to have an appreciation of each of the individual components, and the implications of on-farm decision making for N flows and utilization within those components, but also the need for an overall view of the way in which the total system is operating and influencing N inputs and outputs.

Table 7.2. Origins and sources of N losses within dairy farming systems.

Management stage	Loss 'generator'	Volatilization (NH$_3$) Mode	Importance[1]	Denitrification (N$_2$O/N$_2$) Mode	Importance[1]	Leaching (NO$_3^-$) Mode	Importance[1]
(a) Herbage Forage production	Fertilizer	Direct loss (from urea) by volatilization	**	Direct loss from fertilizer	***	Direct transfer into drainage/surface waters	*
	Soil organic matter (mineralization)	—	—	Reduction of mineralized/nitrified mineral N	*	Leaching of mineralization release at cultivation	**
Grazed swards	Fertilizer + excreta	As above + volatilization from urine	**	As above plus denitrification of mineral N from dung + urine patches	**	As above plus transfer of excess soil mineral N generated from urine	****
(b) Housed animals: milking parlour, hard standing, housing	Collected excreta	Direct volatilization from urine on floors and bedding	***	—	—	—	—
(c) Dirty water	Excreta in washings	Volatilization during collection, storage and disposal	*	Some denitrification of mobile N	*	Transfer of dissolved and mobile organic N	*

(d) Manures and slurries:

	Source						
Storage	Urinary N	Volatilization from lagoon/tank, etc.	*	—	—	—	—
Application	Urinary N	Volatilization during spreading	****	Denitrification of mineral N	**	Direct transfer of mobile N into waterways	*
Residual effects	Mineralization of organic materials	—	—	Denitrification of mineralized N	**	Transfer of excess mineral N released from accumulated organic matter	**

[1] Importance: ranges from * least to **** most influential.

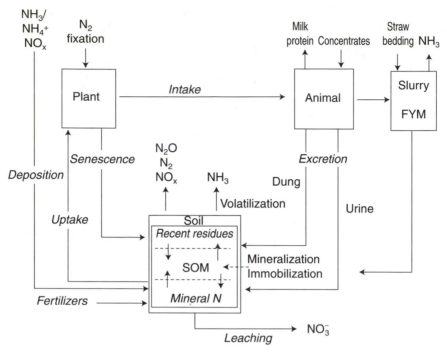

Fig. 7.1.   The nitrogen cycle in dairy farms.

There is therefore need to integrate the impact that on-farm decision-making processes within the dairy farming system (and the external influences governing managerial decisions) have on the way in which N is added, transferred and utilized, or is lost. Grassland systems have an extremely diverse range of managements (i.e. species/breeds/cultivars, grazing/cutting patterns, forms and quantities of inputs and efficiency of resource utilization), environments and soil types. In order to understand more clearly how particular components contribute to the whole and to identify opportunities for future improvements an understanding of the key processes involved is essential. Whitehead (1995) provides a comprehensive review of many of these, but a brief description follows.

Soils

At the heart of any discussion on the sustainability of N is the role that soil processes play. It is here that the majority of the processes which control or release N into forms which are potentially useful for the plant and can then enter the farm cycle (Fig. 7.1) for further transfer and utilization or, alternatively to be potentially available for loss, take place. While some important

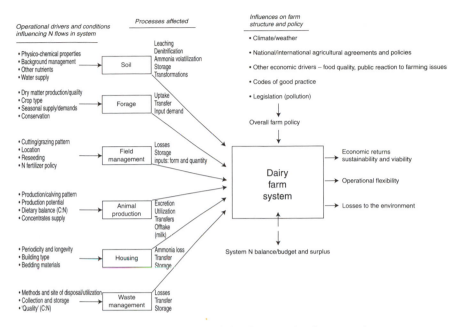

Fig. 7.2. Controls over, and impact of, the flows and utilization of nitrogen in dairy farming systems.

losses occur during housing or waste management phases through ammonia (NH_3) volatilization, the other processes of uptake, mineralization, nitrification, denitrification and leaching are facilitated through the medium of the soil.

Much N is contained within soils (see Table 7.1), a majority of which is held within the organic materials which accumulate over time. Unlike arable systems which, depending on the cultivation practice, more or less attain an equilibrium in soil organic matter (SOM) content and hence background N contents, grassland soils are rarely at equilibrium with respect to SOM and N contents. This makes prediction of the impact of SOM on N behaviour difficult. The other major contrast with arable soils is in the amounts of N present: typically an old pasture soil contains 3–10 t N ha^{-1} compared with 1–4 t N ha^{-1} (to 15 cm) in an arable soil (Archer, 1988). Nitrogen in SOM can be categorized into many different physical and chemical categories (Haynes, 1986), and ranges from materials which have been recently returned or added (e.g. manure, dung, plant residues – Fig. 7.1) to those which may be up to thousands of years old, with a range of degradation/recalcitrance capacities. The relative proportions of these materials is a reflection of past managements and soil/environmental conditions. The current balance of these forms, their physical location and interaction with current environmental/managerial conditions determines the rate of delivery by the action of soil organisms (macro- and micro-) of N into the soil mineral N pool (*mineralization*). There are many specific models of SOM matter turnover which provide arbitrary sub-divisions of the organic materials in order to aid prediction of effects (see de Willigen, 1991). In spite of these, it is, at present, particularly difficult to make prediction of the extent of release of N from SOM in grassland (Jarvis *et al.*, 1996a). This is one particular area in which progress could be made in increasing efficiency because the amounts released from grassland soils into the soil mineral N pool can be substantial and are not fully taken account of in fertilizer recommendation schemes. Recent research has shown that, depending on previous background managements of long-term temperate grassland, more than 300 kg N ha^{-1} can be released during an annual cycle (Hatch *et al.*, 1991; Gill *et al.*, 1995; Clough *et al.*, 1998).

An important component of the SOM pool, which can be regarded as acting as an intermediate ('the eye of the needle through which all organic matter must pass'; Jenkinson, 1990) between SOM and the inorganic pool, is the soil microbial biomass. Nitrogen moving into or from this pool provides the biological mechanism by which mineral N is removed from, in competition with other removal processes, or as the biomass turns over, added to the available pool in the soil (Fig. 7.1).

The soil mineral N pool contains ammonium (NH_4^+) and nitrate (NO_3^-) (Fig. 7.1). Nitrogen can enter as NH_4^+ from degraded organic materials including dead soil microbial biomass and plant residues, urine, farm manures and wastes or directly from some forms of fertilizers. Nitrate N enters either directly from fertilizer addition or after the microbial process of *nitrification,* i.e. the oxidation of NH_4^+ though nitrite (NO_2^-) to NO_3^-. This takes place readily in most soils, but is influenced by aeration and moisture status. It is an important controlling process in most grassland soils because of the large amounts of NH_4^+ added from urine and manure as well as from mineralization. Plants can utilize both forms of N: grassland species, under low temperatures, preferentially utilize NH_4^+ over NO_3^-. Under many circumstances, the natural progression in the soil is to result in release of highly mobile NO_3^- ions into the soil solution. Excess NO_3^- over and above that of the immediate demands for uptake by the crop at any particular time, can create the potential for substantial loss through leaching in excess precipitation (Scholefield *et al.*, 1993) or denitrification. Ammonium ions are not usually mobile because they are bound to cation exchange sites in the soil. However, when, for example, large volumes of slurry or dirty water are added to land or direct losses from fertilizer occur, some movement of NH_4^+ into drainage can take place. There can also be large losses from the soil surface through volatilization when urea in either fertilizer or urine or slurry/manure is added before NH_3 is transformed into ionic form and bonds with exchange surfaces on soil particles.

Nitrate is also 'removed' by transformation into gaseous forms by the *denitrification* process, in the main by microbiological activity. This occurs under anaerobic conditions when the soil is wet and so takes place to the greatest extent in poorly drained soils. In this process, NO_3^- acts as an electron donor to the denitrifying organisms and is transformed into NO_2^- in the first instance and then, sequentially to nitrous oxide (N_2O) and N_2 gases, both of which may be emitted to the atmosphere. Nitrous oxide is an important greenhouse gas and also interacts with ozone and has, therefore, important environmental effects (Granli and Bockman, 1994). Denitrification is an important contributor to the overall losses from a dairy farming system (see Jarvis, 1993: Jarvis *et al.*, 1996b). N_2O is also released during the nitrification process. Nitrite is also an intermediate in both processes and, although generally considered to have a very short half-life, has been found in significant concentrations leaving some Irish soils (Burns *et al.*, 1995) and poses a possible toxic hazard to aquatic fauna.

Other exchanges which may influence directly the pool of inorganic N in the soil are the deposition of atmospheric NH_3 and NH_4^+ (dry and wet, respectively) particularly in down-wind areas from intensive animal production. There may also be significant deposition of NO_x from transport and industrial sources.

The basic requirement for good N management is to attempt to balance flows into the mineral N pool against the demand by the crop, avoiding deficiencies at times of peak growth rate and surpluses at other times. Good fertilizer practice for grass dry matter production aims to do that, but because of the benefits to be had from N fertilizer, use has generally erred towards generosity and excess.

Plants

The growing crop provides another major, albeit temporary, sink for N within the system. Whitehead (1995) has calculated that grass harvested either by cutting or grazing in a temperate environment and producing 8–15 t DM usually contains between 200 and 350 kg N ha^{-1}. If the stubble and roots are also taken into account, then the total plant pool increases to 300–800 kg N ha^{-1}. Plant growth responds in a classical way to N application, although there are some substantial species differences in the characteristics of the growth curve (Doughtey and Rhykerd, 1985) which can be influenced by management (Fig. 7.3a).

In general, the more N that is applied to the soil, the more N there is in the plant. This is not only important from the plant response and herbage quality points of view, but any resultant changes in the C : N balance of cows' dietary intake, may, in turn, influence the retention of N and the extent and mode of excretion by the animal. Plant contents can range between 1% and 5% N (Whitehead, 1995) on a dry weight basis, with

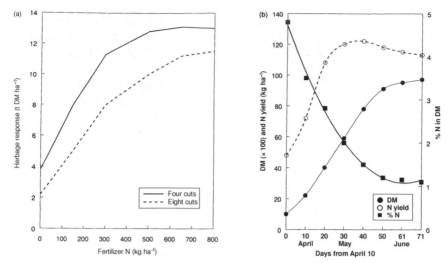

Fig. 7.3. Nitrogen and grass growth. (a) Effects of cutting regimes on herbage dry matter production and response to N: data replotted from Frame *et al.* (1989). (b) Changes in dry matter and N yield (kg ha^{-1}) and N concentration over spring growth period for ryegrass grown in SE England (unpublished data from Green and Corrall, 1965, and replotted from Whitehead, 1995).

distinct annual patterns of decreasing contents with maturity of the sward (Fig. 7.3b): in general, N uptake, for which the capacity for in grasses is high, mirrors the growth pattern (Fig. 7.3b). Managerial influences on the utilization of herbage has substantial effect on the herbage N concentration (Frame, 1992). Grazing pattern also has a considerable influence on the manner of N utilization, however, response to N input as measured by live-stock production criteria also follows a classical response pattern of diminishing returns (Baker, 1986). Under conditions of no constraints, uptake capacity by plants (which is almost entirely in inorganic 'i.e. NH_4^+ and/or NO_3^- form), from the soil solution is high even when only low concentrations are present (Wild *et al.*, 1987). In terms of increasing utilization of N within the system, three concepts relating to the crop are important (Follet and Wilkinson, 1995), i.e. those of (i) N recovery (the N captured in the plant as a proportion of that added), (ii) N requirement and response (dry matter production response to added N) and finally (iii) N use efficiency (the rate of increase in yield per unit of fertilizer applied). Management of the interaction between plant response and N utilization demands a balance of a number of factors: the determination of a realistic yield under the prevailing soil, environmental/climatic and economic conditions, the presence or absence of other limiting factors and the logistics required for the delivery of a good yield potential.

Inputs and transfer of N can also have other effects demanding managerial skills. For example, by influencing botanical composition and possibly affecting the tillering structure of the sward (Wilman, 1965) on some occasions; although when dairy cows grazed perennial ryegrass swards in the Netherlands grown with from 200 to 700 kg N ha^{-1} there were no consistent effects of N on tiller numbers per unit area (Deenen and Lantinga, 1993). Addition of N to mixed legume–grass swards has a major effect on the dynamics of temporal and spatial changes in the balance of white clover and perennial ryegrass in temperate regions.

Species with C$_4$ metabolism are usually more efficient in N utilization than C$_3$ plants (Follet and Wilkinson, 1995): this effect is well illustrated by the performance of maize/corn (*Zea mays* L.), which can produce large quantities of dry matter of relatively low N content. Incorporation of maize into dairy farming managements as an energy rich forage source is an important feature in many environments and the geographic range under which it can be grown has, and is likely to continue to, increased as the result of breeding advances. The use of maize has important implications within a farming system context for the utilization and transfer of N because of a relatively high rate of utilization by the ruminant. There is a large literature describing responses of maize/corn crops to N fertilizer covering a wide range of different effects. The problem of prescribing N fertilizer on a generalized basis for this crop is made difficult because of the wide range of circumstances in which, and for which, it is grown, not the least because of the way in which farm manures and wastes are often used in association with maize growing. Nevertheless, model development is progressing (Cerrato and Blackmer, 1990) and approaches such as a classification system to evaluate a range of factors on crop yield response have been developed which aid the organization and evaluation of the many factors that influence crop yield (Lory *et al.*, 1995). Maize is being increasingly incorporated into new management systems because it may present an opportunity to reduce the overall farm N budget as well as providing an excellent dietary material for high performance dairy cows (Hofman and ten Holte, 1992).

The dairy cow

The major influence of the productive dairy cow on the behaviour of N is as a consumer, processor, excreter and transfer agent, either at grazing or during the housed phases of the production cycle. Large amounts of the ingested N are excreted in dung and urine (Table 7.1). At best, the theoretical maximum efficiency of conversion of N into milk is 40–45% and most usually is less than 25% (van Vuuren and Meijs, 1987). In the typical UK dairy farm case study described earlier (Jarvis, 1993), 19.9% of animal N intake was converted into milk. Consumption, processing and excretion of N as dung and urine therefore have an all important effect in transforming N from a captured, immobile form in forage into other forms which are again accessible, or potentially accessible, for re-utilization or loss (Table 7.2).

As well as these direct effects on N, there are other effects of the animal, which may directly or indirectly affect N transfer or utilization:

- removal of shoots – cutting or grazing – resulting in temporary disruption of the uptake process and removal of sink coupled with changes in resource allocation in the plant;
- increased returns of plant residues with grazing as compared with cutting;
- treading, compaction and poaching of soil – with effects on soil microbial activities involved in mineralization, nitrification, denitrification, through changes in soil structure and porosity;
- physical transfer of N from field to field and to milking parlour and housing and waste store;
- possible impact on sward composition through dietary selection and preference;
- diet rejection through contamination by dung.

Nitrogen excreted indoors and accumulated in manures or slurries represents another large internal N pool within the farming system and a management challenge in being utilized effectively (see later). Before that stage, however, there is opportunity for loss from shed or manure store as ammonia (NH_3). A recent inventory (Pain *et al.*, 1998) of NH_3 loss indicated that over 50% of all UK emissions were derived from cattle production, of which over 40% occurred from housing and manure storage. Ammonia can also be released from grazed swards as well (Bussink, 1996). In general, this loss, as demonstrated with other cattle (Jarvis *et al.*,1989) increases with N input into the system. This effect almost certainly acts though increased dietary N intake and an increased output in urine and Whitehead (1995) has derived a generalized relationship which suggests a linear response between N intake and excretion in urine: thus at 1.5% N in the diet the proportion of the excreted N in urine is c. 45%, which increases to c. 80% of the total at 4% N.

A dairy cow excretes between 10 and 40 l urine day^{-1} (Haynes and Williams, 1993) (the volume of which in large part depends on diet), on 8–12 occasions per day, which contains 2–20 g N l^{-1}. Although NH_3 can be lost directly from urine patches, much of the N in urine quickly infiltrates the soil and enters the soil N transformation cycle. Whitehead (1995) has calculated that 24% of a grazed area is affected by urine, whereas 7% of the area is covered by dung. Typically, dung contains between 1.2 and 4% N on a dry matter basis and is excreted on 7–15 occasions per day to return from 2.5 to 5 kg DM day^{-1}. Contents of mineral N in dung are low and there is little direct loss as NH_3 and the further involvement of the organic N in dung in the recycling processes then depends upon the rate of incorporation into the soil. Incorporation is very much weather dependent and the interaction of moisture, temperature conditions with soil macro fauna and earthworm activity. Under intensive dairy management the rate of disappearance has been recorded as 60 days in spring and 40 days in summer (Bastiman and van Dijk, 1975). Despite these often large returns of excretal N in grazed swards, recorded effects on overall soil N accumulation have been small and variable. Thus whereas Hassink and Neetesen (1991) recorded an overall accumulation of 150 kg ha^{-1} year^{-1} in grazed compared with cut swards, others have found no significant difference (Clement and Williams, 1967).

Manures and slurries

Depending on the management system, dung and urine from the housed animals are usually collected together as semi-liquid slurries containing more or less water depending on mixing with rain water and parlour/yard washings, or more solid manures with straw and other similar bedding materials (farmyard manure (FYM)) during milking and/or the housed phases of the production cycle. The amounts of N excreted per cow under 'typical' management conditions during the winter period (180 days) of a UK production cycle are equivalent to 138 kg N ha^{-1} (Jarvis, 1993). These materials are valuable nutrient reservoirs (Fig. 7.1; Table 7.2) containing much N which should be regarded as a useful resource: in practice, within many dairying systems, because this has to be utilized on the grassland on the farmstead, a number of logistical difficulties can occur. Although much research has been conducted on the nature, utilization and recovery of N supplies, effective use of organic manures still presents a considerable challenge to researcher, advisor, and regulatory bodies (Sims, 1995) in order to integrate existing expertise and knowledge, legislative requirements and nutrient use efficiency. Variability is one of the major problems in making use of these organic materials as efficient suppliers of nutrients.

Superimposed on the variable properties of the manure are then the effects of mode and duration of storage, and method of dispersion on land (often an operational decision determined by windows of opportunity and available land). One of the other major factors influencing the quality of the waste with respect to N context is the opportunity for loss that occurs through varying stages from excretion to spreading (Table 7.2). The losses are, in the main, associated with the change in form from urea to NH_4^+/NH_3 and the release of N as gaseous NH_3. This may occur in the animal house/milking parlour (sufficient urease enzyme is present in most environments to promote the hydrolysis reaction), from the slurry store or manure heap and especially during dispersion on land (Pain *et al.*, 1986).

For one reason or another, manures have a wide range of total N contents and in the ratio of immediately available (NH_4^+) and organic N which will also have a range of half-lives influencing the subsequent rate at which

Table 7.3. Efficiencies (%) of cattle slurry N relative to NH₄NO₃ fertilizer applied in spring (from Pain *et al.*, 1986).

	Winter	Spring	Summer
Apparent N recovery (% N applied)			
Mean	9	22	11
Range	(0–5)	(14–31)	(9–17)
Herbage yield			
Mean	15	31	21
Range	(8–25)	(15–46)	(13–35)

mineralization occurs. This, coupled with changing impact of environmental conditions on uptake by plants at different growth stages means that a wide range of responses and efficiencies of N recoveries has been recorded when slurries, for example, have been applied to grass swards (Table 7.3). There have been two basic approaches to assessing the effectiveness of utilization of organic N (Sims, 1995). The first is to provide decay estimates over time which are usually determined by laboratory mineralization studies and which provide a quantitative estimate of the amount of mineral N that will be mineralized over a period of years and ideally needs multi-year comparisons. The second is to determine a fertilizer equivalent which determines N availability empirically in field trials in comparison with fertilizer N (Sims, 1995). Figure 7.4 shows an example where the amount of total N needed to be supplied from dairy manure to achieve a specific yield or N uptake equivalent to that obtained with fertilizer can be determined. Even with such assessments, the seasonal factors determining grass growth response and NH₃ losses interact to provide a wide range of effects (see Table 7.2).

Within a management system context, consideration of other nutrients must also be made. Build up of P to levels that may exceed soil sorption capacity through manure application has been demonstrated as being important in intensive dairying regions (Sharpley and Menzel, 1987). Even subtle build-ups of P, largely through concentrate input and transfer to organic wastes, may be important in determining P losses and subsequently eutrophication (Haygarth *et al.*, 1998). As well as the effects of surpluses, the N : P : K ratios in manures are usually distinctly different from those required by plants: the K content of slurry is often high relative to that of plant demands, for example. Inappropriate balances can therefore lead to excesses and the potential for enhanced losses of under-utilized N.

N Losses

Over the last 10 years, the extent, forms and pathways of N loss from intensive animal production systems have been much researched and reviewed

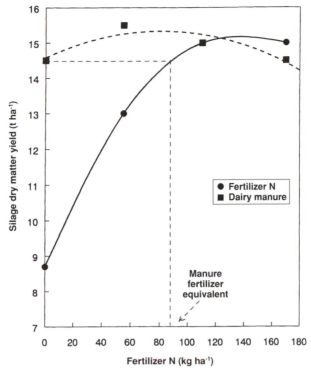

Fig. 7.4. Demonstration of the fertilizer equivalent method to determine the N fertilizer value of added dairy manure applied at 9 t ha^{-1} year^{-1} (Jokela, 1992).

(Haynes and Williams, 1993; Jarvis *et al.*, 1995) and a good deal of information exists on the losses associated with the processes of leaching (Scholefield *et al.*, 1993), denitrification (de Klein *et al.*, 1996) and NH$_3$ volatilization (Bussink, 1996). Relationships between N inputs to the particular system and loss can be established, but because of the complexity and interactive nature of the overall system, it has not always been easy to make reliable predictions of the extent and the ways in which N may be transferred from the system. Within a dairy farm, the opportunities for N loss are many (Table 7.2).

Much of the experimental data has been encapsulated into mathematical models for grassland systems, see for example Scholefield *et al.* (1991) and van de Ven (1996). Although these may differ in the conceptual basis and assumptions that are made to determine the principles on which they operate, all aim to provide a means of predicting production and N flows. By integrating information for all components of the farm management (Tables 7.1 and 7.4) it is possible to define the overall picture as a N budget which demonstrates quite clearly the scale of the losses which can occur

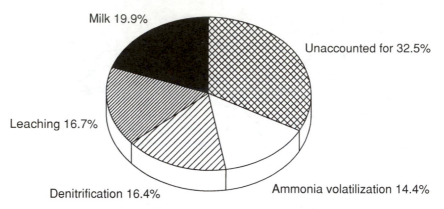

Fig. 7.5. Fate of annual input of N (25.6 t N) to typical UK dairy farm in milk, lost by leaching, denitrification or NH_3 volatilization or unaccounted for (data calculated from Jarvis, 1993).

(Fig. 7.5). In this particular example, the total losses from the modelled farm were equivalent to 47.5% of the annual N input of 25.6 t (337 kg ha^{-1}) from *all* sources. A significant proportion was unaccounted for and, although it is likely that much of this may be sequestered into SOM it is also possible that significant proportions are also lost as emissions to water or atmosphere because of current inadequacies of prediction or measurement.

Responses to Nitrogen and Recommendations

Responses

The general response of herbage to inputs of N is well established and repeatable on a worldwide basis for a wide range of forage types. Broadly speaking, the pattern of response as shown in Fig. 7.6, which is typical for ryegrass in temperate conditions, will be same as that, for example, by bermudagrass and bahiagrass in Florida (Overman *et al.*, 1992). It is largely based on the shape of this response curve that fertilizer recommendation schemes have been developed, in particular the initial steep slope phase of the relationship which is usually dependent upon the background supply of N from SOM. Within the general framework of the curve shown in Fig. 7.6, the effect of SOM plus those of soil, site and environmental account for the very wide range of response characteristics that can be determined across a range of sites (Hopkins *et al.*, 1995). Response trials of this nature have, in the main, been determined for cut swards. When grazing is involved the growth responses are likely to be even more variable through first the difficulty in obtaining meaningful measurements of sward performances under grazing conditions and, second, the further degree of variability

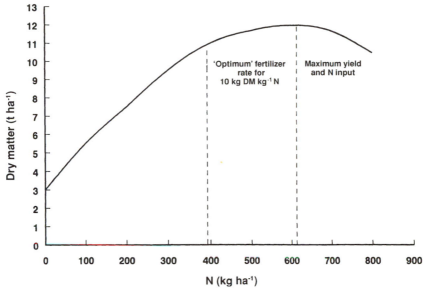

Fig. 7.6. Stylized growth response to added N defining maximum and economic 'optimum' rates of addition: graph redrawn from Holmes (1989).

superimposed by the animal. Nevertheless, some trials have attempted to measure response to N in terms of animal stocking criteria (Baker, 1986).

Other researchers have used final product output as the measure of response to N input. Thus, Holmes (1968) and, more recently Deenen and Lantinga (1993) in the Netherlands have established suggested targets and measured responses, respectively, of milk output per hectare in response to N application rate. There is good agreement between the two data sets in establishing another typical response curve (Fig. 7.7). In the Dutch studies, the response represented c. 16 kg milk kg^{-1} fertilizer N: there were no significant increases in the daily yield of milk per cow or in liveweight gain and the enhanced milk yields therefore were achieved by an increase in stocking rate.

In practical terms, most assessments of N requirements are based on plant responses. Maximum yield is usually of little relevance because of the low returns in terms of dry matter yields per capita of N applied as the maximum is reached. For advice purposes therefore, an economic optimum response has been defined as the point at which herbage response falls to a designated level of herbage production. For example 10 (Holmes, 1989) (Fig. 7.6) or 7.5 (Thomas *et al.*, 1991) kg dry matter kg^{-1} N applied are often used as the criteria.

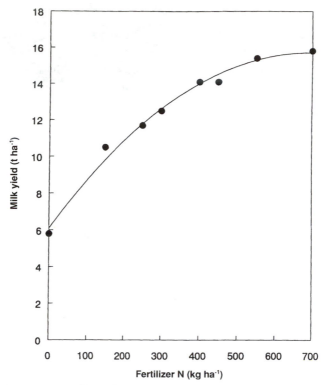

Fig. 7.7. Response in milk production to N fertilizer addition: data taken from Holmes (1968) and Deenen and Lantinga (1993).

Current recommendations

Current recommendations for N fertilizer inputs are therefore based largely on the economic criteria associated with dry matter production. Large numbers of experiments throughout Europe, for example, have established that optima generally lie within the range 200–400 kg N ha^{-1} year^{-1} (Prins *et al.*, 1988). The aims of advice packages have been to provide broad bands of recommendations which cover relatively wide ranges of conditions and circumstances, and because of the high returns obtainable from a relatively cheap product, the inclination has been to target the upper rather than the lower bounds of the range. Current recommended maximum rates in the UK are 300–380 kg N ha^{-1} for grazed grass and from 340 to 420 kg N ha^{-1} for cutting systems (Thomas *et al.*, 1991; MAFF, 1994).

Management criteria are also taken into account, so that recommendations differentiate at least to some extent between cutting and grazing managements, established and new swards, and define broad categories of potential background N supply from the soil (determined largely by immediate past management, the pattern of cutting and/or grazing and in some cases some categorization of the nature of the site, e.g. soil texture × rainfall). Essentially, however, such schemes are broad brush approaches and do not, for example, take true account of inputs of N from all other sources, namely excretal returns from grazing animals, farm manures, and especially the supplies from mineralized SOM. In the past this has not been seen to be a major problem because forage crops generally have appeared to be increasingly efficient at removing high proportions of the N applied. Apparent N recoveries in the Netherlands increased from 60% or less before 1970 to up to 90% in the 1980s (van der Meer *et al.*,1986). This in fact may mask overall effects because efficient recovery by the plant may represent high levels of N in the diet and consequently high levels of recycled N within the system which may be potentially difficult to manage and retain.

Future trends

The increasing drive towards increasing efficiency of N use, with reduced environmental impact primarily in mind, but with improved utilization of a valuable resource as another important issue, has stimulated the development of alternative approaches to the convention of starting with the economic optimum and applying adjustments of the nature of those described above. A major deficiency of this approach is that no account is taken of the major influences of weather patterns on N transformations and the extent of accumulation of mobile mineral N which may not be usefully processed, and therefore lost. There is therefore little or no consideration of the need to link the economic output of the system with the efficiency of utilization, losses and environmental impact. A tactical approach to fertilizer N application has recently been developed which takes account of the amount of mineral N present in the soil before each fertilizer addition which is then applied at a rate pre-determined by a model in order to achieve a given efficiency of N use (Titchen and Scholefield, 1992). Adoption of this approach on experimental field sites on commercial farms in the UK demonstrated the potential of this method, based on soil testing in the field, for maintaining herbage production while reducing both the amount of fertilizer applied and the level of residual mineral N in the soil in the autumn. This approach is being further developed so that the soil testing procedure is more closely integrated with a more sophisticated modelling approach based on a model of the grassland N cycle (Scholefield *et al.*, 1997). This approach will enable links between fertilizer and other N inputs, grassland production and environmental impact to be established in such a way that dairy farmers can understand their own system and manipulate their managements in order to achieve the appropriate balance of economic and environmental goals for their enterprise.

Nitrogen in Dairy Farming Systems

The development of the soil testing approach described above represents the adoption of one particular diagnostic tool to good advantage for one particular component of the total management system. This has potential for lowering inputs and reducing the total N surplus for that system. As our knowledge of the behaviour of N within grassland managements has increased, so it has become more and more apparent that the interactions that occur which affect the transformations, utilization and loss of N through climatic and environmental factors and through managerial decisions and actions, require a comprehension of the total system. It is only through such knowledge that the impact of one action taken in response to a particular issue can be fully accounted for in the remainder of the production cycle. Thus reduced NH_3 volatilization from manures and slurries is likely to have consequences for availability for uptake and for losses by denitrification, for example.

A number of methods have been taken to develop systems understanding and interactions. At the experimental scale, a number of 'farmlet' units have been established to examine changes in N management and their sustainability. Thus Laws *et al.* (1995) have described a replicated farmlet experiment which compared a 'conventional' management for beef cattle with those based on either white clover and injected slurry (as compared with surface spread) with a nitrification inhibitor or on a tactical N fertilizer methodology also with injected slurry. This farmlet experiment operated within a 1 ha experimental unit to integrate elements of cutting, grazing and slurry return. At a slightly larger scale, a comparison of dairy farm N management has recently been described (Peel *et al.*, 1997). This experiment, known as MIDAS, compared three management systems (Table 7.4) which were (i) a conventional management following good practice and based on high output (economic optima for fertilizer additions, slurry broadcast, least cost minimum 18% crude protein in feed: (ii) a reduced loss, high output system (tactical reduction in fertilizer, diluted broadcasted slurry, incorporation of maize silage and no surplus effective rumen degradable protein) and (iii) a minimal loss, reduced intensity system, but with the same total output from a greater total area (planned reduction in fertilizer, slurry injected, no surplus effective rumen degradable protein). All systems aimed to produce 6000 kg milk per cow. The initial results (Table 7.4) demonstrated some significant reductions in leaching and other losses and the inputs were substantially reduced in both alternative systems. Even in the case of system 2, however, it was estimated that the financial margins per hectare were reduced by c. 10% compared with current commercial practice. If achievement of environmental targets carries financial reward then this difference becomes less important.

Table 7.4. Initial nitrogen balance sheets for MIDAS experimental dairy managements (data abstracted from Peel *et al.*, 1997) kg N ha^{-1} year^{-1} (average data for 2 years).

		Management system	
	Conventional	Reduced loss, high output	Minimal loss reduced intensity
Inputs			
Fertilizers	325	190	146
Feed	117	116	90
Atmosphere	30	30	30
Total	472	336	266
Outputs			
Milk	66	64	54
Liveweight gain	6	7	4
Total	72	71	58
Surplus	400	265	208
Measured leaching losses	58	29	24
'Modelled' total losses	175	115	57

Other comparisons of N flows and surpluses in fully operational whole farm systems have also been made: Table 7.5 compares the N surplus in an experimental (De Marke) dairy farm in the Netherlands with those in two conventional farms following good practice. The specific aim of the De Marke system was reduced N inputs, surpluses and losses and as the data in Table 7.5 indicate, the inputs to this system have been reduced to 72% of those for the conventional systems and the surplus N to 65% within the first 2 years of operation. Total output of N in the product (milk) did not change a great deal.

Other total farm balance studies have been conducted on commercial farms (e.g. Bacon *et al.*, 1990, USA; Ferrugia *et al.*, 1997, France) and these provide a valuable insight into the overall nutrient (N) management packages for dairying systems which could not be obtained from conventional small-scale experimental plots examining nutrient dynamics. Relationships are being established between inputs and surpluses within current managements (Ferrugia *et al.*, 1997; Neeteson and Hassink, 1998) which may help to identify trends and outliers and then to establish the background causes for those outliers. Farm nutrient balancing/budgeting by farmers/practitioners should be encouraged and this may be an important on-farm diagnostic for assessment of current N use efficiency, potential pollution and the

Table 7.5. Nitrogen inputs and outputs (kg N ha⁻¹ year⁻¹) from DeMarke experimental system compared with two commercial dairy farms (data taken from Neetesen and Hassink, 1998 and den Boer *et al.*, 1996, respectively).

| | 'Experimental' farm, DeMarke | Commercial farms | |
		Kloosterboer (sand)	Achtenkamp (clay)
Inputs			
Bought in feed	19	15	6
Concentrates	102	114	67
Fertilizers	63	164	213
Atmospheric deposition	49	52	46
Symbiotic fixation	12	4	4
Miscellaneous	2	—	—
Total	247	349	336
Outputs			
Meat	35	14	11
Milk	39	69	53
Miscellaneous	10	15	19
Total	84	98	83
Surplus	163	251	253

development of a range of scenarios for improvement of particular dairy systems.

As an aid to this, a model/systems synthesis approach is required which can be used to produce model predictions and best estimates to define flows and utilization of N under a range of circumstances. This has been attempted for both Dutch and UK conditions (Table 7.6), and is useful first to define typical balances and N pools and flows (Fig. 7.1; Table 7.1) within the total farm context and second to identify opportunities for improving utilization and sustainability. Examination of a number of these desk study options within both Dutch and UK systems (Table 7.6) showed considerable potential for improvement, maintaining production but reducing losses per unit area, and/or per unit product. It is remarkable that despite the substantial differences in stocking rate, purchased fertilizer, use of maize and other management features, surplus N per tonne of milk produced on both conventional managements was very similar.

Conclusions: Opportunities for the Future

It is clear that the perceptions over the use of N within dairy farming systems have changed enormously over the last decade and will continue to

Table 7.6. Desk studies of options to optimize the utilization of N within dairy farming systems in the Netherlands (Aarts *et al.*, 1992) and the UK (Jarvis *et al.*, 1996).

	Netherlands			UK					
	Conventional	A	B	Conventional	A	B	C	D	E
Milk yield (kg cow⁻¹)	5547	7000	8500	5554	5554	5554	5554	5554	5554
Milk production (kg ha⁻¹)	13195	13195	13195	7454	7454	5665	7454	7454	5665
Cows ha⁻¹	2.3	1.9	1.5	1.3	1.3	1.0	1.3	1.3	1.0
Young stock (% cows – LU basis)	?	77	57	62	62	62	62	62	62
Grass (ha)	22	16	16	76	76	56.3	56.3	56.3	56.3
Maize (ha)	3	9	6	–	–	–	19.7	19.7	19.7
Fodder beet (ha)	–	–	3	–	–	–	–	–	–
N inputs (kg ha⁻¹)	370	251	182	337	242	210	269	199	173
Purchased fertilizer N (kg ha⁻¹)	330	159	91	250	155	–	185	115	–
Purchased concentrates (kg ha⁻¹)	4995	2348	1788	1703	1703	1362	1703	1703	1362
Purchased maize silage (kg ha⁻¹)	2135	–	–	–	–	–	–	–	–
Surplus N (kg ha⁻¹)	477	191	122	270	175	156	202	132	120
kg surplus N tonne⁻¹ milk	36	14	9	36	23	27	27	18	21

Options:
The Netherlands: A, more milk per cow, fewer cows, reduced fertilizer, increased maize; B, as for A, adapt housing, reduce fertilizer further, introduce fodder beet. UK: A, tactical fertilizer; injected slurry; B, white clover mixed sward, no fertilizer N; C, introduce maize silage; D, option A + option C; E, option B + option C.

change with the general demand for more sustainable and more acceptable agricultural managements. Intensive agriculture is inherently 'leaky' with respect to N and the opportunities for emission of dissolved or gaseous forms from dairying are numerous (Table 7.2). However, the corollary follows that there are therefore numerous opportunities to reduce these losses and develop more sustainable systems and a number of these have been indicated already. The key to progress is improved diagnostics at a range of scales in combination with the new technologies and approaches that are being developed. Examples of these are discussed below, firstly in terms of the main components of the system and then finally with respect to the whole system.

Management of soil N

The soil provides a very large pool of N, most of it immobile at any one time, but with potential to be released. One of the major difficulties in providing accurate fertilizer recommendations is the inability to predict the background supply of N from the SOM. Because of the substantial differences that different managements can create in patterns of N mineralization/immobilization in soils, information is required at the level of individual paddocks or pastures so that greater precision can be made to adjustments to inputs rather than accepting uniform inputs. Better understanding of the processes involved in the supply of N from organic matter (Jarvis *et al.*, 1996a) and model predictions will allow the possibile development of new tools which can be used to provide the level of diagnosis required. One such possibility is the use of soil thermal units which have an excellent relationship with mineralization rate (Clough *et al.*, 1998). This, coupled with a simple laboratory test, could provide a relatively easy means of assessing the likely mineralization supply at any one time. Such improved information is essential to the new strategies of fertilizer utilization (Scholefield *et al.*, 1997) which integrate modelling and soil diagnosis. The potential for this tactical adjustment of fertilizer in reducing inputs is illustrated in Table 7.6: the current need is to develop the research findings into versatile decision support systems for use by farmers.

Management of plant N

Minimizing luxury uptake of excess N, and maintaining appropriate C : N ratios in plant tissues for animal requirements provides another opportunity to reduce inputs. The inherent differences between the utilization of N in ensiled maize and that of most ensiled grass materials offers one way forward, looking for opportunities through breeding to change the characteristics of other species may be another. Care, however, has to be taken that increased *uptake* efficiency is not the only goal, because of the subsequent processing, excretion and transfer of the N by the dairy cow. Grazing management changes to allow the sward to become a late season sink to remove excess N in the soil have also been proposed (Titchin *et al.*, 1989): again the utilization of the N in that material has to be considered as part of a total package. Soil testing as an diagnostic tool has already been mentioned and plant testing with the same aim is another possibility on which new fertilizer strategies could be based. While plant sap testing has been promoted for a number of years for horticultural crops, the situation is much more difficult for most grassland species because of the problem of determining which specific tissue and form of N to diagnose and the much more complicated effects of management (cutting, grazing) on the morphology and physiology of the plant than most other cropping types. Nevertheless, it is an option worthy of consideration.

Another possibility is greater reliance on fixation as a supplier of N. Table 7.6 shows that overall losses may be reduced when a white clover based management is introduced, but there may be only minimal advantages in terms of N lost per unit of product because of the penalty in livestock carrying capacity with many legume based systems as opposed to fertilizer managements. Legumes are important components of many grasslands: the productivity of much of New Zealand agriculture is still dependent on N supplied initially by white clover (Ledgard, 1991) and production in other climatic zones is equally dependent on N inputs from other legumes. Within UK and many Western European grasslands, clover-based systems are in the minority and although some studies have indicated possible production benefits where animals utilize legume based diets (Wilman, 1989), overall dry matter production is often lower. Clover-based swards have also been viewed with optimism because N losses are often lower than from highly fertilized systems (Ryden *et al.*, 1984). It has been argued that the extent of losses relates to the extent of N input (thus grazed monocultures of white clover lose almost as much N as highly fertilized ones, Macduff *et al.*, 1990), and grass–clover swards lose similar amounts to fertilized ones with comparable inputs of N (Cuttle *et al.*, 1992). Nevertheless, under some environments, and taking into account that there is some compromise with overall production (see Table 7.6), losses per unit area can be reduced with mixed clover swards (Tyson *et al.*, 1997).

Management of animals

Reducing N intake by the animal reduces excretal N and partitions a greater proportion of the excretal N into less mobile forms into dung: both of these effects reduce the opportunities for loss (Table 7.6). The interactions between N supply, intake and utilization are described in Chapter 13, and there may be opportunity for greater manipulation of the form of the N present in dairy cow diets as well as reducing total amounts to increase efficiency. In general, there will be many opportunities for closer specification of the requirements for particular classes of stock by manipulating dietary composition. Some studies have been directed at manipulating diet so that improvements in specific emissions can be achieved. Thus Smits *et al.* (1997) manipulated the supply of rumen protein surplus of dairy cows with the aim of reducing urinary concentrations of urea and consequently reducing the NH_3 losses from the floors and manures generated from housed animals and substantial reductions were achieved. There have also been attempts to manipulate urine volume by providing additional NaCl so that N concentration in the excreted urine was diluted by an increased water intake (van Vuuren and Smits, 1997). In the field, this could be important in reducing the local 'hot spot' accumulation of excessive and vulnerable contents of mobile N in the soil. Prevention of such areas in a grazed pasture is of course difficult but good management should be targeted at providing as uniform as possible distribution of excreta. Although limited by the individual local scales of influence (urine patches for example) involved in grassland compared with arable fields, technology for the precision placement of fertilizer in relation to some infield measurement of growth differences may also become feasible in time to overcome some of the spatial variability created by the nature of animal excreta as well as background natural variability.

Management of stored animal excreta

Nitrogen in the collected excreta and other (e.g. bedding etc.) materials represents a very large pool of mobile N, vulnerable to loss. Options for better management and utilization have been described extensively recently (see Stevens and Laughlin, 1997). Although these manures are important reservoirs of N, farming management logistics and costs often mitigate against easy uptake of some of the methodologies proposed, for example, for improved application of slurries to land (e.g. injection, dilution, acidification, nitrification, inclusion of a nitrification inhibitor etc.). Nevertheless, both legislation, e.g. in the Netherlands, and Codes of Good Practice, e.g. in the UK, are encouraging farmers to reconsider the better use of the nutrients contained in these materials. One of the major problems with this is the variable nature of their composition, both in terms of their total N contents and their composition. Improvements in the existing range of technologies for assessing the N-supplying power of FYM and slurries over both short and longer terms would improve the chances of them being used in a more productive way.

Total system management

It has been clear throughout the whole of this chapter that, while understanding, information, actions and reactions are required at a field level for improved sustainability of N management, these cannot be taken in isolation from each other and from other components of the system, so that their total interactive effects can be taken into account. Table 7.6 illustrates some options which have been examined in desk studies of dairy farms. Understanding the system and its potential for change and the implications of these changes should become an important, larger scale diagnostic aid for the dairy farmer and his advisors. Development of systems description software coupled with N cycling models will be required and will allow the practitioner to develop alternative management strategies incorporating new technologies where these are required. Before this becomes available, simple on-farm, N budgeting/balancing provides an immediate appreciation of the extent of N inputs and surpluses and can help to identify any weaknesses in current management. This is already being encouraged by farm advisory groups in a number of countries.

Finally, there are current debates about the inherent viability of specialist intensively managed stocking units as sustainable units. It has been argued that these are producing not only unacceptable environmental and ecological impact but also economical and social effects which cannot be sustained (Rabbinge, 1992). One option, it is argued, would be for a renaissance of mixed farming systems in which inputs could be reduced, and there would be more efficient utilization of animal manures and a broadening of the crop rotation (Latinga and Rabbinge, 1997). A further development of this argument by some has been to encourage a move towards organically based management. These are some of the options that are under current investigation in terms of understanding their N flows and losses, their productive output and economic margins and their behaviour with respect to N as compared with 'conventional specialist' units. While the results from these developing studies will allow their environmental and production performance impact to be assessed accurately, it will be the national and international economic and other policy drivers, plus (and increasingly important) consumer and environmental 'quality' criteria which will determine longer term trends in dairy farm management and changing N use.

References

Aarts, H.F.M., Biewinga, E.E. and Van Keulen, H. (1992) Dairy farming systems based on efficient nutrient management. *Netherlands Journal of Agricultural Science* 40, 285–299.

Archer, J. (1988) *Crop Nutrition and Fertilizer Use*. Farming Press, Ipswich, 278pp.

Bacon, S.E., Lanyon, L.E. and Schlauder, R.M. (1990) Plant nutrient flow in the management pathways of an intensive dairy farm. *Agronomy Journal* 82, 755–761.

Baker, R.D. (1986) Efficient use of nitrogen fertilizers. In: Cooper, J.P. and Raymond, W.F. (eds) *Grassland Manuring*. British Grassland Society, Reading, pp. 15–27.

Bastiman, B. and van Dijk, J.P.E. (1975) Muck breakdown and pasture rejection in an intensive pasture system for dairy cows. *Experimental Husbandry* 28, 7–17.

Boer, D.J., den, Middlekoop, J.C., van and Bussink, D.W. (1996) *Minimizing Nutrient Losses in Dairy Farming*. NMI, Wageningen, 105pp.

Burg, P.F.J., van and Prins, W.H (1981) Nitrogen and intensification of livestock farming in EEC countries. *Proceedings of the Fertilizer Society, London* 199, 1–78.

Burns, L.C., Stevens, R.J., Smith, R.V. and Cooper, J.E. (1995) The occurrence and possible sources of nitrite in grazed, fertilized grassland soil. *Soil Biology and Biochemistry* 27, 47–59.

Bussink, D.W. (1996) Ammonia volatilization from intensively managed dairy pastures. PhD thesis. Landbouwuniversiteit, Wageningen, 177pp.

Cerrato, M.E. and Blackmer, A.M. (1990) Comparison of models for describing corn yield response to nitrogen fertilizer. *Agronomy Journal* 82, 138–143.

Clement, C.R. and Williams, T.E. (1967) Leys and soil organic matter. 2. The accumulation of nitrogen in soils under different leys. *Journal of Agricultural Science, Cambridge* 69, 133–138.

Clough, T.J., Jarvis, S.C. and Hatch, D.J. (1998) Relationships between soil thermal units, nitrogen mineralization and dry matter production in pastures. *Soil Use and Management* 14, 65–69.

Cuttle, S.P., Hallard, M., Daniel, G. and Scurlock, R.V. (1992) Nitrate leaching from sheep-grazed grass/clover and fertilized grass pastures. *Journal of Agricultural Science, Cambridge* 119, 335–342.

Deenen, P.J.A.G. and Lantinga, E.A. (1993) Herbage and animal production responses to fertilizer nitrogen in perennial ryegrass swards. 1. Continuous grazing and cutting. *Netherlands Journal of Agricultural Science* 41, 179–203.

Dougherty, C.T. and Rhykerd, C.L. (1985) The role of nitrogen in forage-animal systems. In: Heath, M.E., Metcalfe, D.S. and Barnes, R.F. (eds) *Forages: The Science of Grassland Agriculture*, 4th edn, Iowa State University Press, Ames, Iowa, pp. 318–325.

Ferrugia, A., Pichot, L. and Perrot, C. (1997) Mineral balance and farming systems. In: Jarvis, S.C. and Pain, B.F. (eds) *Gaseous Nitrogen Emissions from Grasslands*. CAB International, Wallingford, pp. 405–408.

Follet, R.F. and Wilkinson, S.R. (1995) Nutrient management of forages. In: Barnes, R.F., Miller, D.A. and Nelson, C.J. (eds) *Forages Vol. II. The Science of Grassland Agriculture,* 5th edn. Iowa State University Press, Ames, Iowa, pp. 55–82.

Frame, J. (1992) *Improved Grassland Management*. Farming Press, Ipswich, 351pp.

Frame, J., Harkness, R.D. and Talbot, M. (1989) The effect of cutting frequency and fertilizer nitrogen rate on herbage productivity from perennial ryegrass. *Research and Development in Agriculture* 6, 99–105.

Gibb, M.J., Ivings, W.E., Dhanoa, M.S. and Suttons, J.D. (1992) Changes in body components of autumn-calving Holstein-Friesian cows over the first 29 weeks of lactation. *Animal Production* 55, 339–360.

Gill, K., Jarvis, S.C. and Hatch, D.J. (1995) Mineralization of nitrogen in long-term pasture soils: effects of management. *Plant and Soil* 172, 153–162.

Granli, T. and Bockman, O.C. (1994) Nitrous oxide from agriculture. *Norwegian Journal of Agricultural Sciences, Suppl. No. 12,* 1–128.

Hassink, J. and Neeteson, J.J. (1991) Effect of grassland management on the amounts of soil organic N and C. *Netherlands Journal of Agricultural Science* 39, 225–236.

Hatch, D.J., Jarvis, S.C. and Reynolds, S.E. (1991) An assessment of the contribution of net mineralization to N cycling in grass swards using a field incubation method. *Plant and Soil* 138, 23–32.

Haygarth, P.M., Jarvis, S.C., Chapman, P.J. and Smith, R.W. (1998) Mass balances of phosphorus in grassland systems and potential for transfer to waters. *Soil Use and Management* 14, 160–167.

Haynes, R.J. (1986) The decomposition process: mineralization, immobilization, humus production and degradation. In: Haynes, R.J. (ed.) *Mineral Nitrogen in the Soil Plant System.* Academic Press, London, pp. 152–186.

Haynes, R.J. and Williams, P.H. (1993) Nutrient cycling and soil fertility in the grazed pasture ecosystem. *Advances in Agronomy* 49, 119–199.

Hofman, T.B. and ten Holte, L. (1992) *Field Experiments with Grass, Maize and Fodder Beet.* Research in 1990 at 'De Marke'. Verslag - CAB-DLO 1992 No. 159, 35pp.

Holmes, W. (1968) The use of nitrogen in the management of pasture for cattle. *Herbage Abstracts* 38, 265–277.

Holmes, W. (1989) *Grass: Its Production and Utilization.* Blackwell Scientific, Oxford, 306pp.

Hopkins, A., Murray, P.J., Bowling, P.J., Rook, A.J. and Johnson, J. (1995) Productivity and nitrogen uptake of ageing and newly sown swards of perennial ryegrass (*Lolium perenne* L.) at different sites and with different nitrogen fertilizer treatments. *European Journal of Agronomy* 4, 65–75.

Jarvis, S.C. (1993) Nitrogen cycling and losses from dairy farms. *Soil Use and Management* 9, 99–105.

Jarvis, S.C., Hatch, D.J. and Lockyer, D.R. (1989) Ammonia fluxes from grazed grassland: annual losses from cattle production systems and their relation to nitrogen inputs. *Journal of Agricultural Science, Cambridge* 113, 99–106.

Jarvis, S.C., Scholefield, D. and Pain, B.F. (1995) Nitrogen cycling in grazing systems. In: Bacon, P.E. (ed.) *Nitrogen Fertilization in the Environment.* Marcel Dekker, New York, pp. 381–420.

Jarvis, S.C., Stockdale, E.A., Shepherd, M.A. and Powlson, D.S. (1996a) Nitrogen mineralization in temperate agricultural soils: processes and measurements. *Advances in Agronomy* 57, 187–235.

Jarvis, S.C., Wilkins, R.J. and Pain, B.F. (1996b) Opportunities for reducing the environmental impact of dairy farming managements: a systems approach. *Grass and Forage Science* 51, 21–31.

Jenkinson, D.S. (1990) The turnover of organic carbon and nitrogen in soil. *Philosophical Transactions of the Royal Society, London* B, 329, 361–368.

Jokela, W.E. (1992) Nitrogen fertilizer and dairy cow manure effects on corn yield and soil nitrate. *Soil Science Society, America Journal* 56, 148–154.

Klein, C.A.M. de (1994) Denitrification in grazed grasslands in the Netherlands. PhD thesis, Universiteit Utrecht, 119pp.

Lantinga, E.A. and Rabbinge, R. (1997) The renaissance of mixed farming systems: a way towards sustainable agriculture. In: Jarvis, S.C. and Pain, B.F. (eds) *Gaseous*

Nitrogen Emissions from Grasslands. CAB International, Wallingford, pp. 408–410.

Laws, J.A., Pain, B.F. and Jarvis, S.C. (1995) Grassland management systems to reduce N losses and increase efficiency: a farmlet approach. In: Pollet, G.E. (ed.) *Grassland into the 21st Century: challenges and opportunities.* BGS, Occasional Symposium, No. 29, Harrogate. BGS, Reading, pp. 310–311.

Ledgard, S.F. (1991) Transfer of fixed nitrogen from white clover to associated grasses in swards grazed by dairy cows estimated using ^{15}N methods. *Plant and Soil* 131, 215–223.

Lory, J.A., Russelle, M.P. and Randall, G.W. (1995) A classification system for factors affecting crop response to nitrogen fertilization. *Agronomy Journal* 87, 869–876.

Macduff, J.H., Jarvis, S.C. and Roberts, D.H. (1990) Nitrates leaching from grazed grassland systems. In: Calvet, R. (ed.) *Nitrates – Agriculture – Eau.* INRA, Paris, pp. 405–410.

[MAFF] Ministry of Agriculture Fisheries and Food (1994) *Fertilizer Recommendations for Agricultural and Horticultural Crops* (RB 209). HMSO, London, 112pp.

Neeteson, J.J. and Hassink, J. (1998) Nitrogen budgets of dairy farms in the Netherlands. *European Journal of Agronomy* (in press).

Overman, A.R., Wilkinson, S.R. and Evers, G.W. (1992) Yield response of Bermuda grass and Bahia grass to applied nitrogen and overseeded clover. *Agronomy Journal* 84, 998–1001.

Pain, B.F. Smith, K.A. and Dyer, C.J. (1986) Factors affecting the response of cut grass to the nitrogen content of dairy cow slurry. *Agricultural Wastes* 17, 189–202.

Pain, B.F., Weerden, T.J. van der, Chambers, B.J., Phillips, V.R. and Jarvis. S.C. (1998) A new inventory for ammonia emissions from UK agriculture. *Atmospheric Environment* 32, 309–313.

Peel, S., Chambers, B.J., Harrison, R. and Jarvis, S.C. (1997) Reducing nitrogen emission for a complete dairy system. In: Jarvis, S.C. and Pain, B.F. (eds) *Gaseous Nitrogen Emissions from Grasslands.* CAB International, Wallingford, pp. 383–390.

Prins, W.H., Dilz, K. and Neeteson, J.J. (1988) Current recommendation for nitrogen fertilization within the EEC in relation to nitrate leaching. *Proceedings of the Fertilizer Society* 276.

Rabbinge, R. (1992) Options for integrated agriculture in Europe. In: Lenteren, J.C., Minks, A.K., van and Ponti, O.M.B. (eds) *Proceedings International Conference, Velhoven, Sept 1991.* Pudoc Scientific Publishers, Wageningen, pp. 211–218.

Ryden, J.C., Ball, P.R. and Garwood, E.A. (1984) Nitrate leaching from grassland. *Nature* 311, 50–53.

Scholefield, D., Brown, L., Jewkes, E.C. and Preedy, N. (1997) Integration of soil testing and modelling as a basis for fertilizer recommendations for grassland. In: Lemaire, G. and Burns, I.G. (eds) *Diagnostic Procedures for Crop N Management.* INRA, Paris, pp. 139–147.

Scholefield, D., Lockyer, D.R., Whitehead, D.C. and Tyson, K.C. (1991) A model to predict transformation and losses of nitrogen in UK pastures grazed by beef cattle. *Plant and Soil* 132, 165–177.

Scholefield, D., Tyson, K.C., Garwood, E.A., Armstrong, A.C., Hawkins, J. and Stone, A. (1993) Nitrate leaching from grazed grassland lysimeters: effects of fertilizer input, field drainage, age of sward and patterns of weather. *Journal of Soil Science* 44, 601–613.

Sharpley, A.N. and Menzel, R.G. (1987) The impact of soil and fertilizer phosphorus on the environment. *Advances in Agronomy* 41, 297–324.

Sims, J.T. (1995) Organic wastes as alternative nitrogen sources. In: Bacon, P.E. (ed.) *Nitrogen Fertilization in the Environment.* Marcel Dekker, New York, pp. 487–536.

Smits, M.C.J., Valk, H., Monteny, G.J. and van Vuuren, A.M. (1997) Effect of protein nutrition on ammonia emissions from cow houses. In: Jarvis, S.C. and Pain, B.F. (eds) *Gaseous Nitrogen Emissions from Grasslands.* CAB International, Wallingford, pp. 101–107.

Stevens, R.J. and Laughlin, R.J. (1997) The impact of cattle slurries and their management on ammonia and nitrous oxide emissions from grassland. In: Jarvis, S.C. and Pain, B.F. (eds) *Gaseous Nitrogen Emissions from Grasslands.* CAB International, Wallingford, pp. 233–256.

Thomas, C., Reeve, A. and Fisher, G.E.J. (1991) *Milk from Grass.* BGS, Reading, 112pp.

Thornley, J.H.M. and Verberne, E.L.J. (1990) A model of nitrogen flows in grassland. *Plant, Cell and Environment* 12, 863–886.

Titchen, N.M. and Scholefield, D. (1992) The potential of a rapid test for soil mineral nitrogen to determine tactical application of fertilizer nitrogen to grassland. *Nitrate and Farming Systems, Aspects of Applied Biology* 30, 223–229.

Titchen, N.M., Wilkins, R.J., Phillips, L. and Scholefield, D. (1989) Strategy of fertilizer nitrogen applications to grassland for beef: effects on production and soil mineral N. In: Desroches, R. (ed.) *Proceedings of the XVIth International Grassland Congress, 4–11 October, 1989, Nice, France.* French Grassland Society, Versailles Cedex, France, pp. 123–124.

Tyson, K.C., Scholefield, D., Jarvis, S.C. and Stone, A.C. (1997) A comparison of animal output and nitrogen leaching losses recorded from drained fertilized grass and grass/clover pasture. *Journal of Agricultural Science, Cambridge* 129, 315–323.

Van de Ven, G.W.J. (1996) A mathematical approach to comparing environmental and economic goals in dairy farming on sandy soils in the Netherlands. PhD thesis, Landbouwuniversiteit, Wageningen, 238 pp.

van der Meer, H.G. and van uum-van Lohuyzen, M.G. (1986) The relationships between inputs and outputs of nitrogen in intensive grassland systems. In: Meer, H.G. van der, Ryden, J.C. and Ennick, G.C. (eds) *Nitrogen Fluxes in Intensive Grassland Systems.* Martinus Nijhoff, Dordrecht, pp. 1–18.

Van Vuuren, A.M. and Meijs, J.A.C. (1987) Effects of herbage composition and supplement feeding on the excretion of nitrogen in dung and urine by grazing dairy cows. In: Meer, H.G., van der, Unwin, R.J., Dijk, T.A. van and Ennick, G.C. (eds) *Animal Manure on Grassland and Fodder Crops: Fertilizer or Waste?* Martinus Nijhoff, Dordrecht, pp. 17–25.

Van Vuuren, A.M. and Smits, M.C.J. (1997) Effect of nitrogen and sodium chloride intake on production and composition of urine in dairy cows. In: Jarvis, S.C. and Pain, B.F. (eds) *Gaseous Nitrogen Emissions from Grasslands.* CAB International, Wallingford, pp. 95–99.

Weissbach, F. and Ernst, P. (1994) Nutrient budgets and farm management to reduce nutrient emissions. In: 't Mannetje, L. and Frame, J. (eds) *Grassland and Society,*

Proceedings of the 15th General Meeting of the European Grassland Federation, Wageningen.

Whitehead, D.C. (1995) *Grassland Nitrogen.* CAB International, Wallingford, 397pp.

Wild, A., Jones, L.H.P. and Maduff, J.H. (1987) Uptake of mineral nutrients and crop growth: the use of flowing nutrient solutions. *Advances in Agronomy* 41, 171–219.

Willigen, P. de (1991) Nitrogen turnover in soil-crop systems. Comparison of fourteen simulation models. *Fertilizer Research* 27, 141–149.

Wilman, D. (1965) The effect of nitrogenous fertilizer on the rate of growth of Italian ryegrass. *Journal of the British Grassland Society* 20, 248–254.

Wilman, D. (1989) The growth of white clover (*Trifolium repens* L.) in field swards in Wales. A review. In: *Proceedings of the 16th International Grassland Congress, 4–11 October, 1989, Nice, France.* French Grassland Society, Nice, pp. 1039–1040.

Phosphorus Management and Sustainability

8

B.W. Mathews[1], J.P. Tritschler II[2] and S.C. Miyasaka[3]

[1]College of Agriculture, University of Hawaii at Hilo, 200 W. Kawili St, Hilo, Hawaii 96720-4091, USA
[2]Applied Epidemiology Inc., PO Box 2424, Amherst, Massachusetts 01004, USA
[3]College of Tropical Agriculture and Human Resources, University of Hawaii at Manoa, Beaumont Branch Station, 461 W. Lanikaula St, Hilo, Hawaii 96720-4037, USA

Introduction

Phosphorus (P) is one of the essential nutrients for plants and animals. It is vital for life and in the USA about one third of the human recommended daily allowance for P is supplied by dairy products (Etgen and Reaves, 1978). The most critical role of P in both plants and animals is its involvement in energy storage through the high energy bonds found in adenosine triphosphate (ATP). For this reason, P is essential to all energy transformations. The high energy bonds that hold the nucleic acid sequences of DNA and RNA together are also formed with P (Marschner, 1995).

In addition to the P consumed in grazed herbage, dairy cattle on pasture usually receive P via supplemental feeds and minerals, thereby adding P to the pasture system. Dairy pastures also may receive P inputs from fertilizers and barnyard manure. Excessive importation of P into the pasture system may result in undesired export of this valuable nutrient to environmentally sensitive surface waters (Van Horn *et al.*, 1991; Aarts *et al.*, 1992; Mathews *et al.*, 1996a). Water quality concerns have recently brought P management by dairies into the public limelight to a greater extent than any other mineral. This review of the complex nature of P in the soil–plant–animal system is intended to help place P issues in better perspective for those interested in sustainable dairy pasture management. Important first steps in environmentally friendly and economically viable P management are recognizing the P retention characteristics of the soil and the P requirements of plants and animals.

Relative importance of phosphorus

The total P concentration in the surface 10–20 cm of improved pasture soils may range from 0.4 to 4.0 g kg^{-1} (Walker *et al.*, 1959; Mathews *et al.*, 1996b), but most are probably in the lower half of this range (Stevenson, 1986). Unfortunately, the quantity of total P in soils has little or no relationship to the availability of P to plants or potential for losses to surface waters sensitive to P-induced eutrophication (increased biological productivity of water which may limit its use for drinking, recreation, and industry). Next to N, the second major nutrient limitation to grassland productivity is often P. This is due both to P chemistry in the soil and its importance to plant and animal growth.

Phosphorus is often deficient in the moderately to highly weathered acid soils in both temperate and particularly tropical regions of the world (Plucknett, 1970; Mays *et al.*, 1980; Cornforth and Sinclair, 1982). These low fertility soils often retain large amounts of added fertilizer P in unavailable (fixed) forms. This deficiency limits forage production because of both poor root development and poor plant nutrition. Even when herbage P concentrations are adequate for maximum growth of most forages (generally between 2.0 and 3.0 g P kg^{-1}; Mays *et al.*, 1980; Kelling and Matocha, 1990) they are often insufficient in P for optimal performance of lactating dairy cows and growing heifers which require dietary concentrations between 3.0 and 4.0 g P kg^{-1} (McDowell, 1985; NRC, 1989). Insufficient herbage P for dairy cattle is most frequently a problem with tropical forages (McDowell, 1985; Davison *et al.*, 1986; Minson, 1990). Dietary P supplementation in salt blocks/mixes or concentrates is generally practised as low cost insurance on dairy pastures.

Soil Phosphorus

Forms of soil phosphorus

Most of the naturally occurring soil P is derived originally from primary minerals of the apatite group, in which P is present essentially as tricalcium phosphate, $Ca_3(PO_4)_2$ (Sanyal and DeDatta, 1991). Apatite declines with time as a result of soil weathering and is absent altogether in well developed soils as the P is converted to other inorganic forms and organic P (Stevenson, 1986). Organic P often comprises > 50% of the total soil P and can accumulate at rates between 5 and 15 kg ha^{-1} year^{-1} in intensively managed pastures (Walker *et al.*, 1959; Dalal, 1977). Hence, on a pasture receiving annual phosphate fertilizer additions, the soil organic P concentration also increases but at a much slower rate than inorganic P (Walker *et al.*, 1959). Much of the labile (readily solubilized) P in permanent pasture soils is thought to be present in the soil organic matter and organic residues, in addition to inorganic P forms (Tate *et al.*, 1991; Ibrikci *et al.*, 1994). Phosphate in the soil solution is closely linked to a labile fraction of the soil organic P by means of a rapid P cycling through the soil biomass. Annual P fluxes of 23–32 kg ha^{-1} through soil microbial biomass have been measured under grass pastures in England and New Zealand (Brookes *et al.*, 1984; Perrot and Sarathchandra, 1989). Phosphate released by enzymatic breakdown of decaying organisms (mineralization) may be either re-immobilized into the soil biomass, taken up by plant roots or retained by soil minerals. Figure 8.1 shows a simplified version of the P cycle in a grazed pasture.

The directly bioavailable ionic form of P found in soil solution and surface waters is determined by pH. At pH levels commonly encountered in agricultural soils, the $H_2PO_4^-$ and HPO_4^{2-} orthophosphate ions predominate, with the former being the principal form at pH < 7.2 (Sanyal and DeDatta, 1991). Soil pH (in water) in the range of 5.5–6.8 (depending on the soil colloid chemistry) is considered to be most desirable because of decreased P solubility when soil pH is beyond this range (Sanyal and DeDatta, 1991). The lower half of the range is preferable in highly weathered soils dominated by 1 : 1 clays, amorphous aluminosilicates, oxides/hydrated oxides of Al and Fe or organic soils. In less weathered soils dominated by 2 : 1 clays, the upper half of the range is preferable. Concentrations of phosphate-P in most soil solutions are extremely low (≤ 0.10 mg l^{-1}) and should not lead to environmental problems (Fox, 1979). In order to maintain a constant supply of P to the plant, the P absorbed by plant roots from soil solution must be continually replenished through labile P associated with the soil solid phase (Sanyal and DeDatta, 1991). In addition to the inorganic phosphate ions, a substantial portion (20–90%) of soil solution P is usually present as soluble organic-P compounds (Dalal, 1977). Some of these compounds become readily bioavailable through hydrolysis by phosphatase enzymes produced by plant roots and rhizosphere microorganisms (Tarafdar and Claassen, 1988; Haynes and Williams, 1993).

In long-term pasture soils, there is often a surface layer about 1- to 2-cm thick that consists mostly of organic matter (Haynes and Williams, 1993). Much of the water soluble (passes a 0.45-μm pore diameter membrane filter) reactive P (RP; all reactive forms by the molybdate blue procedure) is often concentrated in this thin layer (Mathews *et al.*, 1997). In some sandy soils (Spodosols) with low P retention throughout the A horizon, high concentrations of water soluble RP (10 mg kg^{-1}) may be found, however, to a depth of 15–20 cm (Graetz and Nair, 1995). The surface 1–2 cm of soil in pastures is also the zone of interaction with runoff and for a given soil type, correlations between extractable or water/dilute salt soluble soil P from this

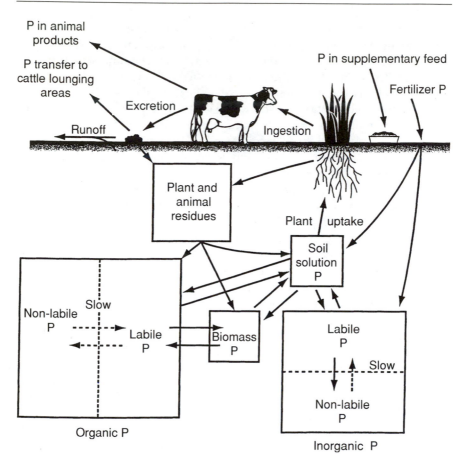

Fig. 8.1. The phosphorus cycle in a dairy pasture.

depth and runoff P concentration are often quite good ($r = 0.72\text{--}0.99$; Sharpley *et al.*, 1978; Pote *et al.*, 1996). Saturation extracts of this soil layer may also be useful in predicting the relative proportions of RP and organic P in surface water drainage/runoff, but further research is needed (Mathews *et al.*, 1994).

Retention of soil phosphorus

Soluble P in soils can be specifically adsorbed by colloid surfaces and often is precipitated in the presence of soluble Ca, Al, and Fe (Sanyal and DeDatta, 1991). The relative proportions of P retained by the soil through adsorption or precipitation are difficult to quantify and transitory in nature.

Retention, by which is meant simply the removal of P from solution, is however, thought to be primarily an adsorption process. This is due to the rapid rate (seconds) at which most of the added P is removed from solution in soils (Ayres and Humbert, 1957; Fox and Searle, 1978; Sanyal and DeDatta, 1991). Retention of P by soil is generally regarded as both a desirable and undesirable soil phenomenon. Retention is desirable in that it minimizes loss of applied P by lateral flow to waterways and leaching to groundwater. It is undesirable insofar as a substantial portion of the P retained is in forms which either are chemically unavailable to plants (non-labile) or which are so slowly available that much of the fertilizer value is lost. Most of these unavailable forms are thought to result from sub-surface diffusive migration of adsorbed P into the structure of soil colloids and aggregates or through occlusion of adsorbed P by the development of fresh mineral coatings on colloid surfaces (Sanyal and DeDatta, 1991). This lack of reversibility of P retention by the soil solid phase is referred to as hysteresis. In addition to pH, factors affecting P retention include colloid chemistry, temperature, organic acids, soil solution ionic strength, cation : anion balance, and water content (Sanyal and DeDatta, 1991).

Acidic soils vary in their phosphate adsorption capacity from very high (> 2000 mg P kg^{-1}) to very low (< 100 mg P kg^{-1}). The magnitude of adsorption based on colloid chemistry for soils in decreasing order is as follows: soils high in amorphous aluminosilicates (allophane and imogolite) and hydrated Al/Fe oxides (i.e. ferrihydrite) > crystalline Al/Fe oxides (gibbsite-goethite) > 1 : 1 clays (kaolinite-halloysite) > 2 : 1 clays (smectites) > quartz and Al-free organic colloids (Ayres and Humbert, 1957; Fox and Searle, 1978; Sanyal and DeDatta, 1991). For example, it would not be unusual for an Andisol (dominated by amorphous aluminosilicates and hydrated Al/Fe oxides) to have a P adsorption capacity in excess of 2000 mg kg^{-1}, while an Ultisol (dominated by crystalline Al/Fe oxides and 1 : 1 clays) and a Mollisol (dominated by 2 : 1 clays) may have adsorption capacities of approximately 750 and 100 mg P kg^{-1}, respectively (Fox and Searle, 1978). Sandy soils with little clay or Al/Fe oxides and hydrated oxides do not adsorb substantial P (< 50 mg kg^{-1}) and considerable leaching of P may occur (Mansell *et al.*, 1991; Graetz and Nair, 1995). Most soils, however, strongly sorb P. In neutral to alkaline soils, P solubility is reduced by reaction with Ca^{2+} and calcareous minerals (Sanyal and DeDatta, 1991).

Assessment of soil phosphorus status

Dilute acid soil test extractants (i.e. Mehlich I, Mehlich III, Bray P$_1$ or modified-Truog procedures) are commonly used to assess availability of labile soil P in the southeastern USA and other humid regions of the world, while the alkaline (pH = 8.5) Olsen extractant is used under a wide range of conditions. Research has shown that depending on the soil texture, mineralogy, and bulk density, it may take anywhere from 2 to 20 kg P ha^{-1} to increase dilute acid or Olsen extractable P in the surface soil (top 15–20 cm) by 1 kg ha^{-1} (Ayres and Humbert, 1957; Kamprath and Watson, 1980; Tamimi and Matsuyama, 1990; Roberts et al., 1994). For rapid laboratory estimation of this relationship, Silva (1985) developed a simple P-extraction curve procedure that entails equilibrating incremental amounts of P with a series of soil samples before extraction. Values in the latter half of the range above are typical for highly weathered and volcanic ash soils (Hydrudands) with high P sorption capacities. Soils of this type often require a very high initial level of P fertilization (500 kg P ha^{-1}) for successful establishment of improved pastures (Plucknett, 1970).

Critical minimum soil test P concentrations for near maximum pasture production are generally thought to be between 20 and 50 mg kg^{-1} depending on the extractant, soil type, and plant species (Kamprath and Watson, 1980; Jones, 1990; Tamimi and Matsuyama, 1990; Roberts et al., 1994; Davison et al., 1997). Lower concentrations (10 mg kg^{-1}) may be adequate for some deep-rooted perennial forages in the presence of substantial subsoil P (Ibrikci et al., 1994). Sandy soils often require greater concentrations than fine textured soils due to slower rates of soil solution P diffusion at a given water potential and lower buffering capacities to replenish P uptake by plants (Kamprath and Watson, 1980). Within a soil texture, low bulk density soils (i.e. 0.5–0.7 g cm^{-3}) require greater concentrations than medium bulk density soils (i.e. 1.0–1.4 g cm^{-3}) to provide an equal amount of 'available P' per unit volume of soil (Tamimi and Matsuyama, 1990). Information on critical soil test P concentrations is best obtained from local agricultural research and extension agencies because extrapolation beyond the area of calibration may lead to erroneous conclusions (Jones, 1990). In addition, it is important to note that dilute acid soil P extractants are often unsuitable for use on calcareous soils or soils that have been fertilized with raw rock phosphate. This is because the dilute acid extractants are often found to dissolve P from relatively insoluble Ca-P minerals or a portion of unreacted P from the rock, resulting in deceptively high values of 'available P' (Sanyal and DeDatta, 1991).

Pasture requirement for phosphorus fertilizer and phosphorus recycling

The dairy pasture requirement for P fertilizer is greatest at the high levels of production established with high N supply (200–400 kg year^{-1}) from fertilizer and legumes (Davison et al., 1997). At low levels of N supply in grass-based pastures, there is often little response to P fertilization. While P fertilizer recommendations based on periodic soil testing and plant analysis are most advisable, research has often indicated that once the soil P concentration for near maximum pasture production has been achieved, annual P inputs of 20–40 kg ha^{-1} are required to maintain long-term optimal productivity (Vicente-Chandler *et al.*, 1974; Middleton and Smith, 1978; Cornforth and Sinclair, 1982; Davison *et al.*, 1997). In contrast, twice as much P is often required in hay systems where P is removed from the system in the harvested herbage (Vicente-Chandler *et al.*, 1974). This is proof that recycled P from dung and plant residues (litter, dead roots, etc.) make a very important contribution to P supply for pasture growth (Blue and Gammon, 1963; During and Weeda, 1973). In a year's time, however, P cycles in a manner to some degree similar (the extent depends on grazing efficiency) to that where herbage is clipped and removed in at least half the pasture area. This is because over half the pasture area is not likely to be influenced by P return in dung (accounts for ≥ 95% of excretal P) in any given year at normal stocking rates for improved pastures (Cornforth and Sinclair, 1982). For example, if we assume a moderate stocking rate of 4 mature cows ha^{-1}, the area of pasture receiving faeces each day is 1.0 m^2 cow^{-1}, and each dung patch affects plant P nutrition in an area about three times its initial area of physical coverage, then approximately 25–40% of the pasture area is influenced by P return in dung per year, depending on the degree of overlap (During and Weeda, 1973; Haynes and Williams, 1993; Mathews *et al.*, 1996a). A major factor affecting P release is the physical breakdown of the dung. It is well established that the physical breakdown of the dung is slower if it dries (Haynes and Williams, 1993). Dickinson and Craig (1990) and Mathews (1992) demonstrated that continuously moist conditions favour the release of P from dung compared with alternate wet and dry conditions, presumably due to enhanced activity of soil organisms.

Approximately 70% of the P intake by lactating dairy cows on pasture is returned to the system (Hutton *et al.*, 1967; Haynes and Williams, 1993). As a practical guide, the P concentration in dung is about twice that of the forage consumed (Minson, 1990). Each mature cow excretes approximately 14–18 kg P year^{-1}, principally (55–75%) in inorganic forms such as calcium phosphate (Safley *et al.*, 1984; Harris *et al.*, 1990). The models of Middleton and Smith (1978) and Cornforth and Sinclair (1982) account for P loss in animal products (6.8 g P kg^{-1} liveweight gain, 1.0 g P kg^{-1} milk); off-pasture losses in dung during stock handling/milking (0.5–3.0 kg P cow^{-1} year^{-1}); and substantial P transfer (10–30% of the total return in dung depending on grazing management and environmental conditions) to small, less productive, pasture zones near shade, water, flat areas, and supplemental feeders where cattle tend to congregate or rest (lounging areas). Phosphorus conversion to unavailable forms in the soil (related to soil mineralogy) and runoff/drainage losses are also considered. In many dairy pastures, it should be noted, however, that P inputs to the system via supplemental feed and minerals are often equal to or greater than the amount removed from dairy pastures in animal liveweight gain and milk production (Mathews *et al.*, 1994; Martz *et al.*, 1996). Therefore, little if any fertilizer P may be needed to maintain adequate soil P in such systems. Careful monitoring, however, of pasture soil/herbage P concentration and spatial distribution is highly advisable (Mathews *et al.*, 1996a) in addition to a balance sheet approach to estimated P inputs into the system (Martz *et al.*, 1996). In a study of four dairies on the Island of Hawai'i, Mathews *et al.* (1996b) found no evidence of soil P depletion below desirable concentrations on pastures that had not received fertilizer P for 10–35 years, but were grazed by dairy cows receiving substantial supplemental feed. Build-up of soil P in pasture systems due to supplemental feeding has been observed (Benacchio *et al.*, 1970) and may be a concern in areas near environmentally sensitive surface waters because of the well documented role of P in accelerated eutrophication (Ward *et al.*, 1990).

Phosphorus fertilization

Phosphorus fertilization of grass-based pastures has generally followed the classical approach of broadcast and incorporation to a depth of 5–10 cm during establishment, followed by annual top-dressings depending on soil test results (Mays *et al.*, 1980). The importance of adequate P on newly-cleared soils cannot be overemphasized when establishing improved pastures (Walker *et al.*, 1959; Plucknett, 1970; Mays *et al.*, 1980). In establishing grass–legume mixtures on soils with moderate to high P retention, however, there is often a definite advantage to banding the legume seed with P fertilizer (Fenster and Leon, 1979). Apparently this creates a more favourable fertility environment for the developing legume seedlings as they expand their root systems than do broadcast applications. Subsequent P applications should be broadcast to promote effective legume growth beyond the banded area. Application of P early in the spring may be most effective in stimulating legume growth in grass–legume mixtures (Haynes, 1980; Mays *et al.*, 1980).

All the commercially available, chemically processed phosphate fertilizers (CPPF) listed below are good, readily soluble sources of P for grass-based pastures and should be considered relatively equal in agronomic effectiveness as sources of P (Withers and Sharpley, 1995). Triple or concentrated superphosphate (TSP) has been a commonly used fertilizer P source since the 1930s. This material contains approximately 20% P (46% P_2O_5) and consists primarily of monocalcium phosphate ($Ca(H_2PO_4)_2$). Single superphosphate (SSP) consists of monocalcium phosphate and gypsum ($CaSO_4$) and contains approximately 9% P (20% P_2O_5) and 12% S. It has been used on pastures since the late 1800s and is an excellent P fertilizer to use on sites that are sensitive to S deficiency in addition to P (Haynes and Williams, 1993). In most situations, however, higher analysis P fertilizers such as TSP or diammonium phosphate (DAP) are used and S is applied through other sources if plant tissue or soil analysis indicates a need for S.

Diammonium phosphate $((NH_4)_2HPO_4)$ and monoammonium phosphate (MAP; $NH_4H_2PO_4$) are also excellent P sources. For many years, however, the production and use of DAP has overshadowed the production and use of MAP. Diammonium phosphate contains 20% P (46% P_2O_5) and 18% N while MAP contains 22% P (51% P_2O_5) and 11% N. Diammonium phosphate is more widely used than any other P fertilizer.

In contrast to the CPPF listed above, soft phosphate rock (SPR) that has been mined and crushed is preferred by 'organic' producers because it is 'natural' (Sharpley and Halvorson, 1994). The P concentration of SPR generally ranges from 11% to 17% (25% to 40% P_2O_5), but readily available P only accounts for about 10–30% of the total P compared with nearly 100% for CPPF. In their review of SPRs as a pasture fertilizer, Bolan *et al.* (1990) concluded that SPR can be as effective as CPPF, per kilogram of P applied, as a 'maintenance' fertilizer on established pastures that have a soil pH < 6.0 (in water) and a mean annual rainfall > 800 mm. The rate of SPR dissolution is too slow on soils with greater pH and lower rainfalls. Even when pH and rainfall are favourable, release of P from SPR is still often slow and soil P retention too rapid for effective SPR use during pasture establishment. In addition, the agronomic effectiveness of SPR is typically less than CPPF on high P retaining soils and this often cannot be overcome with high rates of SPR due to an inverse relationship between the rate of SPR application and the rate of SPR dissolution (Bolan *et al.*, 1990).

Use of animal manure on pastures has recently been reviewed by Edwards (1996). Dairy manure generally contains between 0.6 and 1.2% P and its agronomic effectiveness as a source of P is comparable to that of CPPF (Safley *et al.*, 1984; Eck and Stewart, 1995). Most agricultural best management practices for fertilization with manure are presently based on providing sufficient N, in a timely manner, to meet crop N requirements for a realistic yield. Guidelines have been developed to adjust manure rates according to the expected N availability in the manure and residual effects from previous applications (Van Horn *et al.*, 1991; Eck and Stewart, 1995). Constant manuring of soils, however, to meet crop N requirements creates agronomically excessive concentrations of soil P (Eck and Stewart, 1995). This is due to an unfavourable N : P ratio ($\approx 4 : 1$) in dairy manure and other animal wastes relative to crop and pasture plant N : P ratios ($\approx 10 : 1$). This creates a serious dilemma for many dairy farmers with limited land resources for manure recycling. In addition, manure applications may increase P solubility in both acid and alkaline soils beyond that expected from similar amounts of P applied as CPPF (Lund and Doss, 1980; Eghball *et al.*, 1996; James *et al.*, 1996). While this could often be beneficial from an agronomic standpoint on soils with high P retention capacity (During and Weeda, 1973; Iyamuremye and Dick, 1996), it is certainly undesirable from an environmental standpoint on soils with little capacity to retain P (Allen, 1988; Mansell *et al.*, 1991). Some possible explanations include (i) microbial release from manure of certain mobile, but at present, poorly characterized forms of organic P; (ii) competition for adsorption sites on the soil between inorganic-P and low-molecular-weight (LMW) organic anions (i.e. oxalate, citrate, and malate) released during manure decomposition; and (iii) complexing of LMW organic anions with Ca^{2+}, Al^{3+}, or Fe^{3+}, lowering the activities of these free cations in solution (Eghball et al., 1996; Iyamuremye and Dick, 1996; James et al., 1996). With the latter point, the activity of soluble P from mineral phosphates is increased to satisfy the thermodynamic principle of constant solubility. Hence, manure applications may be useful to enhance P availability from SPRs and CPPF (Iyamuremye and Dick, 1996).

Phosphorus losses to surface waters

Predicting the solubility of P in soils is important in areas with shallow water-tables and sandy soils that discharge into sensitive surface waters (Allen, 1988; Aarts et al., 1992; Breeuwsma et al., 1995; Graetz and Nair, 1995). For instance, in The Netherlands, land application of animal wastes is now regulated based on the percentage of saturation of the P sorption (retention) capacity of surface and subsoils, as estimated by an ammonium oxalate extraction procedure (Breeuwsma et al., 1995):

$$PS = \frac{Pox}{PSC} \times 100\%$$

PS = degree of phosphate saturation (%)

Pox = sorbed P (extractable by ammonium oxalate) (mmol kg^{-1})

PSC = phosphate sorption capacity (mmol kg^{-1})

Measuring P saturation of the P sorption sites both describes the potential of a soil to enrich drainage with P (high degree of P saturation) and also helps to predict how much of the P added in fertilizers or manures will be retained by the soil in a form that is relatively resistant to loss in drainage (low P saturation). The critical value may be as low as 25% for some soils and is defined as the saturation percentage for a given soil type that should not be exceeded to prevent adverse effects on water quality. The reader is referred to Pote et al. (1996) for further information on this procedure. Some states in the USA are now establishing upper, environmentally based 'critical limits' for soil test P because of concerns about pollution of surface waters (Withers and Sharpley, 1995). Typically, these upper soil test limits (120–200 mg P kg^{-1} depending on the extractant) are set at fivefold to tenfold the agronomic critical level that identifies a soil as non-responsive to P fertilization (Withers and Sharpley, 1995). Once a soil exceeds this value, no further additions of P (manures or fertilizer) are recommended. Unfortunately, considerable time is required for noticeable P depletion through removal in crop and livestock products (McCollum, 1991; Aarts et al., 1992).

Therefore, farmers are often left with the difficult task of finding alternative end-uses for animal wastes. This is an overly simplistic approach to a complex problem because drainage water quality will differ from farm to farm under similar pollution abatement practices and soil test P values due to differences in soil physico-chemical properties, terrain, and the resultant solute transport processes (Allen, 1988). It is entirely possible that a soil rated agronomically excessive in P could continue to receive dairy manure applications with no unfavourable environmental effect whatsoever, simply because there is no transport process to carry sufficient amounts of biologically available P to the nearest sensitive surface water.

Phosphorus in runoff from well managed pastures is usually associated with soluble P forms rather than particulate forms (suspended colloidal and organic material) because of limited soil erosion in such systems (Chichester *et al.*, 1979; Sharpley and Halvorson, 1994). The RP and total soluble P (TP; all reactive forms following persulphate oxidation of the sample) concentration in surface water drainage or runoff from dairy pastures is quite variable and is dependent on soil and livestock management. For example, Reese *et al.* (1982) found that RP concentration in runoff from dairy pastures on a Cecil sandy clay loam (clayey, kaolinitic, thermic Typic Kanhapludults) in South Carolina averaged 6–9 mg RP l^{-1} over 6 weeks before returning to background concentrations (0.1–0.5 mg RP l^{-1}) when slurry or dry dairy manure was surface-applied at rates ($\approx$ 10,000 kg DM ha^{-1}) considered adequate for near optimum tall fescue (*Festuca arundinacea* Schreb.) growth over a year. Similar results behaviour would be expected with surface applications of CPPF and even under severe conditions, P loss in pasture runoff is generally less than 5% of that applied (Sharpley *et al.*, 1978; Sharpley and Halvorson, 1994; Edwards *et al.*, 1996). Injection of manure slurries/effluents into the soil would probably decrease runoff P (Allen, 1988; Sharpley and Halvorson, 1994), but may substantially increase soluble P in subsurface horizons of soils with low P retention (Mathews *et al.*, 1996b). No measure of TP was made in the study by Reese *et al.* (1982), but research in Florida has shown that RP generally comprises 80–100% of the TP draining from pastures with the difference being due to organic P (Allen *et al.*, 1988). Mathews *et al.* (1994) found that RP concentration in surface water drainage averaged 1.82 mg l^{-1} (1.96 mg TP l^{-1}) when dairy heifers had access to a pasture swale area on a high P (124 mg P kg^{-1} by Mehlich I extraction), high water table, Chipley sand in Florida. When the heifers had limited access to a swale area, RP concentration averaged from 0.70 mg l^{-1} (0.82 mg TP l^{-1}). Drainage P concentrations may also increase by more than twofold during periods of pasture flooding due to solubilization of reductant soluble P compounds (Sanyal and DeDatta, 1991; Mathews *et al.*, 1994). As P leaves the pasture, often through grassed filter strips and farm water conveyance systems, there is generally a progressive decrease in P load downstream, however, by water dilution and adsorption by sediments (Sharpley and Halvorson, 1994; Edwards *et al.*, 1996). Clearly this must be considered in assessing the impact of P transported in runoff as a function of dairy management on the quality of the receiving surface waters. The greatest potential for short-term high concentrations of P loss occurs during the first few storms following manure or fertilizer P applications to pasture (Reese *et al.*, 1982; Ward *et al.*, 1990). Losses from pastures are generally less than 5 kg P ha^{-1} year^{-1} (Ward *et al.*, 1990).

A general water quality goal of ≤ 0.35 mg TP l^{-1} was established in 1989 for surface waters draining from improved pastures compared to 1.2 mg TP l^{-1} for feedlots in peninsular Florida, USA (SFWMD, 1989). The Netherlands has recently set a target concentration of 0.15 mg TP l^{-1} for livestock operations (Breeuwsma *et al.*, 1995). Whether these are realistic standards for enforcement is subject to much debate and may lead to economic hardship for producers (Canfield and Hoyer, 1988; Withers and Sharpley, 1995). The US Environmental Protection Agency recommends that agricultural runoff entering a stream should not exceed 1.0 mg TP l^{-1} and that TP should not exceed 0.05 mg l^{-1} in any stream at the point it enters a lake or reservoir (USEPA, 1986).

Plant Phosphorus

External requirement of forages for phosphorus and phosphorus uptake

External P requirement for near maximum growth varies markedly between pasture species and also differs between seedlings and established plants. The soil solution concentration of orthophosphate-P in most productive soils is generally in the range of 0.01–0.10 mg l^{-1} although some forages (mostly legumes) require slightly higher concentrations (0.15–0.30 mg l^{-1}) for maximum growth, particularly during establishment (Fox, 1979; Mouat, 1983).

Inorganic P uptake occurs primarily as $H_2PO_4^-$, because it is the predominant ion in the pH range of 4–7 (Sentenac and Grignon, 1985). A $H^+/H_2PO_4^-$ cotransport system is thought to be responsible for the P uptake (Marschner, 1995).

Phosphorus deficiency

Phosphorus deficiency is often characterized by reddening or purpling of leaves and (or) petioles (similar to cold effects) and by reductions in leaf area, number of leaves, shoot branching, and tiller number (Marschner, 1995; Halsted and Lynch, 1996). Phosphorus deficiency also limits forage legume establishment, nodulation, and N_2-fixing activity. Therefore, it is often a major limiting factor in developing successful grass–legume mixtures in pastures. Use of legumes is especially keyed to the P status of the soil because in most cases they are less competitive for P than the associated grasses (Haynes, 1980; Mays *et al.*, 1980). This is because the low mobility of P in soils gives a competitive advantage to grasses with their more extensive root systems consisting of smaller diameter roots and longer root hairs (Caradus, 1980).

Phosphorus interaction with other nutrients

Nutrient balance is one of the most important factors in P response. The interaction of N, P and K are of special significance, often determining whether responses to additions occur and the extent to which they are efficient and profitable (Dibb *et al.*, 1990; Davison *et al.*, 1997). Responses to P and K often only occur when N is adequately supplied. On the other hand, N response is generally enhanced with added P and K if soil availability of these nutrients is limiting. High soil P may limit the uptake of the micronutrients Zn, Fe and Cu, although agronomically these interactions may not be sufficient to affect yield of forages receiving micronutrients from supplemental fertilizers or manures (Blue, 1988; Dibb *et al.*, 1990).

Phosphorus concentration in forages

Phosphorus concentration in forages varies depending on species or genotypes, stage of growth, soil fertility, soil moisture, and climate. Critical P concentration required for near maximum growth ranges from 2.0 to 2.5 g kg^{-1} in lucerne (*Medicago sativa* L.), 0.8 to 3.1 g kg^{-1} in clover (*Trifolium* spp.), 2.2 to 3.0 g kg^{-1} in warm-season grasses, 1.2 to 4.6 g kg^{-1} in subtropical and tropical grasses and 2.0 to 3.4 kg^{-1} in cool-season grasses (Mays *et al.*, 1980; Kelling and Matocha, 1990). In subtropical and tropical legumes, critical P concentrations range from a low of 1.2 g kg^{-1} for stylo (*Stylosanthes guianensis* (Aubl.) Sw.) to a high of 3.0 g kg^{-1} for greenleaf desmodium (*Desmodium intortum* (Mill.) Urb.) (Mays *et al.*, 1980).

Like most minerals, increasing maturity results in a decline of P concentrations in whole forages (Minson, 1990). Phosphorus fertilization increases P concentration of forages, but unlike N and K, most forages do not exhibit luxury P uptake much beyond the plant sufficiency level (Mays *et al.*, 1980; Minson, 1990). Increasing soil moisture reduces tortuosity of the diffusion path for P, extends the depletion profile of P in soil solution adjacent to roots (Gahoonia *et al.*, 1994), and increases P accumulation in plants (Payne *et al.*, 1995).

Significant seasonal variations in P concentration have been reported in pastures (Saunders and Metson, 1971; Reay and Marsh, 1976; Kappel *et al.*, 1985; Minson, 1990; Davison *et al.*, 1997). It is likely that forages grazed by dairy cattle are highly variable in nutrient concentrations, and could be deficient in P, depending on the month of the year. The greatest P concentrations often occur during periods of rapid growth and cannot be conclusively related to changes in soil P availability (Saunders and Metson, 1971; Minson, 1990).

Phosphorus efficiency

The cost of applying fertilizer to correct low soil P can be excessive (Plucknett, 1970; Jones, 1990). An alternative, low-input method of improving forage production is selection or breeding of efficient plant species that will grow well at low levels of available soil P (Fenster and Leon, 1979; Caradus, 1994). Significant differences within the germplasm in response to P have been found (Fig. 8.2) for white clover (*Trifolium repens* L.) and several other important forage legumes (Godwin and Blair, 1991).

There are many definitions for the term 'nutrient efficiency' (Clark, 1990; Gourley *et al.*, 1994) and the ranking of species, cultivars, or genotypes depends on the definition, phenological age of plants, and soil P

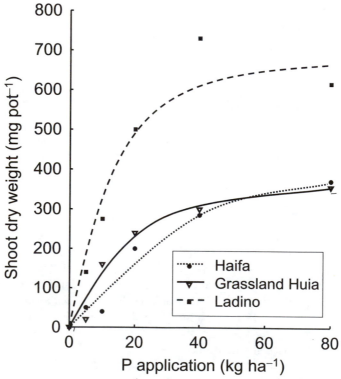

Fig. 8.2. Shoot dry weight (g pot^{-1}) response to added P in several accessions of white clover (adapted from Godwin and Blair, 1991, with permission).

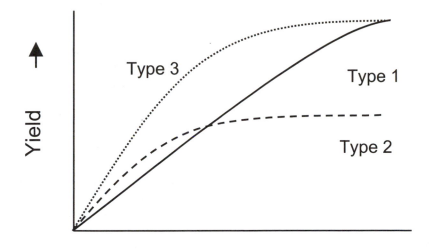

Fig. 8.3. Types of yield response to P in present, required breeding lines and the ideal outcome. Type 1: Present lines, responsive to high phosphorus. Type 2: Genotype required for breeding, responsive to low phosphorus. Type 3: Ideal combination (adapted from Blair, 1993, with kind permission from Kluwer Academic Publishers).

levels (Kemp and Blair, 1991; Blair, 1993). An inappropriate definition for the purpose of selecting P efficient pasture or forage species is the nutrient efficiency ratio, which is defined as total plant biomass produced per unit nutrient absorbed, and is the reciprocal of nutrient concentration (Gourley *et al.*, 1994). Selection based on this definition would result in a lower plant P concentration, whereas a high concentration of P in shoots is needed by animals. Both Blair (1993) and Gourley *et al.* (1994) recommend that P-efficient cultivars should have high yields at non-limiting nutrient availability and near maximum yields at a lower nutrient availability (Fig. 8.3).

Specific plant mechanisms to increase P uptake include increased root growth, longer root hairs, greater rate of P uptake per unit of root, ability to utilize P forms that are relatively unavailable to plants via exudation of organic acids or phosphatases from roots, and vesicular-arbuscular mycorrhizal (VAM) associations (Caradus, 1990; Polle and Konzak, 1990; Gourley *et al.*, 1994). In general, P deficient plants increase partitioning of carbohydrates towards roots, decreasing the shoot : root ratio (Marschner, 1995). Several P efficient species, such as Caribbean stylo (*Stylosanthes hamata* (L.) Taub.) (Smith *et al.*, 1990) or creeping signalgrass (*Brachiaria dictyoneura* (Figure and De Not.) Stapf.) (Halsted and Lynch, 1996), greatly favour root biomass allocation relative to that of the shoot under P stress.

Most plant species form associations with VAM fungi (Howeler *et al.*, 1987). Enhanced P absorption by VAM roots is due to an increased volume of soil explored by fungal hyphae (Li *et al.*, 1991), increased P influx rate per unit length of hyphae (Jakobsen *et al.*, 1992), and production of extracellular acid phosphatases (Tarafdar and Marschner, 1994). For these benefits, there is a cost of 10–20% of net photosynthates from the host to establish and maintain VAM symbiotic associations (Jakobsen and Rosendahl, 1990).

Studies demonstrating the benefit of inoculation with efficient VAM fungi have been carried out in the greenhouse, but there are few studies showing the benefit of VAM fungal inoculation in the field (Powell and Daniel, 1978). Inoculation with efficient VAM fungi are expected to be most beneficial when indigenous VAM fungal propagules are low, such as in eroded soils (Powell, 1980) or under conditions of heavy grazing (Bethlenfalvay *et al.*, 1985).

The potential exists for selection and breeding of forage genotypes that are more P-efficient, given the variability found in response to P. In common bean (*Phaseolus vulgaris* L.), a P-efficient cultivar with large root development was identified in a low P solution culture, and these genetic traits were transferred into an agriculturally useful parent using an inbred backcross line method (Schettini *et al.*, 1987). A long-term programme in New Zealand exists to breed cultivars of white clover that are more P-efficient (Dunlop *et al.*, 1990). Problems have been found, however, with correlation of results from pot studies or solution culture studies to those in the field (Caradus and Snaydon, 1986; Caradus, 1994).

Animal Phosphorus

Phosphorus absorption and metabolism

About 80% of the total P in the animal body is in the bone. In addition to bone formation and its central role in energy metabolism, P is a component of blood, nucleic acids, organic acids, and phospholipids. Phosphorus is also needed by rumen microbes for efficient fermentation and protein synthesis (Minson, 1990; McDowell, 1992).

Phosphorus absorption occurs in the small intestine. Phosphorus absorption is probably an active process (DeLuca, 1979) which is stimulated by 1,25-dihydroxy-vitamin D_3, nevertheless the quantity of P absorbed is directly related to dietary intake. Endogenous P is principally excreted through the faeces, with generally low urinary losses. Excess absorbed P is primarily passed through saliva into the digestive system, thus being excreted in the faeces, the principal route of P excretion in dairy cows.

Apparent P absorption is dependent upon P source, intake level, Ca : P ratio, intestinal pH, animal age, and dietary levels of Ca, Fe, Al, Mn, K, Mg, and fat (Irving, 1964). Overall, the apparent efficiency of P absorption of lactating cows has been demonstrated to be about 40–55% of dietary P (Ward *et al.*, 1972; Hibbs and Conrad, 1983; NRC, 1989). The true P absorptions for clover and bromegrass (*Bromus inermis* L.) by dairy cattle have been observed to be about 45% compared with about 65% for lucerne and tall fescue (Martz *et al.*, 1996). Because of the large contribution of endogenous faecal P, apparent P absorption can vary dramatically depending on the feed (Minson, 1990). When apparent absorption is corrected for endogenous faecal losses, true P availability varies exponentially with P concentration in the feedstuff (Minson, 1990). Unfortunately, there are relatively few data available on true P absorption.

When dietary P was reduced from 0.41% to 0.30% in lactating dairy cows, there was a corresponding decrease of faecal P output by 0.55 g day^{-1} for each 1.0 g day^{-1} decrease in intake (Morse *et al.*, 1992b). More dramatic was the observed increase of 0.8 g day^{-1} for each 1.0 g day^{-1} increase in intake for cows moved to 0.56% P ration. From this it is apparent that P fed above the requirement level is at least 80% excreted in faeces.

Interaction with other nutrients

Phosphorus has been observed to interact with numerous minerals, Ca, S, Zn, Mn, Mo, Fe, Pb, Mg, Cd, and to vary with the protein level in the diet; however under a normal range of pasture-based dietary concentration, these interactions appear to have little effect on P absorption and metabolism (Underwood, 1977; Minson, 1990; McDowell, 1992).

Phosphorus requirements

In revising the ARC (1965), the ARC (1980) reduced the estimated P requirement for most classes of dairy cattle. In contrast, a couple of studies suggest that the requirement should be increased (Ward *et al.*, 1972; NRC, 1989). In light of this, the NRC (1989) points out that the minimum P level is difficult to establish and opted for a 10–22% increase in dietary P requirement over the previous edition (NRC, 1978). In determining the daily P requirements for dairy cattle, the NRC (1989) uses the value of 1.43 g 100 kg^{-1} of live weight for maintenance P based on endogenous P losses. In addition, they conservatively estimate a 50% absorption coefficient. This gives a maintenance level of 2.86 g dietary P 100 kg^{-1} live weight for lactating cows. Both the maintenance value and the absorption coefficient are speculative, and Steenvoorden (1989) noted that France, Germany, The Netherlands, the UK and the USA all use different estimates to determine the P requirements for dairy cattle.

In addition to maintenance, P requirements increase for late gestation or for milk production. For the conceptus, the NRC (1989) uses an adjustment of 0.47 g 100 kg^{-1} of the cow's live weight (1.23 g day^{-1} of conceptus gain). This gives a total late gestation requirement of 3.8 g dietary P 100 kg^{-1} live weight for dry cows. In the lactating cows, P requirement increases in proportion to the total quantity and fat content of the milk produced. As the fat percentage in milk increases the total P requirement increases, but the proportion of P required decreases for each incremental unit in fat. In addition to the maintenance requirement, this results in dietary P needs of 1.68–2.43 g kg^{-1} milk produced with 3.0–5.5 fat %, respectively. For example, the NRC (1989) P requirement for a 650 kg cow producing 30 kg of milk with 4.0 fat % would be 18.6 g P for maintenance plus 59.4 g P for milk production (1.98 g × 30 kg (4.0 fat %) milk) for a total of 78 g day^{-1}. At a consumption of 20 kg DM day^{-1}, this cow would require 0.39% P in her diet. Depending on the cow size and milk production, the P requirement for lactating cows ranges from 0.28% to 0.41% of dietary DM (NRC, 1989). Although it needs to be realized that even if the assumptions upon which the daily requirements are calculated are correct, the potential wide range in intake makes these percentages open to considerable variation.

Effect on production, intake and lactation

Inadequate P intake causes reduced voluntary feed intake, decreased milk production, slower growth rate, impaired reproductive performance, lethargy and unthriftiness (Cohen, 1975; Underwood, 1977; Minson, 1990; McDowell, 1992). The effect on voluntary intake can be very pronounced, with P deficient animals often exhibiting a depraved appetite with craving for objects such as bark, bones, hair or wood (McDowell, 1992).

A number of investigators have shown that milk production and (or) calf weight gains increase, often dramatically, when P deficient cows are supplemented with adequate P (Davison *et al.*, 1986). Generally P concentration in milk does not decrease in P deficient animals, while blood plasma inorganic P declines to below 4 mg dl^{-1}. There are a number of studies that do not demonstrate responses to P supplementation. Possible explanations include: that even when low P may not be the limiting nutrient or that even when P is limiting, the animal's ability to mobilize bone reserves may suffice. For example, improvements in growth and milk production have been observed for several months after P supplementation has ceased (Minson, 1990). Milk production was lower for cows not supplemented with P when they were individually fed a tropical grass–legume mix containing 0.28–0.29% P (Davison *et al.*, 1986). In addition, a decrease in bone Ca and P concentration was observed. Call *et al.* (1987) also observed lower milk production in cows fed 0.24% P than with 0.32 or 0.42% P diets. This decreased milk production was accompanied by decreased feed consumption.

Excess of phosphorus

Under normal conditions P toxicity is implausible. Phosphorus is readily absorbed and excreted making accumulation unlikely. Phosphorus is often supplemented to dairy cattle in excess of requirement without apparent effects on health or performance (Hibbs and Conrad, 1983). Excess P, however, is an economic waste and could potentially have negative environmental impact (Van Horn *et al.*, 1991; Aarts *et al.*, 1992; Tamminga, 1992). Chandler (1996) calculates that P accounts for over 50% of the cost of dairy mineral and vitamin premixes. Although P is proportionally expensive compared with other minerals in premixes, the total cost of mineral supplementation is relatively low. For example, Mathison (1986) estimated the cost of supplementation per Canadian beef cow at only $0.07 per day for Ca, P, Na and Mg. Regardless, high producing dairy cattle are fairly efficient at converting dietary P into milk, with an excess of 30% of the intake being converted to milk P (Steenvoorden, 1989; Van Horn *et al.*, 1994), which is greater than twice the calculated efficiency of utilization of either nitrogen or potassium.

Assessment of phosphorus status

A variety of methods have been used to assess P status. These include:

- forage analysis,
- waste product analysis, principally fecal excretion,
- blood or serum analysis, usually serum,
- tissue analysis, principally bone.

For forage analysis to be an effective method of assessing status, the level of the nutrient in the forage must be indicative of the quantity of the nutrient available to the animal. Similarly, there must be good data available on animal requirements against which the nutrient concentration in the forage can be compared. Although there have been difficulties in determining the nutrient requirements of dairy cattle (see above), detailed requirements are available and these are used extensively in ration formulation (ARC, 1980; NRC, 1989). To some extent one of the difficulties in ration formulations has been the relative lack of data on mineral concentration of feedstuffs, rather than adequate requirement data. Another limitation is that it is often difficult to collect a representative sample of the forage consumed for analysis. For example, P concentration of the fibrous portion of plants is often higher than the more available plant material (Kincaid and Cronrath, 1983;

Montalvo *et al.*, 1987). Nevertheless, forage analysis has the distinct advantage of being non-invasive, therefore it is often most practical to analyse P concentration of the feedstuffs and compare the intake of P with established requirements.

Phosphorus concentration of faecal material could be a reliable indicator of dietary P concentration (Moir, 1960; Cohen, 1974; Holechek *et al.*, 1985), including lactating dairy cows (Morse *et al.*, 1992b). The slope of the regression lines may however, vary with intake and the proportion of endogenous faecal P (Cohen, 1987).

Tissue or fluid analysis are generally better indicators of animal status than forage or soil analysis; however they are only reliable to the extent that they rapidly reflect mineral deficiencies or toxicities. Blood is widely used as an indicator of mineral status because of ease of collection. Deficiencies of P and poor bone mobilization capacities result in decreases in plasma inorganic P to an extent that plasma levels are moderately indicative of P status. In the case of P, serum is a good indicator of status provided stress factors, sampling time, and blood preparation (e.g. haemolysis, time and temperature) are controlled (McDowell, 1985). The critical limit for serum P is 4.5 mg 100 ml^{-1}.

Williams *et al.* (1991) analysed P concentration in a wide variety of tissues and fluids of cattle fed adequate or deficient levels of P. They determined that rib-bone P was the most consistent indicator of dietary status. McDowell (1985) lists the critical levels for P in fat free bone and bone ash as 11.5% and 17.6%, respectively. Recently, Beighle *et al.* (1994) observed a slightly higher P level in bone ash of dairy breeds (18.6%). Although there are routine procedures available (McDowell *et al.*, 1993), obtaining bone samples is still relatively difficult compared with the methods above.

As a practical strategy, the relative ease of obtaining forage or feedstuffs samples, and the observation that P deficiency is generally a gradual, chronic condition would argue for routine forage monitoring. Blood and (or) bone analysis could be used in conjunction with forage analysis to refine or confirm suspected deficiencies.

Feeding strategies

Peeler (1972) ranked the following P sources from least to most available: soft phosphates, bone meal, defluorinated rock phosphate, dicalcium phosphate, monocalcium phosphate, phosphoric acid, and sodium acid phosphate. Oilseed meals and other protein-rich plant by-products, and by-product feeds of animal or fish origin, are high in P. Phytate P is readily available to ruminants, including dairy cows (Morse *et al.*, 1992a). Most roughages are poor sources of P, particularly if the soils upon which they are grown are low in P. Cereal grains are only moderate sources of P, while they are poor sources of Ca. Generally P concentration decreases with forage maturity. There is a tendency for P concentration to decrease as protein concentration decreases.

In a survey of New York dairy farms, Curtis *et al.* (1984) noted that farms were generally over-feeding P. In relation to the NRC (1978) requirements, the ranges were 81–275%, 117–306%, 94–224% and 83–306% for early and late summer, and early and late winter, respectively. Most of these values would still be well in excess of the 10–22% increase in P requirements in the latest NRC (1989). In contrast, a survey of West Virginia dairy management practices found that farms were feeding the NRC (1978) recommended P level on the average, but the range was from 25% to 153% (Varga *et al.*, 1985).

Grazing cattle and forage P concentration

In assessing the mineral concentration of forage grasses and legumes analysed at Pennsylvania State University, Adams (1975) reported for forages that were mainly grass (*n* = 4119) and mainly legume (*n* = 4014) and had P concentrations of 0.23% (range 0.05–0.81%) and 0.29% (range 0.07–0.74%), respectively. The mean coefficient of variation for P analysis in forages was 25%. Minson (1990) illustrated the higher nutritive value of temperate forages compared with tropical forages, by noting that a much higher proportion of temperate forages would satisfy P requirements than tropical forages. For example, approximately 86% of temperate legumes and 54% of temperate grasses would satisfy the minimum requirement for lactating cows, in contrast to only 49% of tropical legumes and 35% of tropical grasses.

Sustainable phosphorus management in the soil–plant–animal system

Recent work suggests that the P concentration of supplementary feeds should be lowered to meet no more than the NRC (1989) dietary allowances (nutrient sufficiency approach), thereby reducing the P excreted while providing adequate amounts for optimal dairy cow performance (Harris *et al.*, 1990; Van Horn *et al.*, 1991). This is not always easy to ensure because animal requirements will vary with season and forage analysis (McDowell, 1985), and there is lack of agreement on P requirements (Steenvoorden, 1989). Further, availability and cost of custom supplements with different P concentrations may be problematic in some areas. Regardless of these limitations, it appears that P is generally over-supplemented (Harris *et al.*, 1990) and expensive compared with other minerals (Chandler, 1996). In conclusion, improved monitoring of dietary nutrition, nutrient cycling and use of grazing management strategies that promote more even distribution of recycled nutrients over dairy pastures (i.e. good distribution of shade, use of portable water and feed supplement units, and perhaps, rotational stocking) will reduce fertilizer requirements and potential for water quality problems (Mathews *et al.*, 1996a, 1996b). Improved P cycling efficiency is a major key to more sustainable pasture systems and additional research that integrates dairy farmers, universities, and environmental regulatory agencies is encouraged.

References

Aarts, H.F.M., Biewinga, E.E. and Van Keulen, H. (1992) Dairy farming systems based on efficient nutrient management. *Netherlands Journal of Agricultural Science* 40, 285–299.

Adams, R.S. (1975) Symposium: new concepts and developments in trace element nutrition. *Journal of Dairy Science* 58, 1538–1539.

Allen, L.H., Jr (1988) Dairy-siting criteria and other options for wastewater management on high water-table soils. *Soil and Crop Science Society of Florida Proceedings* 47, 108–127.

[ARC] Agricultural Research Council (1965) *The Nutrient Requirements of Farm Livestock. No. 2. Ruminants.* Commonwealth Agricultural Bureaux, Farnham Royal.

[ARC] Agricultural Research Council (1980) *The Nutrient Requirements of Ruminant Livestock.* Commonwealth Agricultural Bureaux, Farnham Royal, 125pp.

Ayres, A.S. and Humbert, R.P. (1957) Some aspects of phosphate retention by Hawaiian soils. *Hawaiian Sugar Technologists Reports* 16, 32–36.

Beighle, D.E., Boyazogul, P.A., Hemken, R.W. and Serumaga-Zake, P.A. (1994). Determination of calcium, phosphorus, and magnesium values in rib bones from clinically normal cattle. *American Journal of Veterinary Research* 55, 85–89.

Benacchio, S.S., Baumgardner, M.F. and Mott, G.O. (1970) Residual effect of grain-pasture feeding systems on the fertility of the soil under a pasture sward. *Soil Science Society of America Proceedings* 34, 621–624.

Bethlenfalvay, G.J., Evans, R.A. and Lesperance, A.L. (1985) Mycorrhizal colonization of crested wheatgrass as influenced by grazing. *Agronomy Journal* 77, 233–236.

Blair, G. (1993) Nutrient efficiency – what do we really mean? In: Randall, P.J., Delhaize, E., Richards, R.A. and Munns, R. (eds) *Genetic Aspects of Plant Mineral Nutrition.* Kluwer Academic Publishers, Netherlands, pp. 205–213.

Blue, W.G. (1988) Micronutrients for agronomic and forage crops in perspective. *Soil and Crop Science of Florida Proceedings* 47, 18–29.

Blue, W.G. and Gammon, N., Jr (1963) Differences in nutrient requirements of experimental pasture plots managed by grazing and clipping techniques. *Soil and Crop Science Society of Florida Proceedings* 23, 152–161.

Bolan, N.S., White, R.E. and Hedley, M.J. (1990) A review of the use of phosphate rocks as fertilisers for direct application in Australia and New Zealand. *Australian Journal of Experimental Agriculture* 30, 297–313.

Breeuwsma, A., Reijerink, J.G.A. and Schoumans, O.F. (1995) Impact of manure on accumulation and leaching of phosphate in areas of intensive livestock farming.

In: Steele, K. (ed.) *Animal Waste and the Land-Water Interface*. Lewis Publishers, Boca Raton, Florida, pp. 239–249.

Brookes, P.C., Powlson, D.S. and Jenkinson, D.S. (1984) Phosphorus in the soil microbial biomass. *Soil Biology and Biochemistry* 16, 169–175.

Call, J.W., Butcher, J.E., Shupe, J.L., Lamb, R.C., Boman, R.L. and Olson, A.E. (1987) Clinical effects of low dietary phosphorus concentrations in feed given to lactating dairy cows. *American Journal of Veterinary Research* 48, 133–136.

Canfield, D.E., Jr and Hoyer, M.V. (1988) The eutrophication of Lake Okeechobee. *Lake and Reservoir Management* 4, 91–99.

Caradus, J.R. (1980) Distinguishing between grass and legume species for efficiency of phosphorus use. *New Zealand Journal of Agricultural Research* 23, 75–81.

Caradus, J.R. (1990) Mechanisms for improving nutrient use by crop and herbage legumes. In: Baligar, V.C. and Duncan, R.R. (eds) *Crops as Enhancers of Nutrient Use*. Academic Press, San Diego, California, pp. 253–311.

Caradus, J.R. (1994) Selection for improved adaptation of white clover to low phosphorus and acid soils. *Euphytica* 77, 243–250.

Caradus, J.R. and Snaydon, R.W. (1986) Response to phosphorus of populations of white clover. 3. Comparison of experimental techniques. *New Zealand Journal of Agricultural Research* 29, 169–178.

Chandler, P.T. (1996) Environmental changes as related to animal agriculture – dairy. In: Kornegay, E.T. (ed.) *Nutrient Management of Food Animals to Enhance and Protect the Environment*. Lewis Publishers, Boca Raton, Florida, pp. 7–19.

Chichester, F.W., Van Keuren, R.W. and McGuinness, J.L. (1979) Hydrology and chemical quality of flow from small pastured watersheds: II. Chemical quality. *Journal of Environmental Quality* 8, 167–171.

Clark, R.B. (1990) Physiology of cereals for mineral nutrient uptake, use, and efficiency. In: Baligar, V.C. and Duncan, R.R. (eds) *Crops as Enhancers of Nutrient Use*. Academic Press, San Diego, California, pp. 131–209.

Cohen, R.D.H. (1974) Phosphorus nutrition of beef cattle. 4. The use of faecal and blood phosphorus for estimation of phosphorus intake. *Australian Journal of Experimental Agriculture and Animal Husbandry* 14, 709–715.

Cohen, R.D.H. (1975) Phosphorus and the grazing ruminant. *World Review of Animal Production* 11, 27–43.

Cohen, R.D.H. (1987) Supplementation practices of grazing livestock – macrominerals. In: *Proceedings, Grazing Livestock Nutrition Conference*. University of Wyoming, Laramie, Wyoming, pp. 93–100.

Cornforth, I.S. and Sinclair, A.G. (1982) Model for calculating maintenance phosphate requirements for grazed pastures. *New Zealand Journal of Experimental Agriculture* 10, 53–61.

Curtis, C.R., Erb, H.N., Sniffen, C.J. and Smith, R.D. (1984) Epidemiology of parturient paresis: Predisposing factors with emphasis on dry cow feeding and management. *Journal of Dairy Science* 67, 817–825.

Dalal, R.C. (1977) Soil organic phosphorus. *Advances in Agronomy* 29, 83–117.

Davison, T.M., Isles, D.H. and McGuigan, K.R. (1986) The effect of dietary supplementation with molasses and Christmas Island phosphate on milk yield of cows grazing tropical pastures. *Proceedings of the Australian Society of Animal Production* 16, 179–182.

Davison, T.M., Orr, W.N., Silver, B.A., Walker, R.G. and Duncalfe, F. (1997) Phosphorus fertilizer for nitrogen fertilized dairy pastures. 1. Long term effects on pasture, diet, and soil. *Journal of Agricultural Science, Cambridge* 129, 205–217.

DeLuca, H.F. (1979) The vitamin D system in the regulation of calcium and phosphorus metabolism. *Nutrition Reviews* 37, 161–193.

Dibb, D.W., Fixen, P.E. and Murphy, L.S. (1990) Balanced fertilization with particular reference to phosphorus: interaction of phosphorus with other inputs and management practices. *Fertilizer Research* 26, 29–52.

Dickinson, C.H. and Craig, G. (1990) Effects of water on the decomposition and release of nutrients from cow pats. *New Phytologist* 115, 139–147.

Dunlop, J., Lambert, M.G., van den Bosch, J., Caradus, J.R., Hart, A.L., Wewala, G.S., Mackay, A.D. and Hay, M.J.M. (1990) A programme to breed a cultivar of *Trifolium repens* L. for more efficient use of phosphate. In: El Bassam, N., Damboth, M. and Loughman, B.C. (eds) *Genetic Aspects of Plant Mineral Nutrition*. Kluwer Academic Publishers, Netherlands, pp. 547–552.

During, C. and Weeda, W.C. (1973) Some effects of cattle dung on soil properties, pasture production, and nutrient uptake. I. Dung as a source of phosphorus. *New Zealand Journal of Agricultural Research* 16, 423–430.

Eck, H.V. and Stewart, B.A. (1995) Manure. In: Rechcigl, J.E. (ed.) *Soil Amendments and Environmental Quality*. Lewis Publishers, Boca Raton, Florida, pp. 169–198.

Edwards, D.R. (1996) Recycling livestock manure on pastures. In: Joost, R.E. and Roberts, C.A. (eds) *Nutrient Cycling in Forage Systems*. Potash and Phosphate Institute (PPI) and the Foundation for Agronomic Research (FAR), Manhattan, Kansas, pp. 45–63.

Edwards, D.R., Moore, P.A., Jr. and Daniel, T.C. (1996) Grassed filter strips can reduce losses of nitrogen and phosphorus in runoff. *Better Crops with Plant Food* 80(4), 8–11.

Eghball, B., Binford, G.D. and Baltensperger, D.D. (1996) Phosphorus movement and adsorption in a soil receiving long-term manure and fertilizer application. *Journal of Environmental Quality* 25, 1339–1343.

Etgen, W.M. and Reaves, P.M. (1978) *Dairy Cattle Feeding and Management*. John Wiley, New York, 638pp.

Fenster, W.E. and Leon, L.A. (1979) Management of phosphorus fertilization in establishing and maintaining improved pastures on acid, infertile soils of tropical America. In: Sanchez, P.A. and Tergas, L.E. (eds) *Pasture Production in Acid Soils of the Tropics*. CIAT, Cali, Colombia, pp. 109–122.

Fox, R.L. (1979) Comparative responses of field grown crops to phosphate concentrations in soil solutions. In: Mussel, H. and Staples, R.C. (eds) *Stress Physiology in Crop Plants*. Wiley-Interscience, New York, pp. 81–106.

Fox, R.L. and Searle, P.G.E. (1978) Phosphate adsorption by soils of the tropics. In: Stelly, M. (ed.) *Diversity of Soils in the Tropics*. American Society of Agronomy and Soil Science Society of America, Madison, Wisconsin, pp. 97–119.

Gahoonia, T.S., Raza, S. and Nielsen, N.E. (1994) Phosphorus depletion in the rhizosphere as influenced by soil moisture. *Plant and Soil* 159, 213–218.

Godwin, D.C. and Blair, G.J. (1991) Phosphorus efficiency in pasture species. V. A comparison of white clover accessions. *Australian Journal of Agricultural Research* 42, 531–540.

Gourley, C.J.P., Allan, D.L. and Russelle, M.P. (1994) Plant nutrient efficiency: A comparison of definitions and suggested improvement. *Plant and Soil* 158, 29–37.

Graetz, D.A. and Nair, V.D. (1995) Fate of phosphorus in Florida spodosols contaminated with cattle manure. *Ecological Engineering* 5, 163–181.

Halsted, M. and Lynch, J. (1996) Phosphorus responses of C_3 and C_4 species. *Journal of Experimental Botany* 47, 497–505.

Harris, B., Jr, Morse, D., Head, H.H. and Van Horn, H.H. (1990) Phosphorus nutrition and excretion by dairy animals. Florida Cooperative Extension Service Circular 849, The University of Florida, Gainesville, Florida, 9pp.

Haynes, R.J. (1980) Competitive aspects of the grass–legume association. *Advances in Agronomy* 33, 227–261.

Haynes, R.J. and Williams, P.H. (1993) Nutrient cycling and soil fertility in the grazed pasture ecosystem. *Advances in Agronomy* 49, 119–199.

Hibbs, J.W. and Conrad, H.R. (1983) The relation of calcium and phosphorus intake and digestion and the effect of vitamin D feeding on the utilization of calcium and phosphorus by the lactating dairy cows. Ohio Agriculture Experiment Station Research Bulletin No. 1150, Ohio State University, Wooster, Ohio, 23pp.

Holechek, J.L., Galyean, M.L., Wallace, J.D. and Wofford, H. (1985) Evaluation of faecal indices for predicting phosphorus status of cattle. *Grass and Forage Science* 40, 489–497.

Howeler, R.H., Sieverding, E. and Saif, S. (1987) Practical aspects of mycorrhizal technology in some tropical crops and pastures. *Plant and Soil* 100, 249–283.

Hutton, J.B., Jury, K.E. and Davies, E.B. (1967) Studies on the nutritive value of New Zealand dairy pastures. V. The intake and utilization of potassium, sodium, calcium, phosphorus, and nitrogen in pasture herbage by lactating dairy cattle. *New Zealand Journal of Agricultural Research* 10, 367–388.

Ibrikci, H., Comerford, N.B., Hanlon, E.A. and Rechcigl, J.E. (1994) Phosphorus uptake by bahiagrass from spodosols: modeling of uptake from different horizons. *Soil Science Society of America Journal* 58, 139–143.

Irving, J.T. (1964) Dynamics and functions of phosphorus. In: Comar, C.L. and Bonner, F. (eds) *Mineral Metabolism*, Vol. 2. Academic Press, New York, pp. 149–176.

Iyamuremye, F. and Dick, R.P. (1996) Organic amendments and phosphorus sorption by soils. *Advances in Agronomy* 56, 139–185.

James, D.W., Kotuby-Amacher, J. and Huber, D.A. (1996) Phosphorus mobility in calcareous soils under heavy manuring. *Journal of Environmental Quality* 25, 770–775.

Jakobsen, I. and Rosendahl, L. (1990) Carbon flow into soil and external hyphae from roots of mycorrhizal cucumber plants. *New Phytologist* 115, 77–83.

Jakobsen, I., Abbott, L.K. and Robson, A.D. (1992) External hyphae of vesicular-arbuscular mycorrhizal fungi associated with *Trifolium subterraneum* L. I. Spread of hyphae and phosphorus inflow into roots. *New Phytologist* 120, 371–380.

Jones, R.J. (1990) Phosphorus and beef production in northern Australia. 1. Phosphorus and pasture productivity – a review. *Tropical Grasslands* 24, 131–139.

Kamprath, E.J. and Watson, M.E. (1980) Conventional soil and tissue tests for assessing the phosphorus status of soils. In: Khasawneh, F.E., Sample, E.C. and Kamprath, E.J. (eds) *The Role of Phosphorus in Agriculture*. American Society of

Agronomy, Crop Science Society of America, and Soil Science Society of America, Madison, Wisconsin, pp. 433–469.

Kappel, L.C., Morgan, E.B., Kilgore, L., Ingraham, R.H. and Babcock, D.K. (1985) Seasonal changes of mineral content of southern forages. *Journal of Dairy Science* 68, 1822–1827.

Kelling, K.A. and Matocha, J.E. (1990) Plant analysis as an aid in fertilizing forage crops. In: Westerman, R.L. (ed.) *Soil Testing and Plant Analysis,* 3rd edn. Soil Science Society of America, Madison, Wisconsin, pp. 603–643.

Kemp, P.D. and Blair, G.J. (1991) Phosphorus efficiency in pasture species. VI. A comparison of Italian ryegrass, phalaris, red clover and white clover over time. *Australian Journal of Agricultural Research* 42, 541–558.

Kincaid, R.L. and Cronrath, J.D. (1983) Amounts and distribution of minerals in Washington forages. *Journal of Dairy Science* 66, 821–824.

Li, X.-L., George, E. and Marschner, H. (1991) Extension of the phosphorus depletion zone in VA-mycorrhizal white clover in a calcareous soil. *Plant and Soil* 136, 41–48.

Lund, Z.F. and Doss, B.D. (1980) Coastal bermudagrass yield and soil properties as affected by surface-applied dairy manure and its residue. *Journal of Environmental Quality* 9, 157–162.

McCollum, R.E. (1991) Buildup and decline in soil phosphorus: 30-year trends in a Typic Umbraquult. *Agronomy Journal* 83, 77–85.

McDowell, L.R. (1985) *Nutrition of Grazing Ruminants in Warm Climates.* Academic Press, Orlando, Florida, 443 pp.

McDowell, L.R. (1992) *Minerals in Animal and Human Nutrition.* Academic Press, San Diego, California, 524pp.

McDowell, L.R., Conrad, J.H. and Hembry, F.G. (1993) *Minerals for Grazing Ruminants in Tropical Regions,* 2nd edn. University of Florida, Gainesville, Florida, 77pp.

Mansell, R.S., Bloom S.A. and Burgoa, B. (1991) Phosphorus transport with water flow in acid, sandy soils. In: Jacob, B. and Corapcioglu (eds) *Transport Processes in Porous Media.* Kluwer Academic Publishers, Dordrecht, pp. 271–314.

Marschner, H. (1995) *Mineral Nutrition of Higher Plants,* 2nd edn. Academic Press, New York, 889pp.

Martz, F.A., Weiss, M.F. and Gerrish, J.R. (1996) Macromineral availability and utilization by grazing livestock. In: Joost, R.E. and Roberts, C.A. (eds) *Nutrient Cycling in Forage Systems.* Potash and Phosphate Institute (PPI) and the Foundation for Agronomic Research (FAR), Manhattan, Kansas, pp. 177–192.

Mathews, B.W., Sollenberger, L.E., Nair, V.D. and Staples, C.R. (1994) Impact of grazing management on soil nitrogen, phosphorus, potassium, and sulfur distribution. *Journal of Environmental Quality* 23, 1006–1013.

Mathews, B.W., Sollenberger, L.E. and Tritschler, J.P., II. (1996a) Grazing systems and spatial distribution of nutrients in pastures: Soil considerations. In: Joost, R.E. and Roberts, C.A. (eds) *Nutrient Cycling in Forage Systems.* Potash and Phosphate Institute (PPI) and the Foundation for Agronomic Research (FAR), Manhattan, Kansas, pp. 213–229.

Mathews, B.W., Tritschler, J.P., II, Douglas, N.P. and Madsen, D.L. (1996b) Some chemical characteristics of soils and manures at Hawai'i Island dairies. *Journal for Hawaiian and Pacific Agriculture* 7, 21–35.

Mathews, B.W., Tritschler, J.P., II, and Miyasaka, S.C. (1997) Sustainable phosphorus management in pasture-based dairy systems. *Journal for Hawaiian and Pacific Agriculture* 8, 27–58.

Mathews, C.B. (1992) Cattle manure characterization and decomposition and its potential impact on water quality. MSc thesis, The University of Florida, Gainesville, Florida.

Mathison, G.W. (1986) Supplementing minerals in the cow-calf operation. Managing in Changing Times. *IMC Conference Proceedings*, International Minerals and Chemical Corporation, Saskatoon, pp. 113–126.

Mays, D.A., Wilkinson, S.R. and Cole, C.V. (1980) Phosphorus nutrition of forages. In: Khasawneh, F.E., Sample, E.C. and Kamprath, E.J. (eds) *The Role of Phosphorus in Agriculture*. American Society of Agronomy, Crop Science Society of America and Soil Science Society of America, Madison, Wisconsin, pp. 805–846.

Middleton, K.R. and Smith, G.S. (1978) The concept of a climax in relation to the fertiliser input of a pastoral system. *Plant and Soil* 50, 595–614.

Minson, D.J. (1990) Phosphorus in ruminant nutrition. In: *Forage in Ruminant Nutrition*. Academic Press, San Diego, California, pp. 230–264.

Moir, K.W. (1960) Nutrition of grazing cattle. II. Estimation of phosphorus and calcium in pasture selected by grazing cattle. *Queensland Journal of Agricultural Science* 17, 373–379.

Montalvo, M.I., Veiga, J.V., McDowell, L.R., Ocumpaugh, W.R. and Mott, G.O. (1987) Mineral content of dwarf *Pennisetum purpurem* under grazing conditions. *Nutrition Reports International* 35, 157–165.

Morse, D., Head, H.H. and Wilcox, C.J. (1992a) Disappearance of phosphorus in phytate from concentrates in vitro and from rations fed to lactating dairy cows. *Journal of Dairy Science* 75, 1979–1986.

Morse, D., Head, H.H., Wilcox, C.J., Van Horn, H.H., Hisseem, C.D. and Harris, B., Jr. (1992b) Effects of concentration of dietary phosphorus on amount and route of excretion. *Journal of Dairy Science* 75, 3039–3049.

Mouat, M.C.H. (1983) Phosphate uptake from extended soil solutions by pasture plants. *New Zealand Journal of Agricultural Research* 26, 483–487.

[NRC] National Research Council (1978) *Nutrient Requirements of Dairy Cattle*, 5th edn. National Academy Press, Washington, DC, 105pp.

[NRC] National Research Council (1989) *Nutrient Requirements of Dairy Cattle*, 6th edn. National Academy Press, Washington, DC, 157pp.

Payne, W.A., Hossner, L.R., Onken, A.B. and Wendt, C.W. (1995) Nitrogen and phosphorus uptake in pearl millet and its relation to nutrient and transpiration efficiency. *Agronomy Journal* 87, 425–431.

Peeler, H.T. (1972) Biological availability of nutrients in feeds: availability of major mineral ions. *Journal of Animal Science* 35, 695–712.

Perrott, K.W. and Sarathchandra, S.U. (1989). Phosphorus in the soil microbial biomass of New Zealand soils under established pasture. *New Zealand Journal of Agricultural Research* 32, 409–413.

Plucknett, D.L. (1970) Productivity of tropical pastures in Hawaii. In: Norman, M.J.T. (ed.) *Proceedings of the XI International Grassland Congress*. University of Queensland Press, St. Lucia, Queensland, pp. A38–A49.

Polle, E.A. and Konzak, C.F. (1990) Genetics and breeding of cereals for acid soils and nutrient efficiency. In: Baligar, V.C. and Duncan, R.R. (eds) *Crops as Enhancers of Nutrient Use*. Academic Press, San Diego, California, pp. 81–130.

Pote, D.H., Daniel, T.C., Sharpley, A.N., Moore, P.A., Jr, Edwards, D.R. and Nichols, D.J. (1996) Relating extractable soil phosphorus to phosphorus losses in runoff. *Soil Science Society of America Journal* 60, 855–859.

Powell, C.L. (1980) Mycorrhizal infectivity of eroded soils. *Soil Biology and Biochemistry* 12, 247–250.

Powell, C.L. and Daniel, J. (1978) Growth of white clover in undisturbed soils after inoculation with efficient mycorrhizal fungi. *New Zealand Journal of Agricultural Research* 21, 675–681.

Reay, P.F. and Marsh, B. (1976) Element composition of ryegrass and red clover leaves during a growing season. *New Zealand Journal of Agricultural Research* 19, 469–472.

Reese, L.E., Hegg, R.O. and Gantt, R.E. (1982) Runoff water quality from dairy pastures in the Piedmont region. *Transactions of the American Society of Agricultural Engineers (ASAE)* 25, 697–701.

Roberts, A.H.C., Morton, J. and Edmeades, D.C. (1994) *Fertilizer Use on Dairy Farms.* Dairying Research Corporation Ltd., Hamilton, 36pp.

Safley, L.M., Jr., Barker, J.C. and Westerman, P.W. (1984) Characteristics of fresh dairy manure. *Transactions of the American Society of Agricultural Engineers (ASAE)* 27, 1150–1162.

Sanyal, S.K. and DeDatta, S.K. (1991) Chemistry of phosphorus transformations in soil. *Advances in Soil Science* 16, 1–120.

Saunders, W.M.H. and Metson, A.J. (1971) Seasonal variation of phosphorus in soil and pasture. *New Zealand Journal of Agricultural Research* 14, 307–328.

Schettini, T.M., Gabelman, W.H. and Gerloff, G.C. (1987) Incorporation of phosphorus efficiency from exotic germplasm into agriculturally adapted germplasm of common bean (*Phaseolus vulgaris* L.). In: Gabelman, H.W. and Loughman, B.C. (eds) *Genetic Aspects of Plant Mineral Nutrition.* Martinus Nijhoff, Boston, Massachusetts, pp. 559–568.

Sentenac, H. and Grignon, C. (1985) Effect of pH on orthophosphate uptake by corn roots. *Plant Physiology* 77, 136–141.

[SFWMD] South Florida Water Management District (1989) *Surface Water Improvement and Management (SWIM) plan for Lake Okeechobee.* South Florida Water Management District, West Palm Beach, Florida.

Sharpley, A.N. and Halvorson, A.D. (1994) The management of soil phosphorus availability and its impact on surface water quality. In: Stewart, B.A. and Lal, R. (eds) *Soil Processes and Water Quality.* Lewis Publishers, Boca Raton, Florida, pp. 7–90.

Sharpley, A.N., Syers, J.K. and Tillman, R.W. (1978) An improved soil sampling procedure for the prediction of dissolved inorganic phosphate concentrations in surface runoff from pasture. *Journal of Environmental Quality* 7, 455–456.

Silva, J.A. (1985) Procedure for developing phosphorus extraction curves. In: Silva, J.A. (ed.) *Soil-Based Agrotechnology Transfer.* College of Tropical Agriculture and Human Resources, University of Hawaii, Honolulu, Hawaii, pp. 241–242.

Smith, F.W., Jackson, W.A. and van den Berg, P.J. (1990) Internal phosphorus flows during development of phosphorus stress in *Stylosanthes hamata. Australian Journal of Plant Physiology* 17, 451–464.

Steenvoorden, J.H.A.M. (1989) Manure management and regulations in the Netherlands. In: *Proceedings from the Dairy Manure Management Symposium.*

Northeast Regional Agricultural Engineering Service, Cornell University, Ithaca, New York, pp. 15–22.

Stevenson, F.J. (1986) *Cycles of Soil: Carbon, Nitrogen, Phosphorus, Sulfur, Micronutrients.* Wiley-Interscience, New York, New York, 380pp.

Tamimi, Y.N. and Matsuyama, D.T. (1990) Long-range management of fertility in Hawaiian soils. *Hawaiian Sugar Technologists Reports* 49, 38–43.

Tamminga, S. (1992) Nutrition management of dairy cows as a contribution to pollution control. *Journal of Dairy Science* 75, 345–357.

Tarafdar, J.C. and Claasen, N. (1988) Organic phosphorus compounds as a phosphorus source for higher plants through the activity of phosphatases produced by plant roots and microorganisms. *Biology and Fertility of Soils* 5, 308–312.

Tarafdar, J.C. and Marschner, H. (1994) Efficiency of VAM hyphae in utilization of organic phosphorus by wheat plants. *Soil Science and Plant Nutrition* 40, 593–600.

Tate, K.R., Spier, T.W., Ross, D.J., Parfitt, R.L., Whale, K.N. and Cowling, J.C. (1991) Temporal variations in some plant and soil P pools in two pasture soils of different P fertility status. *Plant and Soil* 132, 219–232.

Underwood, E.J. (1977) *Trace Elements in Human and Animal Nutrition,* 4th edn. Academic Press, New York, 545pp.

[USEPA] United States Environmental Protection Agency (1986) *Quality Criteria for Water,* EPA 440/5-86-001. United States Environmental Protection Agency, Washington, DC.

Van Horn, H.H., Nordstedt, R.A., Bottcher, A.V., Hanlon, E.A., Graetz, D.A. and Chambliss, C.F. (1991) Dairy manure management: strategies for recycling nutrients to recover fertilizer value and avoid environmental pollution. *Florida Cooperative Extension Service Circular* 1016, The University of Florida, Gainesville, Florida, 16pp.

Van Horn, H.H., Wilkie, A.C., Powers, W.J. and Nordstedt, R.A. (1994) Components of dairy management systems. *Journal of Dairy Science* 77, 2008–2030.

Varga, G.A., Hoover, W.H. and Dailey, R.A. (1985) Survey of nutritional management practices and metabolic disorders in West Virginia dairy herds. *Journal of Dairy Science* 68, 1507–1512.

Vicente-Chandler, J., Abruña, F., Caro-Costas, R., Figarella, J., Silva, S. and Pearson, R.W. (1974) Intensive grassland management in the humid tropics of Puerto Rico. *Puerto Rico Agricultural Experiment Station Bulletin* 233. University of Puerto Rico, Mayagüez, 164pp.

Walker, T.W., Thapa, B.K. and Adams, A.F.R. (1959) Studies on soil organic matter: 3 Accumulation of carbon, nitrogen, sulfur, organic and total phosphorus in improved grassland soils. *Soil Science* 87, 135–140.

Ward, G., Dobson, R.C. and Dunham, J.R. (1972) Influence of calcium and phosphorus intakes, vitamin D supplement, and lactation on calcium and phosphorus balance. *Journal of Dairy Science* 55, 768–776.

Ward, J.C., O'Connor, K.F. and Wei-bin, G. (1990) Phosphorus losses through transfer, runoff, and soil erosion. In: *Proceedings of a Symposium on Phosphorus Requirements for Sustainable Agriculture in Asia and Oceania.* International Rice Research Institute, Los Baños, Laguna, Philippines, pp. 167–182.

Williams, S.N., McDowell, L.R., Lawrence, L.A., Wilkinson, N., Ferguson, P.W. and Warnick, A.C. (1991) Criteria to evaluate bone mineralization in cattle: II. noninvasive techniques. *Journal of Animal Science* 69, 1242–1254.

Withers, P.J. and Sharpley, A.N. (1995) Phosphorus fertilizers. In: Rechcigl, J.E. (ed.) *Soil Amendments and Environmental Quality*. Lewis Publishers, Boca Raton, Florida, pp. 65–107.

Grass Silage

P. O'Kiely[1] and R.E. Muck[2]

[1]Teagasc, Grange Research Center, Dunsany, Co. Meath, Ireland
[2]USDA-Agricultural Research Service, US Dairy Forage Research Center, 1925 Linden Drive West, Madison, Wisconsin 53706, USA

Introduction

Significance of silage

The science and technology of making and feeding silage have advanced considerably since the 1960s, enabling the rapid expansion in the quantities of forages conserved by ensilage. Most silage is currently made in Europe, the Russian Federation and North America, and a wide range of crops are now ensiled, with grass being the most common, followed in order by maize (*Zea mays* L.), legumes (mainly lucerne (*Medicago sativa* L.)), whole crop cereals, sorghum (*Sorghum bicolor* (L.) Moench) and beet (*Beta vulgaris* L.) tops (Wilkinson and Bolsen, 1996). Silage has superseded hay as the primary form of conserved forage in some parts of the world (Table 9.1), particularly in climates unsuited to reliable hay making.

Advances facilitating silage making

Various factors have contributed to farmers adopting silage making. These include the development of flail, double-chop and precision-chop harvesters and the big round baler, the availability of polythene covers, evolving options regarding additives and their applicators on the one hand or partial field wilting on the other, and innovations relating to storage (e.g. increasing variety of silo types) and feeding (e.g. self-feeding; easier mechanization

Table 9.1. Estimated quantities (million tonnes dry matter) of hay and silage produced in 1994.

	Hay	Silage
Western Europe[1]	60.3	91.1
Eastern Europe[2]	31.6	16.4
Russian Federation	34.0	52.9
USA	123.0	39.3
Canada	40.9	7.4

[1]European Union (15) plus Norway and Switzerland.
[2]Fifteen countries. (Source: update of Wilkinson and Bolsen (1996).)

relative to hay). Furthermore, improved crop husbandry practices that result in high yields of nutritious, ensilable forage have helped make silage an economically attractive feedstuff (O'Kiely *et al.*, 1997a). Finally, the scientific knowledge quantifying the factors affecting crop yield and quality, conservation efficiency and ruminant nutrition has given farmers an understanding of the steps they have to take to produce and utilize an economically attractive feedstuff.

Silage in farm systems

The type of farming system determines the philosophy underpinning the silage making and feeding systems practised. On the North Island of New Zealand, grass grows throughout the year, albeit in a seasonal pattern, and low-cost dairying systems are synchronized to produce milk in parallel with the supply of grass for grazing. In the winter season, grass growth tends to be adequate to maintain non-lactating cows. Grass is grown to be grazed, and is not generally managed in a planned manner for making into silage. Instead, it is harvested for ensilage if its growth exceeds herd demand for a sufficient period to create a surplus that is difficult to utilize efficiently by grazing. Consequently, silage making tends to be opportunistic, with the area of grass to be harvested, together with its yield and nutritive value, being variable.

In countries such as Ireland, the annual pattern of milk production also reflects the pattern of grass growth. However, very little grass growth occurs in winter, and cows are normally housed for between 2 and 5 months each winter. In this grassland management system, predetermined proportions of grassland are allocated for grazing or silage making during the first half of the grass growing season, based on the estimated yield and nutritive value of the grass on the projected silage harvest dates. After one or two cuts of grass have been harvested for ensiling, and the required supply of winter forage is conserved, the regrowth of grass is grazed.

In areas such as the Po valley in Northern Italy, cows are often fed in confinement systems and not grazed. Here, specific crops (e.g. *Lolium multiflorum* Lam.) are grown under irrigation expressly for ensiling. The yield and quality of grass for ensilage should be more predictable under such conditions.

The optimal quality for silage within any grass-based cattle system will vary depending on factors such as (i) the level and quality of animal production required, (ii) the duration for which silage will be offered and its proportion in the diet, (iii) intake and conversion efficiency characteristics of the forage, (iv) the value of the animal produce, (v) ensiling method, (vi) the cost and supply of silage, (vii) the cost, quality and availability of alternatives, and (viii) the costs of land, buildings and labour. The optimal quality for silage can therefore change with changing economic forces.

Grass for Silage

Principles of ensilage

The aim of ensiling is to retain as much of the animal production potential of the harvested crop as feasible. This includes minimizing quantitative losses and controlling qualitative changes. The main factors determining the preservation of grass as silage are outlined schematically in Fig. 9.1. The primary requirement for successful ensilage is the rapid achievement of anaerobic conditions, and its maintenance thereafter. This terminates plant

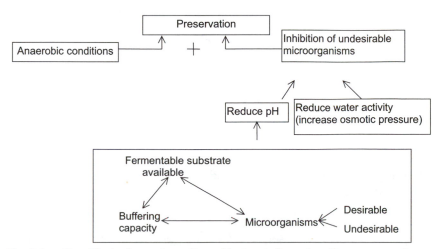

Fig. 9.1. Simplified schematic outline of factors influencing silage preservation (temperature excluded).

respiration, inhibits the growth of aerobic microorganisms, restricts temperature rise and creates conditions favourable for the growth and development of lactic acid bacteria (Pitt *et al.*, 1985). The second requirement is to create conditions within the ensiled forage that inhibit the activity of undesirable anaerobic microorganisms. This is normally achieved by reducing pH sufficiently and quickly, but may be assisted by simultaneously reducing the water activity (e.g. wilting) (McDonald *et al.*, 1991). Both low pH and undissociated organic acid molecules may be responsible for inhibitory effects depending on the specific microbial species, and a fall in the pH of silage increases the proportion of undissociated acids such as lactic and acetic acids (Baird-Parker, 1980).

When grass is mown or harvested, enzyme systems in the plant, including those controlling respiration, are still active. Once the harvested crop is sealed from air, plant respiration removes the trapped oxygen, thereby creating an anaerobic environment. Microbial respiration may also contribute to attaining anaerobic conditions. In the absence of oxygen, plant cells lyse, releasing their contents. This results in rapid proteolysis and enzymatic breakdown of carbohydrates as well as the provision of substrates for anaerobic microbial growth (i.e. fermentation) (Pitt *et al.*, 1985). The quicker that both anaerobic conditions are achieved and pH is reduced, the sooner the activities of aerobic and undesirable anaerobic microorganisms, and respiratory and proteolytic plant enzymes, are inhibited, and consequently the greater the probability of achieving a successful preservation of grass as silage (Woolford, 1984; Charmley, 1995).

Silage fermentation under farm conditions is not a controlled process, particularly when compared with industrial fermentation processes. Most aspects of silage-making systems therefore contribute to stacking the probabilities in favour of lactic acid-dominant fermentation, but the precise characteristics of the silage are still prone to considerable variability. Furthermore, although mean values are quoted for the composition of grass silage, it is a heterogeneous feedstuff (Spoelstra, 1990) with the possibility for distinctly contrasting micro-niches within (Jonsson, 1989).

Ensilability characteristics

As indicated in Fig. 9.1, the reduction in silage pH is a function of three principal factors: fermentable substrate (primarily soluble carbohydrates), buffering capacity (BC) of the plant and the microorganisms present on the crop at ensiling.

Water soluble carbohydrates (WSC)

The primary fermentable substrates in temperate grasses are cold water soluble carbohydrates such as glucose, fructose, sucrose and fructans (Smith, 1973). Because microorganisms utilize these sugars in an aqueous medium, and because the water content in grass may vary considerably, the fermentable substrate concentration is expressed in the aqueous phase (i.e. $g\ l^{-1}$ or $g\ kg^{-1}$ grass juice) as an index of ensilability. The values presented are therefore a product of both the water and WSC contents in the grass.

The content and composition of WSC in grass varies with grass type and plant part. In a comparison of *Lolium perenne* L. and *Lolium multiflorum* conducted during a 2-year interval, McGrath (1988) noted that fructans accounted for about 70% of the soluble carbohydrates (range 0–90%). Fructans were a more prominent component of stem than of leaf, and the WSC content (DM basis) was at least 50% higher in stem than in leaf. *Lolium multiflorum* had less fructans than *Lolium perenne*, but more of other sugars. In a 4-year comparison of sward types managed under an intensive cutting regime, Wilson and Collins (1980) showed that there is almost a twofold range in the average WSC content among commonly found sward types and grass genera (Table 9.2).

The concentration of WSC in grass DM and the proportion of fructans in WSC increase as the growth stage advances (McGrath, 1988), reflecting increased concentrations in both leaf and stem and an increase in the ratio of stem to leaf. Furthermore, there appears to be a seasonal effect, with the concentration of WSC ($g\ kg^{-1}$ DM) tending to peak in May or June, and decreasing gradually thereafter to about half the peak value in late September (Jones *et al.*, 1965).

Fertilizer is applied to ensure satisfactory yields of silage. Applied N can reduce grass DM and WSC concentrations and increase buffering capacity (Table 9.3), the effects being greater with increasing rates of N addition and as the interval after N application decreases. In contrast, application of

Table 9.2. Effect of sward type on ensilability indices (source: Wilson and Collins (1980)).

	DM ($g\ kg^{-1}$)	WSC ($g\ l^{-1}$)	Buffering capacity (mEq kg DM^{-1})	NO$_3$ (mg l^{-1} juice)
Lolium multiflorum Lam.	203	36	235	311
L. multiflorum + Trifolium pratense L.	203	35	247	91
Poa trivialis L.	207	33	264	308
L. perenne L.	201	29	272	428
L. perenne + Trifolium repens L.	200	28	302	19
Festuca rubra L.	205	24	258	230
Phleum pratense L.	192	22	257	232
Agrostis tenuis Sibth.	200	22	244	290
Old permanent pasture[1]	194	20	265	255
Dactylis glomerata L.	182	17	246	289

[1]Indigenous mixture.
DM, dry matter; WSC, water soluble carbohydrates.

Table 9.3. Changes in grass ensilability indices over time where a series of rates of N fertilizer were applied on March 31 (source: O'Kiely *et al.* (1997b)).

		kg N ha^{-1} on March 31			
	Sampling date	0	50	100	150
Dry matter (g kg^{-1})	May 6	212	180	165	153
	May 20	255	195	186	170
	June 2	208	176	173	158
WSC (g l^{-1})	May 6	37	28	22	19
	May 20	29	22	15	15
	June 2	28	22	24	17
Buffering capacity	May 6	241	264	297	311
(mEq kg^{-1} DM)	May 20	220	237	262	275
	June 2	175	169	184	205

WSC, water soluble carbohydrates.

P (O'Kiely and Tunney, 1997) or K (Keady and O'Kiely, 1996) has minor effects on ensilability indices. Cattle or pig slurry has a less negative effect on grass ensilability than an equivalent level of fertilizer N. The main negative effects of slurry seem to be from the nutrients it supplies to grass, although occasionally microbial contamination effects occur (O'Kiely *et al.*, 1994; Rammer, 1996). Where slurry is applied in an even and timely manner, at an appropriate rate, and with the inorganic fertilizer input modified to take account of the probable nitrogen contribution from slurry, silage fermentation is unlikely to be compromised. However, application of solid manures during the grass growing season causes serious problems to silage fermentation, increasing the risk of a clostridial fermentation (Rammer, 1996).

Weather has considerable direct and indirect effects on DM and WSC concentrations, affecting the balance between WSC being produced and utilized. Increasing light intensity increases both DM and WSC (g kg^{-1} DM) concentrations, increasing temperature increases DM and reduces WSC concentrations, and increasing water supply reduces DM and WSC concentrations (Deinum, 1966). Over a series of days, these factors can have a considerable cumulative effect of the content of WSC available. Diurnal variation in both DM and WSC concentrations in grass is marked (Wilkinson *et al.*, 1994), reflecting both the drying of the grass (e.g. evaporation of dew) and the production of WSC by photosynthesis.

Starch is the main non-structural polysaccharide in the vegetative tissue of tropical grasses; starch is stored mainly in the leaf rather than the stem (Smith, 1973). Because most lactic acid bacteria active during silage fermentation do not metabolize starch directly (Sneath *et al.*, 1986), tropical grasses can be deficient in fermentable substrate and therefore difficult to preserve properly (Catchpoole and Henzell, 1971). Furthermore, they tend to be lower in protein (Lyttleton, 1973) and digestibility (Minson and McLeod, 1970) than temperate grasses.

The concentration of WSC decreases during ensilage. The concentration of residual WSC plus fermentation products in silage normally exceeds that of WSC in fresh herbage, implying that other sources of fermentable substrate become available such as structural carbohydrates, amino acids and organic acids. Hemicellulose is likely to be the primary source of substrate (e.g. pentoses), initially undergoing hydrolysis by plant enzymes and later acid hydrolysis (Dewer *et al.*, 1963).

Buffering capacity (BC)

The BC of grass for ensilage is defined as the equivalents of acid per unit DM required to lower crop pH from 6 to 4, and can be crudely used as an index of the amount of sugar required for an adequate natural lactic fermentation. The BC of *Lolium multiflorum* and *Trifolium pratense* L. has been attributed primarily to organic acid salts, orthophosphates, sulphates, nitrates and chlorides, with only about 10–20% resulting from the action of plant proteins (Playne and McDonald, 1966). In studying two permanent pastures of indigenous grasses, Muck *et al.* (1991a, b) found organic acids to account for only 15–30% of the BC, but reported a strong correlation for BC with both crude protein content and *in vitro* DM digestibility.

Typical BCs in grasses range from 150 to over 500 mEq kg^{-1} DM (McDonald *et al.*, 1991; Muck *et al.*, 1991a; O'Kiely *et al.*, 1994), suggesting that the sugar requirement for adequate preservation of grass silages should vary by fourfold. BC varies by grass species (Table 9.2) and usually is higher in legumes and lower in maize than grasses (McDonald *et al.*, 1991). The BC decreases with advancing maturity in grass except for autumn regrowth (Muck *et al.*, 1991a) and is increased by high or late application of N fertilizer (Table 9.3). In lucerne, moisture stress reduces buffering capacity (Muck and Walgenbach, 1985). On the same basis as for WSC, a strong argument can be made for expressing BC on a liquid phase basis when using it as an index of ensilability. To date, this has not been the convention.

Microorganisms

The populations of microorganisms normally found on growing grass are usually dominated by obligate aerobes and facultative anaerobes, with smaller proportions of obligate anaerobes (spores) often present. These populations include pseudomonads, enterobacteria, actinomycetes, filamentous fungi, yeasts, acetic acid bacteria, homofermentation and heterofermentative lactic acid bacteria, propionic acid bacteria and spores of clostridia (Dellaglio and Torriani, 1991).

Lactic acid bacteria are Gram-positive, non-sporing and microaerophilic (Lindgren, 1991). The main genera active during ensilage are *Lactobacillus, Pediococcus, Enterococcus, Lactococcus, Streptococcus* and

Leuconostoc. Those with homofermentative activity ferment 1 mole glucose or fructose to 2 moles lactic acid, whereas the heterofermenters produce lactic acid plus products such as acetic acid, ethanol or mannitol (McDonald *et al.*, 1991). Most strains of lactic acid bacteria associated with silage are capable of fermenting sucrose and the main hexoses and pentoses available (Sneath *et al.*, 1986), but relatively few can ferment the fructans of grass (Muller and Lier, 1994). However, fructans are almost completely hydrolysed by plant enzyme activity during ensilage (Henderson *et al.*, 1972).

The numbers and types of lactic acid bacteria on grass vary considerably (Fehrmann and Muller, 1990; Moran *et al.*, 1991), with lower counts being obtained on the leaf, followed by the upper stem, than on other plant parts (Moran *et al.*, 1991). The numbers of lactic acid bacteria increase quickly within the silo once anaerobic conditions are established and substrate becomes available, with peak numbers of approximately 10^9 colony forming units (CFU) g^{-1} herbage being obtained within days (Dellaglio, 1985). These gradually decline to about 10^6 g^{-1} in lactic acid dominant silages but can remain much higher where preservation is unsatisfactory (Lindgren, 1991; Fig. 9.2). Within the lactic acid bacteria population, a generalized change in predominant species occurs during the ensilage period, with leuconostocs and lactococci initially dominant, followed rapidly by homofermentative pediococci and lactobacilli, and with heterofermentative lactobacilli finally dominating under low pH conditions (Seale, 1986).

Clostridia are Gram-positive, spore-forming, rod-shaped anaerobic bacteria whose main habitat is soil, as well as decaying vegetation and animal excreta (Cato *et al.*, 1986). Their growth is inhibited in the presence of oxygen and at acidic pH, low temperature or low water activity (McDonald *et al.*, 1991). In wetter silages, clostridia often result in a poor fermentation that reduces intake potential for cattle. Saccharolytic clostridia ferment mainly sugars and lactic acid during ensilage, producing butyric acid and other products, while proteolytic clostridia ferment mainly amino acids either by deamination or decarboxylation, producing amines, ammonia and other products (Woolford, 1984). Growth limiting factors for saccharolytic clostridia are pH < 4.2–4.5, water activity (a_w) < 0.95 and temperature $< 10°C$, with corresponding values for proteolytic clostridia of < 4.7–5.0, < 0.95–0.97 and $< 10°C$ (Lindgren, 1991). These factors interact so that, for example, wilting a crop directly lowers water activity but also raises the critical pH for inhibiting clostridial growth (Leibensperger and Pitt, 1987). Hypothetical numbers of spores under good and bad ensilage conditions are shown in Fig. 9.2. High numbers of clostridial spores on forage at ensiling need not cause bad preservation, but a badly preserved silage (i.e. high pH, butyric acid and NH_3-N) is likely to have large numbers of clostridia (Rammer, 1996).

Weissbach *et al.* (1976) found that the WSC : BC ratio and the DM content were useful in determining if a forage would be susceptible to

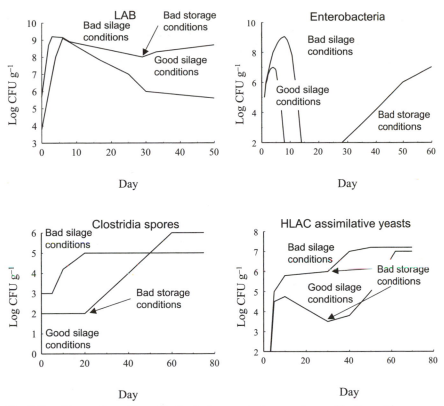

Fig. 9.2. Illustration of hypothetical changes in the numbers of lactic acid bacteria (LAB), enterobacteria, clostridia spores and lactic acid (HLAC) assimilating yeasts under good (lactic acid dominant fermentation) and bad (lactic acid deficient fermentation) silage conditions and during bad storage (i.e. non-anaerobic) conditions (source: Lindgren (1991)). (CFU: colony forming units.)

clostridial fermentation (Fig. 9.3). The lower the WSC : BC ratio the higher the DM content necessary to reduce the possibility of a clostridial silage. Such a relationship just produces general guidelines because it does not take into account factors affecting the release of plant sugars, the speed of the lactic acid bacterial fermentation, and the relative numbers of clostridia and lactic acid bacteria at ensiling. Consequently, a silage with a satisfactory WSC : BC ratio may have a clostridial fermentation and *vice versa*. Nevertheless, Fig. 9.3 is useful in demonstrating the interrelationship between WSC, BC and DM content on the probability of a clostridial fermentation.

Enterobacteria in silage are Gram-negative, non-sporing, rod-shaped and facultatively anaerobic (Beck, 1978). They are found in soil and animal excreta as well as in herbage. They can ferment carbohydrates to short-chain organic acids, particularly acetic acid, or butanediol. They have weak proteolytic activity but can deaminate and decarboxylate some amino acids,

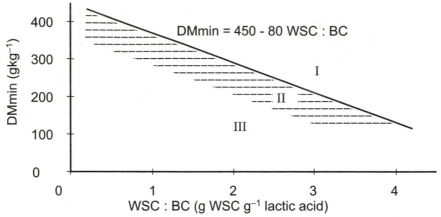

Fig. 9.3. Minimum dry matter content for anaerobically stable silages. Silage quality: I – good, II – uncertain, III – poor. Ratio of WSC : BC is determined from water soluble carbohydrates (WSC) as expressed as g WSC per 100 g DM and buffering capacity (BC) as measured as the g lactic acid per 100 g DM required to lower unensiled crop pH to 4.0 (adapted from Weissbach *et al.* (1976)).

and most species can reduce nitrate to ammonia via nitrite and some can also produce nitrous oxide (McDonald *et al.*, 1991). This latter activity has both positive and negative aspects. Nitrite can inhibit clostridial activity, but nitrogen oxide gases are potentially hazardous to farmers, especially in tower silos where gases may be trapped for long periods. The number of enterobacteria on herbage varies widely. Enterobacteria may increase in the early stages of ensilage (Fig. 9.2), but should decrease rapidly if pH falls quickly and sufficiently (Muck, 1991). Growth limiting factors include pH < 4.5–5.0, a_w < 0.95 and temperature < 8°C (Lindgren, 1991).

Yeast are budding eukaryote organisms of low proteolytic activity (Jonsson, 1989), some of which have a high fermentative ability for sugars (producing mainly ethanol) and others that have a high respiration capacity for lactic acid. These latter yeasts are normally considered the most important for initiating aerobic spoilage of grass silages (Woolford, 1984). Possible changes in total counts during ensilage are shown in Fig. 9.2, and growth limiting factors are pH < 2.0, a_w < 0.62–0.85 and temperature < 0–5°C (Lindgren, 1991). Moulds such as *Aspergillus*, *Fusarium* and *Penicillium* can develop on silage under aerobic conditions. They significantly reduce the quantity and quality of feed available as well as posing health risks to livestock and humans (Kalac and Woolford, 1982).

Changes During Conservation

The manner by which the principles controlling the preservation of grass as silage are fulfilled in farm practice varies between (Table 9.4) and within

Table 9.4. Mean DM concentrations (g kg^{-1}) of grass silages in European countries (1994) (source: Wilkinson *et al.* (1996)).

	Mean DM concentration (g kg^{-1})		Mean DM concentration (g kg^{-1})		Mean DM concentration (g kg^{-1})
Portugal	195	Spain	266	Switzerland	350
Finland	200	Sweden	275	Slovenia	350
Ireland	211	France	275	Germany	350
Norway	220	Serbia + Montenegro	290	Austria	370
Poland	230	Romania	300	Hungary	375
Russian Fed.	240	Denmark	300	Netherlands	400
Italy	240	Czech Rep.	320	Luxemburg	410
Great Britain	250	Belgium	330	Iceland	450

countries. Considerations relating to climate, farm structure and economics have led to a wide range in the relative importance of grass silage, the types of sward used, their DM content at harvesting, additive use (if any) and the systems of harvesting, storage and feeding employed (Wilkinson *et al.*, 1996). The contributory causes of both the quantitative and qualitative changes occurring during harvesting, storage and feeding therefore vary between countries and within countries, and there can be considerable variation between cuts, years and locations.

The losses and changes in forage quality during harvesting and storage have been reviewed by Rotz and Muck (1994), with typical values shown in Table 9.5. A more detailed summary (Table 9.6) of the energy losses during ensilage and of their causing factors has been estimated by Zimmer (1980). Opportunities exist to modify some of these losses by altering the management system used for silo filling, sealing or feedout, by quickly pre-wilting the crop, by the strategic and appropriate use of silage additives, etc.

Factors Affecting Nutritive Value

The nutritive value of silage is mediated through its intake, digestibility and supply of lipogenic, glycogenic and aminogenic nutrients, and ultimately on how it quantitatively and qualitatively influences milk production and animal growth and health (Van Vuuren *et al.*, 1995). Several factors may cause variation in these characteristics, but they are determined mainly by the characteristics of the fresh forage at harvest and by the modifications undergone during harvesting and conservation (Dulphy and Demarquilly, 1991). Generally, the nutrient content and intake potential of silage are lower than those of the crop immediately before harvesting.

Table 9.5.   Typical DM losses and quality changes during hay and silage production from grass.

Type of loss	Dry matter loss (g kg^{-1} DM)		Change in nutrient concentration (g kg^{-1} DM)		
	Range	Normal	CP	NDF	TDN
Respiration[1]	20–80	50	8	32	−18
Rain damage[1]					
5 mm	10–30	20	−2	9	−5
25 mm	40–140	80	−13	53	−30
50 mm	80–270	150	−27	110	−60
Mowing/conditioning	10–20	10	0	0	0
Tedding	10–30	10	−2	4	−4
Swath inversion	10–30	10	0	0	0
Raking	10–200	50	−3	5	−6
Baling					
small bale	20–60	40	−5	9	−10
round bale	30–90	60	−10	18	−20
Chopping	10–80	30	0	0	0
Hay storage					
inside	30–90	50	−13	32	−18
outside	50–220	120	0	80	−48
Silo storage					
sealed	60–140	80	8	9	−37
stave	70–170	100	12	22	−47
bunker	100–160	120	15	36	−56

[1]Respiration loss includes plant and microbial respiration for crop cured without rain damage. Rain damage includes leaf loss, nutrient leaching, and microbial respiration resulting from rain damage. (Source: Rotz and Muck (1994).)
CP, crude protein; NDF, neutral detergent fibre; TDN, total digestible nutrients.

Quality of the standing crop

Studies in which the digestibility of grass was compared with well conserved silage made from the same grass showed that the effects of ensilage on digestibility were small (McDonald and Edwards, 1976; Dulphy and Demarquilly, 1991). Similar comparisons have shown a good relationship between the intake of grass and the resultant well conserved silage, but with a reduced intake of the silage (Dulphy and Demarquilly, 1991).

As grass plants mature physiologically, the proportion of cell walls and their constituent fractions increase and the cell content fraction decreases. The corresponding decline in forage quality is mainly a reflection of a decrease in the leaf : stem ratio and a greater rate of decline in the digestibility of the stem compared with the leaf (Mowat *et al.*, 1965) due to

Table 9.6. Energy losses during ensilage and causing factors (source: Zimmer (1980)).

Process	Classified as	Approx. losses ($J kJ^{-1}$)	Causing factors
Residual respiration	Unavoidable	10–20	Plant enzymes
Fermentation	Unavoidable	20–40	Microorganisms
Effluent	Mutually	50–>70	DM content
or		or	
Field losses by wilting	Unavoidable	0–>50	Weather, technique, management, crop
Secondary fermentation	Avoidable	0–>50	Crop suitability, environment in silo, DM content
Aerobic deterioration during storage	Avoidable	0–>100	Filling time, density, silo, sealing, crop suitability
Aerobic deterioration after unloading (heating)	Avoidable	0–>150	As above, DM content, silage unloading technique, season
		Total 70–>400	

Table 9.7. Winter or spring defoliation of a grass sward and its effects on *in vitro* DM digestibility ($g kg^{-1}$) at first-cut harvesting in May (source: O'Kiely and O'Riordan (1994, 1995).

Dec 2	Uncut (U)			Cut to 10 cm stubble (H)			Cut to 5 cm stubble (L)		
March 16	U	H	L	U	H	L	U	H	L
May 18	754	786	818	824	824	815	821	813	832

lignification. The digestibility of grass harvested for ensiling decreases with advancing growth stage, particularly once ear emergence has commenced (Minson *et al.*, 1960). The date of ear emergence varies between geographical locations and years due to climatic factors (Cooper, 1952), while cell wall digestibility can be decreased by high temperatures (Moir *et al.*, 1977). Grass digestibility is also influenced by species (Haggar, 1976) and cultivar (Green *et al.*, 1971). The digestibility of grass harvested for first-cut silage can be altered significantly by the extent and timing of defoliation the previous winter and/or spring (O'Kiely and O'Riordan, 1994, 1995) (Table 9.7). Furthermore, the rate of decline in forage digestibility around ear emergence can temporarily increase by a factor of two or three where heavily fertilized, high yielding crops of leafy grass lodge badly during wet weather,

Table 9.8.   Effects of stage of maturity and period of ensilage on forage composition, feed intake, milk production and nutrient utilization (source: Cushnahan *et al.* (1996)).

Treatment	Stage of maturity		Duration of ensilage (weeks)			
	Early	Late	0	3	9	52
Chemical composition of forage						
DM (g kg^{-1})	183	176	174	176	185	182
pH	4.3	4.1	5.3	4.0	3.8	3.7
Total N (g kg^{-1} DM)	29.1	20.7	24.6	25.3	25.0	24.6
Soluble N (g kg^{-1} DM)	15.0	10.0	4.5	14.2	15.2	15.5
Ammonia N (g kg^{-1} N)	58.6	57.5	11.3	55.7	70.9	94.3
n-Butyric acid (g kg^{-1} DM)	2.5	1.4	0.0	1.0	4.2	2.6
Lactic acid (g kg^{-1} DM)	75.0	89.3	18.8	92.1	99.2	106.7
Animal performance						
DM intake (kg day^{-1})	14.6	11.9	15.3	14.5	13.3	9.9
Milk yield (kg day^{-1})	19.0	15.5	18.6	18.4	17.3	14.8
Butterfat (g kg^{-1})	36.4	37.9	38.5	37.5	37.3	35.2
Protein (g kg^{-1})	30.6	29.7	32.2	30.4	29.9	28.0
DM digestibility	0.816	0.761	0.765	0.783	0.803	0.809

and where the lodged crop is left lying under these wet conditions (O'Kiely *et al.*, 1987).

The rate and extent of forage digestibility determines the clearance rate of forage from the rumen, which in turn will affect forage intake and, subsequently, milk production and growth (Paterson *et al.*, 1994). Although the intake and performance response of cattle to silage digestibility varies considerably (Thomas, 1980), high digestibility in well preserved silage usually increases intake and milk production with dairy cows (Gordon 1980a, b; Steen and Gordon, 1980; Thomas *et al.*, 1981; Castle, 1982; Rohr, 1991; Bosch *et al.*, 1992; Cushnahan *et al.*, 1996), and is frequently paralleled by a decrease in milk fat concentration and an increase in milk protein concentration (Table 9.8). However, a season effect may also be important.

Both Castle and Watson (1970) and Peoples and Gordon (1989), when comparing well preserved spring- and autumn-harvested silages of relatively similar digestibility, reported lower intakes and consequently lower milk production with the autumn-harvested forage.

Preservation quality

In providing cattle with grass or silage, the nutrients in these feeds are firstly presented to the rumen microflora before their products of metabolism, together with components avoiding digestion in the rumen, are potentially available to the animal. The most striking chemical differences between grass and silage are the replacement of WSC by fermentation products and the substantial breakdown of protein. These, and possibly other changes, generally result in the intake of grass silage being lower to a variable extent than for the corresponding green forage (Dulphy and Demarquilly, 1991; Mayne and Cushnahan, 1995), with the production of milk or body tissue being at least as variable. Cushnahan *et al.* (1996) ensiled unwilted ryegrass without additive at two stages of maturity and found (Table 9.8) that as the duration of ensilage continued to week 52, an extensive lactic acid dominant fermentation progressively developed; this was paralleled by decreases in silage intake, milk yield, and milk fat and protein concentrations. Cushnahan and Mayne (1995) compared the nutritive value of fresh grass (zero grazed) with silage made following treatment with either an inoculant (*Lactobacillus plantarum*) additive or an additive applying high rates of carboxylic acids, to achieve extensive and restricted fermentations, respectively. Both of these silages should have had very rapid rates of pH decrease after ensiling and were excellently preserved. Neither intake nor milk production were depressed significantly following ensilage, although the extensively fermented silage had reduced concentrations of butterfat and protein (Table 9.9). Keady and Murphy (1995) compared the nutritive value of fresh grass (zero grazed) with unwilted formic acid treated silage offered alone or supplemented with sucrose or sucrose plus fishmeal, and found that ensiling had no effect on forage DM intake relative to the parent herbage, but significantly decreased milk yield and milk protein concentration. Supplementation of the silage with sucrose did not overcome the drop in milk yield relative to the fresh herbage but sucrose plus fishmeal increased the output of milk and milk solids.

Extensive breakdown of protein during ensilage generally results in reduced intake and animal performance. These silages may also have high concentrations of volatile fatty acids. Whereas the water content of grass can influence silage quality through its effects on factors such as fermentation and effluent losses, the water content *per se* in silage is unlikely to be a critical factor affecting intake and performance. A review by Waldo and Jorgansen (1981) suggested that the intake of digestible organic matter would be much lower for direct-cut silages made without additive than for conserved forages of higher DM concentration, but that improvements in preservation due to the use of additives such as formic acid should lead to substantial improvements in intake. The data collated by Parker and Crawshaw (1982) indicate an increase in intake in response to formic acid when silage made without additive was imperfectly preserved while there was no benefit from formic acid when the control silage was well fermented. A high water content in grass at ensiling can predispose the forage to high in-silo losses and reduced nutritive value. However, the use of high-sugar crops, the judicious use of N fertilizer, the strategic use of

Table 9.9. Effects of ensiling and pattern of silage fermentation on the chemical composition of forage as offered (g kg^{-1} DM, unless otherwise stated) (source: Cushnahan and Mayne (1995)).

		Treatments	
	Fresh grass	Extensively fermented silage	Restricted fermented silage
Dry matter (DM) (g kg^{-1})	149	171	183
Total N	31.8	31.3	29.4
pH	5.14	3.81	4.08
Ammonia-N (g kg^{-1} N)	13.8	36.0	46.2
Buffering capacity (mEq kg^{-1} DM)	405	967	557
Ethanol	ND	9.0	23.8
Acetic acid	ND	11.1	4.9
Propionic acid	ND	1.2	2.8
i-Butyric acid	ND	0.5	0.2
Lactic acid	ND	124.6	27.0
Gross energy (MJ kg^{-1} DM)	18.6	19.0	18.9
Acid detergent fibre	259	291	290
Neutral detergent fibre	588	531	570
Hemicellulose	329	241	280
Acid detergent lignin	68	62	65
Cellulose	191	229	225
Ash	90	98	106
Water soluble carbohydrate	183	36	100
Silage DM intake (kg day^{-1})	17.3	16.7	17.5
Milk yield (kg day^{-1})	20.7	21.2	21.8
Milk constituents (g kg^{-1})			
Butterfat	40.7	36.1	39.6
Protein	32.6	30.0	31.5
Lactose	47.0	46.6	46.3
Yield of milk constituents (kg day^{-1})			
Butterfat	0.83	0.77	0.86
Protein	0.67	0.64	0.69
Lactose	0.97	0.99	1.01
Liveweight (kg)	582	583	592

appropriate additives and pre-wilting the crop, can overcome many of the problems associated with high water content.

Overall, Mayne and Cushnahan (1995) concluded that as the standard of silage preservation and conservation improved, the depression in its nutritive value due to ensilage became progressively less. This continued to a point where for silage whose fermentation was restricted by a high input of carboxylic acids, intake and the ability to support milk yield and composition were comparable to the pre-ensiled herbage. They postulated that the key factors influencing silage intake and animal performance are related to protein and fibre fractions in the silage, and in particular the relative rates and extent of digestion of these components within the animal. In contrasting the higher intakes and milk yields achieved by cows grazing well managed grassland compared with consuming typical good quality grass silage, they apportioned most of the improvement to the superior digestibility of the grazed herbage (3–4 weeks' regrowth) compared with the grass silage (7–8 weeks' regrowth) rather than to ensilage *per se*. They largely attributed the effects of ensilage to progressive changes in the nitrogenous components. This, it was suggested, offers considerable potential for grass silage if proteolytic activity during ensilage is reduced, either by ensuring rapid pH reduction in the ensiled crop or by attaining rapid increases in herbage DM before ensiling.

Chamberlain and Choung (1993) state that because silage is low in readily available energy for rumen microbes, and because ammonia is quickly released by rumen degradation of silage nitrogenous constituents, the efficiency of rumen microbial fixation of ammonia is low. This results in a much reduced input of total amino acids into the small intestine, within a digesta relatively deficient in some amino acids. Chamberlain and Choung (1993) suggest a broad classification of silages into two or three groups depending on the concentrations of lactic acid and residual sugars, and postulate on the mechanisms by which these silages undergo rumen digestion and subsequent metabolism, and ultimately are consumed and support the production of animal produce. The authors propose possible strategies of energy and protein supplementation of these different silages. A case is made that high inputs of organic acid silage additives, by restricting fermentation during ensilage, could increase microbial protein supply to the ruminant and thereby improve animal performance. However, it is also acknowledged that other factors can be influential – for example, bacterial inoculants can increase animal performance without having markedly shifted the balance between lactic acid and residual sugars in silage.

Oxygen

Exposure to oxygen during silo storage and feedout is a major factor in determining the difference in crop quality before and after ensilage. The primary effect of oxygen is to permit plant and microbial respiration, leading to the loss of digestible material and the production of carbon dioxide, water and heat.

During silo filling, plant respiration is normally the dominant means by which oxygen is removed from the silo (McDonald *et al.*, 1991). This respiration consumes the plant sugars, the principal substrate for the lactic acid bacteria. The slower the silo is filled, the more sugar is lost and unavailable for fermentation. Thus, in direct-cut or lightly wilted grass silages, slow filling increases the probability of not reaching a stable pH at which clostridial fermentation is inhibited. In heavily wilted silages, the heating produced by excessive respiration is of greater concern. If respiration raises silage temperatures to 35°C or 40°C, Maillard or browning reactions are promoted; these polymerize amino acids and sugars and cause additional heating that may result in permanent heat damage to the silage. Regardless of the crop DM content, slow filling promotes the growth of fungi, aerobic bacteria and enterobacteria (Ohyama *et al.*, 1975; Pahlow, 1985). This may predispose a silage to heating at feeding and/or possibly promote mycotoxin and endotoxin production, which could affect animal performance and health.

Pitt and Muck (1993) found that DM losses at the silo face during emptying were related primarily to the linear feedout rate and porosity of the silage, both measures of oxygen exposure. Aerobic microbial numbers and temperature were very important to the stability of the silage as it is fed, and to the subsequent effects of heating on reducing silage intake. Inherent factors related to the chemical composition of the silage can also be important (O'Kiely and Muck, 1992).

Chop length

Chop length and type can impact indirectly on silage nutritive value through their effects on forage compaction, cell content release, speed and direction of fermentation, and aerobic stability, and directly through their effects on intake, rate of passage, digestibility and animal performance. Furthermore, the type of machinery used and the silage-making, storing and feeding systems within which it is used, can interact with the above factors through their effects on speed of filling and feed-out, soil contamination, forage temperature, ease of eating to appetite, etc. Caution is therefore needed when reviewing literature to separate experiments studying the effects of chop length on conservation and/or nutritive value from those comparing conservation systems wherein a number of variables are changing.

Seale *et al.* (1982) compared ensiling unwilted grasses unchopped (long), chopped (25 mm) or minced (4 mm screen) in laboratory silos and found faster rates of pH decline and reduced concentrations of N present as ammonia-N with minced compared with unchopped forages. Marsh (1978) confirmed many other comparable published results. However, after reviewing the information available, Marsh (1978) concluded that machinery type is unlikely to alter silage fermentation in farm-scale silos except where fine chopping improves consolidation of heavily wilted silages (> 400 g DM kg^{-1}) or where the harvesting system interacts with the ability to achieve good silage-making practices. For example, a forage pick-up wagon produces chop lengths in the order of 100 mm or longer. Such long and unlacerated chop lengths may not be a problem with a leafy, immature grass but may be more difficult to consolidate during silo filling with stiffer, overly mature or high DM crops.

Fineness of chop may affect animal performance. Although information reviewed by Marsh (1978) and reported by Castle *et al.* (1979) and Petit *et al.* (1993) showed improvements in intake and milk yield with finer chopping, many of the differences could be explained, at least in part, by the poorer preservation of the longer forage. In contrast, where the unwilted silages being compared were equally well preserved, intake and milk production could be similar across a range of chop lengths within the chop length ranges being studied (i.e. from 14 mm nominal precision chop to flail harvested (unchopped) forage; Gordon, 1982, 1986). Finer lengths of chop than this can increase intake by cattle. From an animal perspective, chop length can be too small. North American forage harvesters are typically set to shorter theoretical lengths of cut than comparable harvesters in Europe and may be set as low as 6 mm. At such short chop lengths, saliva production and rumination can be affected, and dairy cows are more susceptible to rumen acidosis and displaced abomasums. McCullough (1991) suggests that silage that contains 15–20% of forage particles over 38 mm will meet the need for a forage mat and good rumen function. Armentano and Pereira (1997) propose the use of an effectiveness factor to provide an improved measure of the fibre value (i.e. neutral detergent fibre (NDF) concentration) of a forage. A physical effectiveness factor that is influenced by forage particle size is fundamental to this.

Overall, except for overly-fine chop length, it appears that equally good preservation, intake and dairy cow performance can be achieved across a range of forage chop lengths, provided that there is good silo management, particularly filling rapidly and packing well, and good feeding practices.

Wilting

As indicated in Table 9.4, the mean DM content of grass silage varies substantially by country. This mainly reflects the prevailing climatic conditions at harvest and the opportunities to wilt grass before chopping. Wilting reduces or eliminates effluent production and clostridial fermentation, both of which can reduce nutritive value of silage. However, wilting is weather-dependent and, if attempted during unsuitable weather conditions, may result in severe quantitative and qualitative losses (Rotz and Muck, 1994). Additionally, prolonged wilting can restrict regrowth (Schukking and Overvest, 1980), while the drier, more porous silage is more susceptible to aerobic losses (Pitt and Muck, 1993).

Many earlier experiments compared wilting with a poorly preserved unwilted silage as the control treatment (Marsh, 1979). Under such circumstances, distinct milk production benefits could accrue from wilting. Even where unwilted silage was not badly preserved, wilting led to an improvement or restriction in silage fermentation, and a resultant increase in intake. However, where unwilted silage was well preserved, often due to the use of formic acid as an additive, the increased intake (4–6%) in response to wilting was not reflected in an increase in milk production (–2 to –3%) (Rohr and Thomas, 1984; Gordon, 1989). Consequently, wilting, separate from improving preservation, appears to have little beneficial effect on nutritive value.

The decision to wilt should thus be based primarily on other factors. Milk output per hectare grass cut for silage is likely to be higher with direct-cut than with comparable two-stage (i.e. cut and pick-up separately) harvesting systems (Gordon, 1986) mainly because of lower field losses with the direct-cut system. Furthermore, the consumption of silage effluent by livestock (Steen, 1986; Patterson and Kilpatrick, 1991) can further improve the output of animal product per hectare with the direct-cut system. However, direct-cut systems must absorb the costs of handling large quantities of water in harvesting and feeding, possibly the costs associated with additives such as formic acid, as well as the collection and safe utilization (normally by landspreading) of silage effluent and the costs of the corrosion it causes to concrete yards and silos. Successful wilting can eliminate the production of silage effluent which can be a source of significant in-silo losses (Table 9.6). Finally, the risk of a poor fermentation in the direct-cut system must be weighed versus losses in harvesting and storage in the wilted system. Because of the complexity of such factors, programs such as *DAFOSYM* (Rotz *et al.*, 1993) are valuable in accounting for the interactions between weather patterns and the harvesting system and their effects on farm profitability.

Additives

Silage additives fall into several main groups: acids, sugars, bacterial inoculants, enzymes, sterilants and non-protein nitrogen. All but non-protein nitrogen additives are commonly used on grass silages. The effects of these additives have been reviewed (McDonald *et al.*, 1991; Muck and Bolsen, 1991; Spoelstra, 1991; Harrison *et al.*, 1994).

Good quality silage can often be made without additive treatment. As various additive types have different modes of action, no one additive is ideal for all circumstances. Moreover, from a farmer's perspective, the purpose of using an additive is to make a profit from the investment. Consequently, a farmer must decide if the technical or biological benefits from the use of an additive, i.e. reduced conservation losses or improved animal performance or a combination of both factors, result in an economic advantage.

Acids and sugars are used primarily to address the problem of insufficient sugars in the crop leading to a clostridial silage. By adding an acid, formic acid being the most common, pH is immediately lowered and the amount of fermentation required to reach a stable pH is reduced. Rates of formic acid addition of approximately 2.5 l t^{-1} reduce grass pH directly from 6.0 to about 4.9, with lactic acid-dominant fermentation, which develops after a lag, hopefully reducing the pH to about 3.9 with unwilted forage. Higher rates of formic acid addition (e.g. 4 l t^{-1}), as are common in some counties, will initially reduce the pH further, relying on a subsequently more restricted fermentation. Castle (1975) has shown that where formic acid altered fermentation from being dominated by clostridia to lactic acid bacteria, milk production improved in response to increased intake. Alternatively, both Parker and Crawshaw (1982) and Mayne and Steen (1989/90) suggest that in situations where good fermentations were obtained with untreated silage, formic acid treatment had little effect on animal performance. Sugars, primarily molasses, provide additional substrate for the lactic acid bacteria to facilitate the attainment of a low pH. Of these two approaches, acid is easier (although more hazardous) to apply uniformly, and the immediate pH drop helps preserve more true protein in the silage. However, the main reason for using either additive type is to avoid a clostridial fermentation. Therefore these additives are mainly used in making direct-cut and lightly wilted (DM < 220 g kg^{-1}) silages.

Bacterial inoculants nominally aim to apply homofermentative lactic acid bacteria to a crop at target rates that range from 10^5 to 10^6 bacteria g^{-1} crop. The homofermentative lactic acid bacteria, if successful, should promote a fast and efficient fermentation that is high in lactic acid and low in acetic acid, ethanol and carbon dioxide. If the inoculant bacteria dominate the fermentation, they should improve dry matter recovery from the silo by 2–3% and lower pH (Muck and Bolsen, 1991). Because the inoculant bacteria must compete with the natural population, they are not always successful. A recent literature survey has shown that inoculants were successful at improving silage fermentation in two-thirds of published experiments (Muck and Kung, 1997), whereas animal performance was improved in approximately half of published experiments (Kung and Muck, 1997). Such surveys assume that the published experiments are a representative sample of all experiments conducted. Farmers may improve the chances of success by using a product that supplies a high number of bacteria per unit of crop. Furthermore, they are more likely to obtain a worthwhile biological response if the inoculants are applied to grass with high, rather than marginal or insufficient, concentrations of available WSC.

Enzyme additives are products that typically contain a mix of cell-wall-degrading enzymes (cellulases, hemicellulases, pectinases) and sometimes amylases and glucose oxidase. They aim to provide additional sugars quickly for fermentation in the silo. Some have successfully reduced fibre content (Muck and Kung, 1997), particularly in grasses, thereby providing additional sugars. Unfortunately, improvements in animal performance have not necessarily followed from the pre-digestion of forage fibre (Kung and Muck, 1997). However, the reduction in fibre has also increased the consolidation of the crop in the silo. This is an advantage in drier silages (DM > 300 g kg^{-1}), reducing porosity and subsequently lowering aerobic losses (Muck and Kung, 1997). However, in unwilted and lightly wilted crops, increased consolidation has led to higher effluent losses (Muck and Kung, 1997). These results suggest that enzyme products are not particularly advantageous in providing sufficient sugars quickly enough to prevent clostridial fermentation in wetter crops. Currently the main opportunity for a farmer to recover the cost of an enzyme additive is from improved dry matter recovery in grass silages made at 300–400 g DM kg^{-1}.

Health problems

Animal health problems associated with silage-fed cattle, and related to undesirable microorganisms (e.g. moulds, *Listeria*, clostridia), undesirable chemicals (e.g. mycotoxins, nitrogenous compounds) and metabolic disorders (e.g. rumen acidosis, lameness, displaced abomasum, ketosis, impaired fertility, hypomagnesaemia, mineral deficiencies), have been discussed by Stark and Wilkinson (1988) and Van Vuuren *et al.* (1995). Some of these are mentioned elsewhere in this chapter.

Health problems related to silage contaminants are most commonly associated with moulds (hyphal fungi) and *Listeria monocytogenes,* although toxins from clostridia and enterobacteria are possible (McDonald *et al.*, 1991). Moulds may produce a variety of mycotoxins that can cause poor performance, abortions and susceptibility to disease in cattle and sheep. Because moulds are aerobic microorganisms, mycotoxin production is unlikely in a well sealed silo and is more likely to occur before ensiling or at unloading. Factors affecting mycotoxin production are not well understood.

Listeria monocytogenes is the aerobic bacteria responsible for listeriosis, a disease of both humans and ruminants. It does not grow well below pH 5.0. In general, mould and listeria activities are impaired significantly where anaerobic conditions are achieved quickly and maintained and where the forage pH is reduced quickly to the level appropriate to the DM concentration of the forage (McDonald *et al.*, 1991; Muck *et al.*, 1992).

Other

The ability to realize the potential nutritive value of any silage depends on the animals to which it is offered having genuine *ad libitum* access to it. In general, the higher the proportion of the diet that a range of silages constitute, the greater the effects of differences in their nutritive value on animal productivity. In theory, mixing rations to achieve a better balance between ingredients could increase total intake and may result in a more synchronous release of energy and nitrogenous components in the rumen, thereby increasing microbial protein synthesis and nitrogen utilization. However, the guidelines to achieve these improvements in practice are not conclusively proven.

High quality silage is a prerequisite in many feeding systems if limitations of energy intake are not to restrict milk yields excessively. The formulation of the concentrate supplement to silage in such cases should aim to minimize the substitution rate of silage while providing substrates that encourage efficient microbial utilization of silage nitrogen in the rumen. The amount and type of carbohydrate, lipid and protein in the supplement are all critical considerations (Chamberlain *et al.*, 1989).

References

Armentano, L. and Pereira, M. (1997) Measuring the effectiveness of fiber by animal response trials. *Journal of Dairy Science* 80, 1416–1425.

Baird-Parker, A.C. (1980) Organic acids. In: Silliker, J.H. (ed.) *Microbial Ecology of Foods*, Academic Press, London, pp. 126–135.

Beck, Th. (1978) The microbiology of silage fermentation. In: McCullough, M.E. (ed.) *Fermentation of Silage – a Review*. National Feed Ingredients Association, West Des Moines, Iowa, pp. 61–116.

Bosch, M.W., Tamminga, S., Post, G., Leffering, C.P. and Muylaert, J.M. (1992) Influence of stage of maturity of grass silages on digestion processes in dairy cows. 1. Composition, nylon bag degradation rates, fermentation characteristics, digestibility and intake. *Livestock Production Science* 32, 245–264.

Castle, M.E. (1975) Silage and milk production. *Agricultural Progress* 50, 53–60.

Castle, M.E. (1982) Feeding high quality silage. In: Rook, J.A.F. and Thomas, P.C. (eds) *Silage for Milk Production*. Technical Bulletin No. 2, NIRD, UK and Hannah Research Institute, UK, pp. 127–150.

Castle, M.E., Retter, W.C. and Watson, J.N. (1979) Silage and milk production: comparisons between grass silage of three different chop lengths. *Grass and Forage Science* 34, 293–301.

Castle, M.E. and Watson, J.N. (1970) Silage and milk production: a comparison of grass silages made with and without formic acid. *Journal of British Grassland Society* 25, 65–71.

Catchpoole, V.R. and Henzell, E.F. (1971) Silage and silage-making from tropical herbage species. *Herbage Abstracts* 41, 213–221.

Cato, E.P., George, W.L. and Finegold, S.M. (1986) Genus *Clostridium*. In: Sneath P.A.H., Mair, N.S. and Sharpe, E. (eds) *Bergy's Manual of Systematic Bacteriology*, Volume 2. Williams and Wilkins, Baltimore, Maryland, pp. 1141–1200.

Chamberlain, D.G. and Choung, J.-J. (1993) The nutritional value of grass silage. In: O'Kiely, P., O'Connell, M. and Murphy, J. (eds) *Silage Research 1993, Proceedings of Tenth International Silage Conference*. Dublin City University, pp. 131–136.

Chamberlain, D.G., Martin, P.A. and Robertson, S. (1989) Optimising compound feed use in dairy cows with high intakes of silage. In: Haresign, W. and Cole, D.J.A. (eds) *Recent Advances in Animal Nutrition.* Butterworths, London, pp. 175–193.

Charmley, E. (1995) Making the most of silage proteins. *Feed Mix* 3, 28–31.

Cooper, J.P. (1952) Studies on growth and development in *Lolium.* 3. Influence of season and latitude in ear emergence. *Journal of Ecology* 40, 352–379.

Cushnahan, A. and Mayne, C.S. (1995) Effects of ensilage of grass on performance and nutrient utilization by dairy cattle. 1. Food intake and milk production. *Animal Science* 60, 337–345.

Cushnahan, A., Mayne, C.S., Goodall, E.A. and Unsworth, E.F. (1996) Effects of stage of maturity and period of ensilage on the production and utilization of grass silage by dairy cows. In: Jones, D.I.H., Jones, R., Dewhurst, R., Merry, R. and Haigh, P.M. (eds) *Proceedings of Eleventh International Silage Conference,* IGER Aberystwyth, pp. 78–79.

Deinum, B. (1966) Chemical composition and nutritive value of herbage in relation to climate. In: Riley, H. and Skjelvag, A. (eds) *The Impact of Climate on Grass Production and Quality. Proceedings of the 10th General Meeting of the European Grassland Federation.* Norwegian State Research Station, Oslo, pp. 338–350.

Dellaglio, F. (1985) Lactic acid bacteria in silage fermentation. *Microbiologie Aliments Nutrition* 3, 91–104.

Dellaglio, F. and Torriani, S. (1991) Microbiological methods. *Landbauforschung Volkenrode, Sonderheft* 123, 206–217.

Dewar, W.A., McDonald, P. and Whittenbury, R. (1963) The hydrolysis of grass hemicelluloses during ensilage. *Journal of the Science of Food and Agriculture* 14, 411–417.

Dulphy, J.P. and Demarquilly, C. (1991) Digestibility and voluntary intake of conserved forage. In: Pahlow, G. and Honig, H. (eds) *Forage Conservation Towards 2000, Proceedings of European Grassland Federation. Landbauforschung Volkenrode,* Sonderheft 123, 140–160.

Fehrmann, E. and Muller, Th. (1990) Seasonal changes in epiphytic micro-organisms on a grassland plot. *Das Wirtschaftseigene Futter* 36, 66–78.

Gordon, F.J. (1980a) The effect of silage type on the performance of lactating cows and the response to high levels of protein in the supplement. *Animal Production* 30, 29–37.

Gordon, F.J. (1980b) The effect of interval between harvests and wilting on silage for milk production. *Animal Production* 31, 35–41.

Gordon, F.J. (1982) The effects of the degree of chopping grass for silage and method of concentrate allocation on the performance of dairy cows. *Grass and Forage Science* 37, 59–65.

Gordon, F.J. (1986) The influence of the system of harvesting grass for silage on milk output. *Jubilee Report of Agricultural Research Institute of Northern Ireland,* pp. 13–22.

Gordon, F.J. (1989) Effect of silage additives and wilting on animal performance. In: Haresign, W. and Cole, D.J.A. (eds) *Recent Advances in Animal Nutrition.* Butterworths, London, pp. 159–174.

Green, J.O., Corrall, A.J. and Terry, R.A. (1971) *Grass Species and Varieties. Relationship between Stage of Growth, Yield and Forage Quality.* Technical Report No. 8. Grassland Research Institute, Hurley, 81pp.

Haggar, R.J. (1976) The seasonal productivity, quality and response to nitrogen of four indigenous grasses compared with *Lolium perenne. Journal of the British Grassland Society* 31, 197–207.

Harrison, J.H., Blauwiekel, R. and Stokes, M.R. (1994) Fermentation and utilization of grass silage. *Journal of Dairy Science* 77, 3209–3235.

Henderson, A.R., McDonald, P. and Woolford, M.K. (1972) Chemical changes and losses during the ensilage of wilted grass treated with formic acid. *Journal of the Science of Food and Agriculture* 23, 1079–1087.

Jones, D.I.H., ap Griffith, G. and Walters, R.J.K. (1965) The effect of nitrogen fertilizers on the water-soluble carbohydrate content of grasses. *Journal of Agricultural Science, Cambridge,* 64, 323–328.

Jonsson, A. (1989) The role of yeasts and clostridia in silage deterioration – identification and ecology. PhD dissertation. Swedish University of Agricultural Sciences, Department of Microbiology, Report 42.

Kalac, P. and Woolford, M.K. (1982) A review of some aspects of possible associations between the feeding of silage and animal health. *British Veterinary Journal* 138, 305–320.

Keady, T.W.J. and Murphy, J.J. (1995) An evaluation of ensiling *per se* and addition of sugars and fishmeal on the rate of forage intake and performance of lactating dairy cattle. *Irish Journal of Agricultural and Food Research* 34, 96–97.

Keady, T.W.J. and O'Kiely, P. (1996) Effects of potassium and nitrogen fertilisation and formic acid treatment on silage composition, in-silo losses and feeding value. In: Jones, D.I.H., Jones, R., Dewhurst, R., Merry, R. and Haigh, P.M. (eds) *Proceedings of Eleventh International Silage Conference.* IGER Aberystwyth, pp. 128–129.

Kung, L., Jr and Muck, R.E. (1997) Animal response to silage additives. In: *Silage: Field to Feedbunk, NRAES-99.* Northeast Regional Agricultural Engineering Service, Ithaca, New York, pp. 200–210.

Leibensperger, R.Y. and Pitt, R.E. (1987) A model of clostridial dominance in ensilage. *Grass and Forage Science* 42, 297–317.

Lindgren, S. (1991) Microbial dynamics during silage fermentation. In: Lindgren, S. and Pettersson, K.L. (eds) *Proceedings of the Eurobac Conference.* Swedish University of Agricultural Sciences, Uppsala, pp. 135–145.

Lyttleton, J.W. (1973) Proteins and nucleic acids. In: Butler, G.W. and Bailey, R.W. (eds) *Chemistry and Biochemistry of Herbage.* Academic Press, London, pp. 63–103.

McCullough, M.E. (1991) Applied hay and silage feeding for dairy cattle. In: Bolsen, K.K., Baylor, J.E. and McCullough, M.E. (eds) *Field Guide for Hay and Silage Management in North America.* National Feed Ingredients Association, West Des Moines, Iowa, pp. 81–91.

McDonald, P. and Edwards, R.A. (1976) The influence of conservation methods on digestion and utilization of forages by ruminants. *Proceedings of the Nutrition Society* 35, 201–211.

McDonald, P., Henderson, A.R. and Heron, S.J.E. (1991) *The Biochemistry of Silage.* Chalcombe Publications, Marlow, Bucks, 340pp.

McGrath, D. (1988) Seasonal variation in the water-soluble carbohydrates of perennial and Italian ryegrass under cutting conditions. *Irish Journal of Agricultural Research* 27, 131–139.

Marsh, R. (1978) A review of the effects of mechanical treatment of forages on fermentation in the silo and on the feeding value of the silages. *New Zealand Journal of Experimental Agriculture* 6, 271–278.

Marsh, R. (1979) The effects of wilting on fermentation in the silo and on the nutritive value of silage. *Grass and Forage Science* 34, 1–10.

Mayne, C.S. and Cushnahan, A. (1995) The effects of ensilage on animal performance from the grass crop. *68th Annual Report of the Agricultural Research Institute of Northern Ireland*, pp. 30–41.

Mayne, C.S. and Steen, R.W.J. (1989/90) Recent research on silage additives for milk and beef production. *63rd Annual Report of the Agricultural Research Institute of Northern Ireland*, pp. 31–42.

Minson, D.J. and McLeod, M.N. (1970) The digestibility of temperate and tropical grasses. In: Norman, M.J.T. (ed.) *Proceedings of the XIth International Grassland Congress*. Surfers Paradise, pp. 719–722.

Minson, D.J., Raymond, W.F. and Harris, C.E. (1960) Studies in the digestibility of herbage. VIII. The digestibility of S37 cocksfoot, S23 ryegrass and S24 ryegrass. *Journal of the British Grassland Society* 15, 174–180.

Moir, K.W., Wilson, J.R. and Blight, G.W. (1977) The *in vitro* digested cell wall and fermentation characteristics of grass as affected by temperature and humidity during their growth. *Journal of Agricultural Science, Cambridge* 88, 217–222.

Moran, J.P., O'Kiely, P., Wilson, R.K. and Crombie-Quilty, M.B. (1991) Lactic acid bacteria levels on grass grown for silage in Ireland. *Landbauforschung Volkenrode, Sonderheft* 123, 283–286.

Mowat, D.N., Fulkerson, R.S. Tossell, W.E. and Winch, J.E. (1965) The *in vitro* digestibility and protein content of leaf and stem portions of forages. *Canadian Journal of Plant Science* 45, 321–331.

Muck, R.E. (1991) Silage fermentation. In: Zeikus, J.G. and Johnson, E.A. (eds) *Mixed Cultures in Biotechnology*. McGraw Hill, New York, pp. 171–204.

Muck, R.E. and Bolsen, K.K. (1991) Silage preservation and silage additive products. In: Bolsen, K.K., Baylor, J.E. and McCullough, M.E. (eds) *Field Guide for Hay and Silage Management in North America*. National Feed Ingredients Association, West Des Moines, Iowa, pp. 105–126.

Muck, R.E. and Kung, L., Jr (1997) Effects of silage additives on ensiling. In: *Silage: Field to Feedbunk, NRAES-99*. Northeast Regional Agricultural Engineering Service, Ithaca, New York, pp. 187–199.

Muck, R.E. and Walgenbach, R.P. (1985) *Variation in Alfalfa Buffering Capacity*. ASAE Paper No. 85-1535. American Society of Agricultural Engineers, 17pp.

Muck, R.E., O'Kiely, P. and Wilson, R.K. (1991a) Buffering capacities in permanent pasture grasses. *Irish Journal of Agricultural Research*, 30, 129–141.

Muck, R.E., Wilson, R.K. and O'Kiely, P. (1991b) Organic acid content of permanent pasture grasses. *Irish Journal of Agricultural Research* 30, 143–152.

Muck, R.E., Spoelstra, S.F. and van Wikselaar, P.G. (1992) Effects of carbon dioxide on fermentation and aerobic stability of maize silage. *Journal of the Science of Food and Agriculture* 59, 405–412.

Muller, M. and Lier, D. (1994) Fermentation of fructans by epiphylic lactic acid bacteria. *Journal of Applied Bacteriology* 76, 406–411.

Ohyama, Y., Morichi, T. and Masaki, S. (1975) The effect of inoculation with *Lactobacillus plantarum* and addition of glucose at ensiling on the quality of aerated silages. *Journal of the Science of Food and Agriculture* 26, 1001–1008.

O'Kiely, P. and Muck, R.E. (1992) Aerobic deterioration in lucerne and maize silages – effects of yeasts. *Journal of the Science of Food and Agriculture* 59, 139–144.

O'Kiely, P. and O'Riordan, E.G. (1994) Effects of defoliating grass in winter or spring on first- and second-cut yield and on grass and silage quality. In: *Proceedings of the 4th Research Conference of the British Grassland Society*. British Grassland Society, University of Reading, Reading, pp. 55–56.

O'Kiely, P. and O'Riordan, E.G. (1995) Effects of defoliating grass swards in winter on subsequent silage yield and quality. *Irish Journal of Agricultural and Food Research* 34, 81.

O'Kiely, P. and Tunney, H. (1997) Silage conservation characteristics of grass that received a range of rates of phosphorus fertilizer. *Irish Journal of Agricultural and Food Research* 36, 104.

O'Kiely, P., Flynn, A.V. and Wilson, R.K. (1987) New concepts in silage making. *Irish Grassland and Animal Production Association Journal* 21, 38–50.

O'Kiely, P., Carton, O.T. and Lenehan, J.J. (1994) *Effect of Time, Method and Rate of Slurry Application to Grassland Grown for Silage*. F.A.O. Network on Animal Waste Utilization, Bad Zwischenahn, Germany, 11pp.

O'Kiely, P., Moloney, A.P., Killen, L. and Shannon, A. (1997a) A computer programme to calculate the cost of providing ruminants with home-produced feedstuffs. *Computers and Electronics in Agriculture* 19, 23–36.

O'Kiely, P., O'Riordan, E.G. and Moloney, A.P. (1997b) Grass ensilability indices as affected by the form and rate of inorganic nitrogen fertilizer and the duration to harvesting. *Irish Journal of Agricultural and Food Research* 36, 93.

Pahlow, G. (1985) O_2-abhängig Veränderungen der Mikroflora in Silagen mit Lactobacterienzusatz. [Oxygen-dependent changes in the silage microflora in response to added lactic acid bacteria]. *Landwirtschaftliche Forschung* 37 (Kongressband for 1984), 630–639.

Parker, J.W.G. and Crawshaw, R. (1982) Effects of formic acid on silage fermentation, digestibility, intake and performance of young cattle. *Grass and Forage Science* 37, 53–58.

Paterson, J.A., Belyea, R.L., Bowman, J.P., Kerley, M.S. and Williams, J.E. (1994) The impact of forage quality and supplementation regimen on ruminant animal intake and performance. In: Fahey, G.C., Collins, M., Mertens, D.R. and Moser, L.E. (eds) *Forage Quality, Evaluation, and Utilization*. American Society of Agronomy, Madison, Wisconsin, pp. 59–114.

Patterson, D.C. and Kilpatrick, D.J. (1991) Effluent from grass silage for finishing pigs. *Journal of Agricultural Science, Cambridge* 116, 119–124.

Peoples, A.C. and Gordon, F.J. (1989) The influence of wilting and season of silage harvest and the fat and protein concentration of the supplement on milk production and food utilization by lactating cattle. *Animal Production* 48, 305–318.

Petit, H.V., Tremblay, G.F., Savoie, P., Tremblay, D. and Wauthy, J.M. (1993) Milk yield, intake and blood traits of lactating cows fed grass silage conserved under different harvesting methods. *Journal of Dairy Science* 76, 1365–1374.

Pitt, R.E. and Muck, R.E. (1993) A diffusion model of aerobic deterioration at the exposed face of bunker silos. *Journal of Agricultural Engineering Research* 55, 11–26.

Pitt, R.E., Muck, R.E. and Leibensperger, R.Y. (1985) A quantitative model of the ensilage process in lactate silages. *Grass and Forage Science* 40, 279–303.

Playne, M.J. and McDonald, P. (1966) The buffering constituents of herbage and of silage. *Journal of the Science of Food and Agriculture* 17, 264–268.

Rammer, C. (1996) Manure in grass silage production – effects on silage fermentation and its hygienic quality. PhD dissertation, Swedish University of Agricultural Sciences, Agraria 2.

Rohr, K. (1991) Effect of forage quality on milk and meat. *Landbauforschung Volkenrode, Sonderheft* 123, 161–176.

Rohr, K. and Thomas, C. (1984) Intake, digestibility and animal performance. *Landbauforschung Volkenrode, Sonderheft* 69, 64–70.

Rotz, C.A. and Muck, R.E. (1994) Changes in forage quality during harvest and storage. In: Fahey, G.C., Collins, M., Mertens, D.R. and Moser, L.E. (eds) *Forage Quality, Evaluation, and Utilization*. American Society of Agronomy, Madison, Wisconsin, pp. 828–868.

Rotz, C.A., Pitt, R.E., Muck, R.E., Allen, M.S. and Buckmaster, D.R. (1993) Direct-cut harvest and storage of alfalfa on the dairy farm. *Transactions of the American Society of Agricultural Engineers (ASAE)* 36, 621–628.

Schukking, S. and Overvest, J. (1980) Direct and indirect losses caused by wilting. In: Thomas, C. (ed.) *Forage Conservation in the 80's*. Occasional Symposium No. 11 of British Grassland Society, pp. 210–213.

Seale, D.R. (1986) Bacterial inoculants as silage additives. In: Bateson, M., Benham, C.L. and Skinner, F.A. (eds) *Journal of Applied Bacteriology Symposium Supplement*. Blackwell Scientific, Malden, Massachusetts, pp. 9S–26S.

Seale, D.R., Quinn, C.M., Whittaker, P.A. and Wilson, R.K. (1982) Microbiological and chemical changes during the ensilage of long, chopped and minced grass. *Irish Journal of Agricultural Research* 21, 147–158.

Smith, D. (1973) The non-structural carbohydrates. In: Butler, G.W. and Bailey, R.W. (eds) *Chemistry and Biochemistry of Herbage*. Academic Press, London, pp. 105–155.

Sneath, P.A.H., Mair, N.S. and Sharpe, E. (eds) (1986) *Bergy's Manual of Systematic Bacteriology*, Volume 2. Williams and Wilkins, Baltimore, 552pp.

Spoelstra, S.F. (1990) Comparison of the content of clostridial spores in wilted grass silage ensiled in either laboratory, pilot-scale or farm silos. *Netherlands Journal of Agricultural Science* 38, 423–434.

Spoelstra, S.F. (1991) Chemical and biological additives in forage conservation. *Landbauforschung Volkenrode, Sonderheft* 123, 48–70.

Stark, B.A. and Wilkinson, J.M. (1988) *Silage and Health*. Chalcombe Publications, Chelmsford, 57pp.

Steen, R.W.J. (1986) An evaluation of effluent from grass silage as a feed for beef cattle offered silage-based diets. *Grass and Forage Science* 41, 39–45.

Steen, R.W.J. and Gordon, F.J. (1980) The effect of type of silage and level of concentrate supplementation offered during early lactation on total lactation performance of January/February calving cows. *Animal Production* 30, 341–354.

Thomas, C. (1980) Conserved forages. In: Broster, W.H., Johnson, C.L. and Taylor, J.C. (eds) *Feeding Strategies for Dairy Cows*. Agricultural Research Council, London, pp. 8.1–8.4.

Thomas, C., Daley, S.R., Aston, K. and Hughes, P.M. (1981) Milk production from silage. 2. Influence of the digestibility of silage made from the primary growth of perennial ryegrass. *Animal Production* 33, 7–13.

Van Vuuren, A.M., Huhtanen, P. and Dulphy, J.P. (1995) Improving the feeding and health value of ensiled forages. In: Journet, M., Grenet, E., Farce, M-H., Theriez, M. and Demarquilly, C. (eds) *Recent Developments in the Nutrition of Herbivores, Proceedings of 4th International Symposium on the Nutrition of Herbivores.* INRA Edition, Paris, pp. 279–307.

Waldo, D.R. and Jorgensen, N.A. (1981) Forages for high animal production: nutritional factors and effects of conservation. *Journal of Dairy Science* 64, 1207–1229.

Weissbach, F., Schmidt, L. and Hein, E. (1976) Method of anticipation of the run of fermentation in silage making, based on the chemical composition of green fodder. In: *Proceedings of the XII International Grassland Congress, 11–20 June, 1974, Moscow, USSR.* Izd-vo, MIR, Moscow, Vol. III, Part II, pp. 663–673.

Wilkinson, J.M. and Bolsen, K.K. (1996) Production of silage and hay in Europe and North America. In: Jones, D.I.H., Jones, R., Dewhurst, R., Merry, R. and Haigh, P.M. (eds) *Proceedings of Eleventh International Silage Conference.* IGER, Aberystwyth, pp. 42–43.

Wilkinson, J.M., Price, W.R., Russell, S.R. and Jones, P. (1994) Diurnal variation in dry matter and sugar content of ryegrass. In: *Proceedings of 4th Research Conference of the British Grassland Society.* British Grassland Society, University of Reading, Reading, pp. 61–62.

Wilkinson, J.M., Wadephul, F. and Hill, J. (1996) *Silage in Europe – a Survey of 33 Countries.* Chalcombe Publications, Chelmsford, 153pp.

Wilson, R.K. and Collins, D.P. (1980) Chemical composition of silages made from different grass genera. *Irish Journal of Agricultural Research* 19, 75–84.

Woolford, M.K. (1984) *The Silage Fermentation.* Microbiology Series No. 14. Marcel Dekker, New York, 350pp.

Zimmer, E. (1980) Efficient silage systems. In: Thomas, C. (ed.) *Occasional Symposium No. 11 of British Grassland Society.* British Grassland Society, The Grassland Research Institute, Hurley, Berkshire, pp. 186–200.

Grass Baleage

10

C. Ohlsson

*Department of Forage Crops and Potatoes, Danish Institute of
Agricultural Science, Research Center Foulum, PO Box 21,
DK-8830, Tjele, Denmark*

Introduction

Baling grass for silage has become increasingly popular. Compared with
other ensiling techniques, grass baleage, i.e. baled silage, may have greater
flexibility with respect to handling, quality, and feeding, and can compete
favourably with respect to cost in some circumstances. In Europe, ensilage
has increased at the expense of hay. Wilkinson *et al.* (1996) estimated that
56% of conserved grass, or 21 million ha, was ensiled in 33 European countries. Ensiling grass in bales has steadily increased and presently comprises
10–25% of all forage harvested for silage in 15 Western European countries
(Wilkinson *et al.*, 1996). Pfimlin (1993) reported that grass baleage continues to increase in France on small to medium sized farms, involving the use
of equipment for haymaking. In northern Europe, grass baleage has become popular because unstable weather makes conservation of hay risky.
In Iceland, for example, 80% of all silage is ensiled in bales (Wilkinson *et
al.*, 1996). In 1993, 7.2 million tons of dry matter (DM) were ensiled in
England and Wales, corresponding to more than 20 million bales. Thus, the
UK is probably the greatest producer of baled silage in the world.

In spite of its popularity, relatively little technical and scientific information on the ensiling characteristics of big-bale silage is published (Nicholson
et al., 1991). In this chapter, the desired quality characteristics of baled
crops, various processes in ensiling, storing, feeding baled silage, and economics, will be discussed. Although baled grass is used in most industrial-
ized nations, this chapter will mainly refer to grass ensiled in bales under

European conditions because of greater availability of data compared with other parts of the world.

Advantages and disadvantages of baled silage

Baled silage is one system that has certain characteristics that are attractive under certain circumstances, but not under others. But because of its extensive use, it probably no longer should be considered a system of marginal use.

The primary advantages of grass baleage compared with conventional silage systems are as follows:

- Reduced aerobic losses by quickly achieving and maintaining anaerobic conditions and because of quick consumption of silage at feed-out.
- Reduced effluent production because wilting is an inherent part of the system.
- Fits in with grazing management systems, whereby surplus paddocks of grass may be readily removed by baling resulting in reduced wastage. The need of harvesting alternative sites for grass baleage may also be reduced.
- Greater flexibility with regard to harvest date and desired forage quality and to quickly bale and wrap small quantities of grass.
- Greater flexibility for ration formulation (Lindberg and Larsson, 1993).
- Does not require as extensive wilting as hay and, therefore, less weather dependent.
- Marketable and transportable commodity.
- Opportunity for good feed value by greater ability to harvest the crop at an appropriate time (Lindberg and Larsson, 1993).
- Reduced capital investment for equipment and storage of forage (Halvarsson, 1991; Forristal et al., 1995).

The main disadvantages with grass baleage are as follows:

- Difficult to bale in wet weather or with no wilting.
- Flimsy plastic film prone to damage and consequently oxygen ingress.
- Not always suitable system at large, intensively managed farms because of high labour requirements at feed-out.
- Not suitable in total mixed rations unless investment is made in specialized equipment.
- Because of long chop-length, wastage of silage at feeding may be high due to pulling of feed by cattle.
- Variable costs may be high (Forristal et al., 1995).
- Large use of polyethylene because silage in bales has six to eight times larger surface area than silage in clamps (Forristal et al., 1995).
- Wrapped bales stored in fields may be 'unaesthetic' (Lingvall, 1997).

General practices in making, storing and feeding baled silage

Ensiling of grass in bales is a three-step procedure. The first step is to mow the grass and put it in a swath, as there are no balers available commercially for direct harvest. Grass is then often wilted to 300–500 g kg^{-1} DM to avoid production of effluent (Ganeau, 1981; Bosma and Verkaik, 1987) and to produce well preserved silage with less than 100 g kg^{-1} ammonia-N expressed on a total N basis (Haigh, 1990, 1995). The duration of wilting varies from a few hours to 3–4 days depending on weather, degree of crimping or tedding, the desired DM concentration of the herbage, and farm traditions.

The second step is to bale the herbage. In the advent of grass baleage, mainly round bales were used. Presently, the round balers with fixed chamber-size are most common in Europe (A. Riis, Kverneland-Taarup A/S, Denmark, 1997, personal communication). But round balers with a variable chamber capable of making bales with variable diameter, as well as large square balers, are used. The fresh weight of a round bale ranges between 300 kg and 800 kg (Howe, 1987; Ohlsson, 1996, 1997) depending on speed at baling, type of baler, DM concentration of the herbage and stage of maturity at baling. The aim should be to produce bales that have a volume weight of 180–200 kg DM m^{-3}. Lighter bales are uneconomical to produce and heavier bales are difficult to handle.

The third step is to enclose the bales in plastic. It is advisable to seal the bales at the site of storage, whether this be in the field or close to farm buildings, to avoid damage to the fragile plastic from excessive handling after bales have been wrapped. When the technique of baled silage was first developed, round bales were placed in an approximately 115 µm thick plastic sack (Ganeau, 1981) that was tied at the flat end of the bale. Square bales were typically stacked in groups of 10–20 bales and covered with one or more sheets of plastic, similar to the type of plastic used in bunker silos or clamps.

In the mid 1980s, stretch film was introduced to seal round bales, and quickly replaced bagged round bales or stacked square bales. Wrapping with stretch film reduced labour and reduced storage losses compared with bagging. Initially, wrappers only were available for round bales, but lately wrappers also have been produced for square bales.

The guidelines for wrapping vary from country to country. It is recommended that grass bales in Scandinavia be wrapped in a minimum of six layers of stretch film with a 50% overlap between layers. In the UK, four layers of stretch film are recommended. The film should have a minimum thickness of 25 µm and undergo a prestretch of 70%, which means that the film becomes 1.7 times longer as it is wrapped on the bale (Lingvall, 1995). The stretching of the film allows the plastic to be tightly wrapped to the bale and each layer of plastic stick to one another. Because of the tight wrap, condensation is rarer in wrapped bales than in bagged bales.

Composition of Grass and Cereals before and after Ensilage

In principle, the minimum requirements for quality of grass ensiled in bales should be the same as grass ensiled in other types of silos. In practice, however, baled grass has a longer chop length and bales a larger surface area covered with plastic. These factors alone may result in lower density, delayed or different fermentation, and increased risk of fungal growth in baled grass.

Grass for quality silage should be mown in boot to early heading stages (Pettersson, 1988), whereas a later harvest may be more suitable when feeding silage at maintenance to cattle. Similar grass ensiled in bales or in conventional silos will undergo different fermentation patterns. But baled grass is often ensiled at higher DM concentrations, resulting in higher residual concentrations of water-soluble carbohydrate, and lower concentrations of fermentation acids.

Whole crop cereals have become an important silage source in some of the European countries such as Germany, the UK and Denmark. While most whole-crop cereal silage is ensiled in a bunker or tower silo, there has been a desire to ensile cereals in bales. The general recommendation is to harvest winter wheat (*Triticum aestivum* L.) at 400–450 g kg^{-1} DM and barley (*Hordeum vulgare* L.) at 350 g kg^{-1} DM (Skovborg *et al.*, 1979; Kristensen and Skovborg, 1991). At these stages of maturity, the starch concentration is approximately 300 g kg^{-1} of DM. Ensilage at earlier stages of maturity results in a completely different type of silage with a high sugar, but low starch concentration. Little information is available considering optimal maturity stage for ensilage of whole-crop cereals in bales. Ohlsson (1996) did not find a great advantage of harvesting spring barley at 250–300 g kg^{-1} DM on silage quality and aerobic stability compared with barley harvested at 350–400 g kg^{-1} DM.

Ensilage of whole-crop cereals in bales requires extra care because there is a high proportion of stem in relation to ear, 450 g kg^{-1} or higher. Nutrients of whole crops are heterogeneously distributed, the stem containing few nutrients that are usable in fermentation. Further, the stem is stiff and difficult to compress, thereby making ensilage conditions less than ideal (Kim *et al.*, 1995). If whole crop cereals are harvested later than recommended, there also is a risk of losing kernels at baling. Ohlsson (1996) reported that starch concentration in ensiled barley was 16 g kg^{-1} lower for herbage that was cut inside the bale chamber of a big-round baler compared with an uncut control, indicating that excessive physical handling of the crop results in grain losses.

An additional problem of ensiling whole crops, and particularly in bales, is that cereals contain lower concentrations of nitrate than forage grass (Weissbach and Haacker, 1988), thereby increasing the risk for growth of *Clostridium*. In fact, baled whole crop barley had unacceptably high spore counts, > 1000 colony forming units (CFU) per gram fresh silage, unless proper silage additives were used (Ohlsson, 1996; Thøgersen *et al.*, 1996).

Harvest of Crop

The time of cutting is important to obtain a silage of high nutritional value. It may also affect the ability of the baler to compress herbage adequately and produce dense and well-shaped bales. This is especially important for whole-crop cereals that have stiff straw of low nutritional value. Unfortunately, weather conditions have a considerable influence on harvest management. The risk of damage from moulds is generally higher in baled compared with conventionally made silage (Fenlon, 1986) because big bale silage generally is conserved at higher concentrations of DM and sugar (Anderson, 1985). It is not known, however, if the system of grass baleage results in more growth of yeast and mould compared with conventional silage systems.

Wilting

Wilting of baled grass also is essential in producing good, firm bales and to avoid effluent and to improve silage quality by reducing the concentration of ammonia-N, butyric acid and the number of clostridia spores (Jonsson, 1989; Jonsson *et al.*, 1990).

The width of the pick-up of the baler ranges from 1.4 to 2 m. Consequently, the width of the swath may have to be reduced by mounted plates at the rear of the mower. A conveyor belt that is mounted at a 90° angle in relation to the flow of the herbage and that throws the forage to the side may also reduce the width of the swath. A rake mounted in the front of the tractor may also be used to reduce the width of the swath.

From a hygiene point of view, it is preferable to reduce the width of the swath at mowing, without raking the forage before baling. This reduces the risk of soil contamination and consequent growth of clostridia and fungi in the silage compared with one or several turns of the forage by a rake. However, in countries where grass is cut infrequently and at high DM yields, it may be difficult to justify making a narrow, thick swath because wilting will be slow. In such situations, repeated tedding may be necessary.

If herbage is wilted for 24 h or less, dry matter losses normally range from 10 to 30 g kg^{-1} (Zimmer, 1980), whereas prolonged wilting may result in DM losses as high as 130 g kg^{-1} (Zimmer and Wilkins, 1984). Even under unfavourable weather conditions, herbage should not be left on the swath for more than 72 h. Especially under damp and warm conditions, the risk of fungal growth is high, making herbage unsuitable for silage. Timeliness and speed of operation when wilting the herbage is consequently essential. The cooperation among farmers or reliance on contractors improves the chance of getting high quality silage.

The degree of wilting of grass for ensilage in bales varies depending on weather conditions in the various countries, and the preferences and traditions among farmers. Generally, wilting ranges from 300 to 500 g kg^{-1} DM. The advantage of wilting in the upper range is that costs will be lower as more DM may be packed into a single bale and that feed intake generally increases. The disadvantages are that the pore volume of the bale increases with increased DM (Fig. 10.1) and that dry silage may be aerobically unstable. Wilting to 350–400 g kg^{-1} DM rather than 450–500 g kg^{-1} DM before ensilage in bales generally results in a more desirable fermentation (Nicholson *et al.*, 1991).

Silage quality is also affected by the degree of wilting. Loss of water soluble carbohydrates during fermentation decreases with increasing DM concentration of the herbage, as well as the concentration of lactic acid and acetic acid decreasing and pH increasing (Shin, 1990; Petit *et al.*, 1993).

Balers and Baling Techniques

Round bales of grass generally weigh 170–200 kg DM and it is possible to produce 20–50 bales per hour (Gaillard, 1990). The weight of square bales is in the same range, but density is generally higher. Weight and density of the bales depend largely on the DM concentration of herbage at baling, however (Fig. 10.1).

In Denmark round balers cost between US$15,000 and US$39,000, whereas square balers range from US$39,000 to US$104,000 (Landbrugets Maskinoversigt, 1997). As a consequence of high costs, the investment often is not justified unless the farmer is able to use the baler for operations other than grass baleage at his farm.

Round balers

Round balers are divided into three major groups, i.e. fixed-sized bale chambers, variable-sized bale chambers, and a hybrid, which combines the principle of fixed and variable-sized chambers. The fixed-chamber baler (Fig. 10.2) is most common in Europe, while the hybrid is the newest, but least used type.

Fixed chamber
The volume of the fixed-chamber baler is constant, and the bale is rotated by shafts, belts or chains around the periphery. Herbage is not compressed

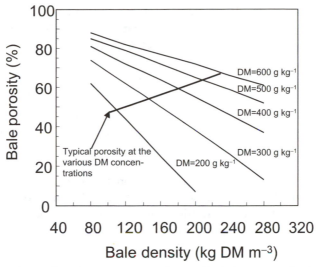

Fig. 10.1. Relationships between pore volume of baled grass and herbage density at different dry matter concentrations. Herbage density has been assumed to be 1.45 g cm^{-3} (data from Thylén and Nilsson (1993)).

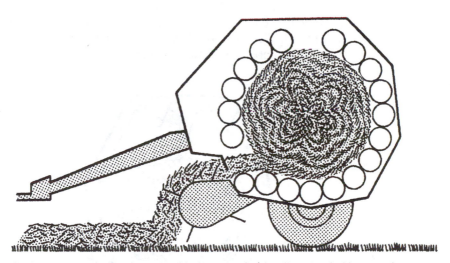

Fig. 10.2. Fixed-chamber round baler. Graphic kindly provided by Trioplast AB, Smålandsstenar, Sweden.

until the bale chamber is full. As a consequence, the core of the bale will be relatively loose. The periphery of the bale becomes more densely compacted, however. The power requirement ranges from 25 to 70 kW. The power requirement increases as the compression phase progresses.

The fixed chamber baler produces bales of fixed diameter and height. Most balers produce bales that are 120 cm tall and 120 cm in diameter. Depending on manufacturer, the diameter of the bales varies from 125 to 155 cm, however.

Variable chamber

Because the chamber volume varies, it is possible with the variable-chamber baler (Fig. 10.3) to produce bales with variable diameter. The belt systems that make the bale rotate compress the herbage immediately upon arrival in the bale chamber. Bales are 120 cm high or higher, and have a diameter from 90 to 180 cm, depending on brand of baler (Landbrugets Maskinoversigt, 1997). The diameter of the bale is determined by setting the pressure inside the chamber to trigger the release of the bale.

The power requirement for round balers ranges from 40 to 70 kW, which is slightly higher than that of a fixed-chamber baler. The power requirement is constant throughout the compression phase.

Hybrid

The hybrid baler (Fig. 10.4) is relatively new on the market and not widely used. During the first part of the compression, the hybrid baler works as a fixed-chamber baler, although the chamber volume is considerably smaller. Like the fixed-chamber baler, the bale is not compressed until the chamber has been filled. After the 'smaller' chamber has been filled, the baler principally works as a variable-chamber baler. Belts, chains, or a combination of

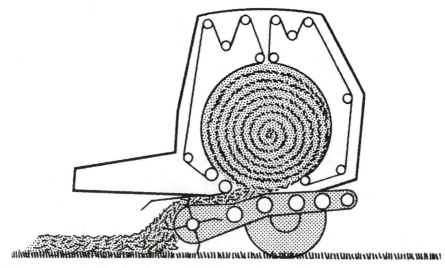

Fig. 10.3. Variable-chamber round baler. Graphic kindly provided by Trioplast AB, Smålandsstenar, Sweden.

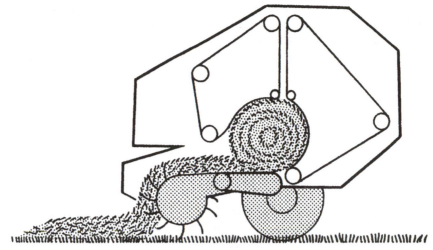

Fig. 10.4. Hybrid round baler. Graphic kindly provided by Trioplast AB, Smålandsstenar, Sweden.

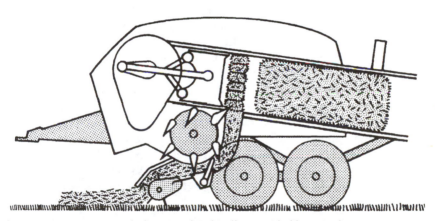

Fig. 10.5. Large square baler. Graphic kindly provided by Trioplast AB, Smålandsstenar, Sweden.

both rotates the bale. The end result is that the core of the bale is slightly looser than that in a variable chamber and that the bale diameter is slightly smaller than that in a fixed-chamber baler.

Large square balers

Most large square balers (Fig. 10.5) work according to a similar set of principles. Herbage is fed in sections from below the bale chamber. A large piston

inside the bale chamber then compresses each section. Depending on manufacturer, the width of the bale is fixed at 80 cm or 120 cm. Most balers are
capable of producing bales of variable length, 80–250 cm. Some balers,
however, have a fixed bale length of 250 cm or 280 cm (Landbrugets
Maskinoversigt, 1997).

Because large square balers are capable of making larger and denser
bales than round balers, power requirement also is greater. The power
requirement ranges from 55 to 164 kW, but for optimal use of the baler, no
less than 100 to 125 kW should be used. Some of the largest, self-propelled
square balers are not suitable for ensilage, but are predominantly used for
straw or hay.

Driving techniques

Careful and skilful driving at baling is necessary to obtain well-shaped bales
of high density. If the swath is much narrower than the width of the pick-up
on the baler, the operator should alternatively drive to the left, straight, to
the right, straight, and to the left. Zigzagging should be avoided because
there is a risk that the diameter of the bale will vary.

There should be a steady flow of herbage into the baler to produce
dense and uniformly shaped bales and this is most easily achieved by
driving slowly. Generally, it is advised not to drive faster than 6 km h^{-1}.
There is generally a negative correlation between high speed and the production of well shaped, dense bales (Table 10.1). Thus, in spite of balers
having a high baling capacity, i.e. number of bales per hour, it is wise to
slow down, especially when baling in fields from high-yielding crops.
Herbage has a heterogeneous structure and it takes some time to compress
it. Because operators often are paid a fixed price per bale, it is not in their
best economic interest to drive slowly, however.

Driving technique of fixed-chamber round balers

As the compression of the bale does not start until the bale chamber is full,
it is possible to drive relatively fast until the later stages of the baling phase.
On most balers, there is a display indicating when the chamber is full. When
the chamber is full, the speed should be reduced until the bale is released
from the chamber. A complement to this technique is to stop and rotate the
bale briefly before continuing the completion of the bale.

There should be sufficient power available to rotate the bale inside the
chamber at the final phase of compression. Certain brands of fixed-chamber
balers allow a continuation of the compression even after the display has
shown that the bale is completed. Compression beyond this phase may
increase the mechanical wear of the baler, however.

Table 10.1. The influence of dry matter concentration and speed at baling on bale density. Data compiled from Lingvall (1995) and Forristal *et al.* (1996).

Dry matter concentration (g kg⁻¹ of fresh wt)	Bale density (kg DM m⁻³)	
	Low speed[1]	High speed[2]
300	140	134
360	181	167
570	182	176

[1]Speed at baling = 6.0–6.4 km h⁻¹.
[2]Speed at baling = 8.0–8.8 km h⁻¹.

Driving technique of variable-chamber round balers

A relatively low speed from the start of baling is important because compression of herbage starts immediately. Unlike the fixed-chamber baler, the chamber of the variable-chamber baler becomes full quickly. Variable-chamber balers generally have a display showing the pressure of the periphery and centre of the chamber, to make it easier for the operator to feed herbage into the sides and centre of the bale chamber. This is obviously more important if the width of the swath is narrower than the width of the pick-up of the baler.

Experience has shown that incorrect driving technique is more likely to produce poorly shaped bales with a variable-chamber round baler than with a fixed-chamber round baler.

Driving technique of square balers

This baler should also be fed a steady flow of herbage. In addition, however, it is essential that the baler be properly adjusted to get well-shaped square bales. Adjustment for length and pressure of each twine should be calibrated to avoid getting 'banana-shaped' bales. Also, it is important to get relatively sharp corners, i.e. 90° angles along the sides of the bale, to avoid slippage of the stretch film at wrapping.

Choice of baler

Most balers are capable of producing bales in excess of 180 kg DM m⁻³ when herbage is wilted to 35% or higher (Fig. 10.1), ensuring good compression of the bales. Therefore, the choice of baler is virtually a matter of individual need.

The fixed-chamber round baler often is more durable and solid than variable-chamber balers, but produces bales that are slightly lighter and with a looser core. Petit *et al.* (1993) reported that bales from a fixed-chamber baler were 8% lighter than bales from a variable chamber baler. A loose core of the bale increases the risk for spoilage, should the plastic film fail to continually exclude oxygen.

Square bales require less space and are easier to handle when stored compared with round bales. Furthermore, square bales consist of small sections of silage that easily separate as twines are cut. Round bales either have to be rolled out or cut by special saws at feeding. Generally, the cost of producing square bales is higher and special wrappers for square bales still are not easily available.

Twine or net to secure bales?

Round bales may be secured by either net or twine. The use of net saves time and avoids the risk of twine being trapped between layers of plastic at wrapping, potentially allowing air to enter the bale. Unless at least four layers of nylon net are used to secure the bale, twine is better to maintain the diameter of the bale. Net is more convenient to remove than twine at feed-out. The net also helps shape the bale, possibly helping good wrapping.

On square bales, twine is the only option for securing the bale. Because square bales have a high density, it is particularly important to choose twine of high quality.

Silage additives

The choice of additives is not as dependent on which silage-making system is used as it is on the composition of the crops at ensilage. Under most circumstances there should be little need for additives in baled grass because of the degree of wilting. But because herbage in bales has a long chop length and a large surface area : volume ratio, fermentation may be slow in addition to silage having many clostridia spores and being aerobically unstable (Lättemäe and Lingvall, 1996; Thøgersen *et al.*, 1996).

Quality of baled grass may be improved and DM losses reduced by adding formic acid and other organic acids in low DM crops (Braithwaite and Jones, 1987; Kjus and Randby, 1987). Silage fermentation in baled wilted crops is generally improved by using additives, including lactic acid bacteria and fibre-degrading enzymes (Lingvall, 1993; Ohlsson, 1996; Braithwaite and Jones, 1987). Ohlsson (1996) reported that application of silage additives to whole-crop barley ensiled in bales increased bale weight by 2–8%, conserved sugar, reduced alcohol, and generally reduced butyric acid and the number of clostridia (Table 10.2). Recent work by Lättemäe

Table 10.2. Quality data on whole barley[1] that was ensiled at Silstrup Research Station Denmark on 27 July, 1995. Bales were cored on 12 March, 1996 to determine silage quality. Data from Ohlsson (1996).

Additive[2]	Bale weight[3] (kg DM m^{-3})	ME (MJ kg^{-1} DM)	pH	Dry matter (g kg^{-1} of fresh wt)	Quality (g kg^{-1} DM)				Clostridia (log CFU 100 g^{-1})
					WSC	Alcohol	Lactic acid	Butyric acid	
Ensimax	230	9.98	4.87	416	146	7	12	1.8	5.93
Kofasil	231	9.52	5.07	445	123	3	13	0.0	4.22
Propionic acid	243	9.92	5.01	466	153	9	5	4.7	5.61
Control	226	9.66	5.52	458	101	15	12	2.1	6.08
Average	232	9.77	5.12	446	131	9	10	2.2	5.46
LSD$_{0.05}$	11.7	0.271	0.141	10.4	14.0	5.8	4.2	3.67	0.464

[1]Barley was harvested at medium to hard dough stage and had the following quality (g kg^{-1} DM): DM = 441, crude protein = 7.2, starch = 24.7, WSC = 10.5 and ME (MJ kg^{-1} DM) = 10.7.

[2]Ensimax (25% formic acid, 25% acetic acid, and 50% lignosulfonate) applied at 4 litres t^{-1}, Kofasil liquid (main ingredient was sodium nitrite) applied at 4.5 litre t^{-1}, and propionic acid applied at 5 litres t^{-1}.

[3]Weight of silage in bales after 7.5 months of storage. Each round bale had a volume of 1.36 m^3.

ME, metabolizable energy; WSC, water soluble carbohydrates; Clostridia expressed as colony forming units (CFU) per gram fresh silage.

(1997) indicates that activity of clostridia and fungi in unchopped wilted grass may be inhibited by using appropriate doses of sodium benzoate or a mixture of sodium nitrite and hexamethylene tetraamine. A concern of using sodium benzoate or sodium nitrite is their potential carry-over effect on rumen microorganisms. Driehuis *et al.* (1996) reported that *Lactobacillus buchneri* produces large quantities of acetic acid that has the potential of inhibiting the activity of fungi. A more thorough description of additives and their mode of action is given in Chapter 9.

When used, silage additives in liquid form should be applied by using a pump that is capable of maintaining a steady pressure at the nozzles. Nozzles are best mounted slightly above and in front of the pick-up of the baler. The additive should be distributed on the whole width of the swath. Often it is better to use two nozzles than one nozzle, because it is easier to remain at steady pressure and steady flow of additives. Powdered additives are less commonly used than liquids because of the difficulty of distributing them evenly into the herbage. A hopper with a mounted auger often is used to distribute the powder inside the baler.

It is more difficult to distribute the additives evenly in crops with long rather than short chop-length and in high-yielding crops. Therefore, many additives may not have the intended effects when applied to baled grass.

Chop-length of forage

Big-bale silage is generally chopped at ensiling. Chop-length of herbage is either reduced by knives mounted in front of the pick-up of the baler or inside the chamber of the baler. The chop-length for round bales generally is 7 cm and 4.5 cm for square bales. Less commonly, 'grass processed' round balers may be used to chop herbage by flails. After chopping, herbage is blown into the bale chamber.

Silage compaction may be improved by reduced chop-length of herbage, with denser bales resulting in lower costs of handling and transport. However, the effects on fermentation are relatively small where proper ensiling techniques are achieved.

Reduced chop-length of baled grass potentially increases DM losses as it is easier for short particles to escape between belts or metal rollers of round balers. Bales also may lose shape when chop-length is reduced. Reduced chop-length in baled whole-crop barley resulted in a 9% increase in density compared with uncut herbage. Further, cut herbage resulted in slightly lower DM losses, improved lactic acid fermentation, but a slightly higher loss of starch (Table 10.3). Other studies have reported 8–26% increase in bale density after reducing chop-length (Weddell, 1990; Jones, 1995).

Table 10.3. Reduced chop-length of ensiled whole-crop barley. A variable-chamber round baler with knives mounted inside the bale chamber was used at Silstrup Research Station, Denmark in 1996. Each treatment mean is an average of 30 bales. Data from a study conducted by Ohlsson (1997).

Chop-length	Silage density (kg DM m^{-3})	DM loss (g kg^{-1})	pH	Quality (g kg^{-1} of DM) DM	Lactic acid	Starch
Cut	192	70	4.38	381	33	215
Uncut	176	83	4.53	376	29	220
Average	184	76	4.46	378	31	218
LSD$_{0.05}$	5.9	19.1	0.154	4.8	5.4	11.6

Hygiene of baled grass

Baled grass is susceptible to spoilage from mold, yeast and clostridia. Part of the susceptibility in baled grass compared with conventional silage is related to high DM at ensilage, long chop-length, large surface area in relation to amount of feed, and a thin cover of stretch film. Dangerous organisms to animals and humans, such as listeria (*Listeria monocytogenes*) and *Clostridium botulinum*, have been identified in baled silage and silage in conventional silos (Ricketts *et al.*, 1984; Fenlon *et al.*, 1989; Ruxton and Gibson, 1995). In spite of these findings, frequency of these dangerous organisms is relatively rare in well-fermented silage.

Presence and survival of *Listeria* is inversely related to quality of silage. The survival and growth of *Listeria* is greatest in a wet season when the concentration of water soluble carbohydrate is low and when a low DM crop is ensiled (Fenlon and Wilson, 1990). Donald *et al.* (1995) reported that oxygen concentrations above 1.0% sustained growth of *Listeria*, whereas less than 0.1% oxygen was lethal to the bacteria. In silage of very poor quality, however, *Listeria* is capable of surviving even under anaerobic conditions (Donald *et al.*, 1995). Presence of *Clostridium botulinum* is also related to poor-quality silage. Hogg *et al.* (1990) reported a suspected incidence of botulism in cattle that originated from feeding baled silage contaminated with poultry carcasses.

Contamination by soil, manure, or deteriorated organic matter increases the risk of obtaining baled silage of poor hygienic quality. Increased stubble height at mowing may alleviate some of the problems because spore-forming anaerobic bacteria may be numerous close to the soil surface (Pedersen and Guttormsen, 1975).

Handling of bales

Bales should not be stored in the field because it is difficult to check for damage to the film and to protect from damage by animals. Questions regarding the aesthetics of storing white or black bales in fields have also been raised. Consequently, tests with 'aesthetically acceptable colours' on bales stored in the field have been conducted (Lingvall, 1997).

Unless the distance to the wrapper is short, it is generally time saving to load four to six bales on a wagon, compared with transporting single bales on the front loader of the tractor. At loading, bales should be gently lifted to avoid damage to the bale. There are specially designed loaders with two to three hydraulically manoeuvred forks that gently squeeze the mantle of the round bale, the curved side of the bale. The same type of loader may be used for square bales.

Timeliness in transportation is important because bales should be wrapped within 2–6 h of being baled (Jones, 1995; Lingvall, 1995). A longer period between baling and wrapping increases the risk of heated herbage. In general, herbage with more than 200–250 g kg^{-1} DM provides an excellent environment for aerobic microbes (Gregory *et al.*, 1963). Sung *et al.* (1985) reported that delayed sealing of grass in round bales resulted in a temperature rise from 26°C to 55°C. Heat in herbage is a result of plant respiration, microbial metabolism and chemical reactions (Koegel and Bruhn, 1971).

Sealing of bales

When wrapping the bales, it is important to select plastic of good quality, to use proper wrappers and to handle properly the bales after wrapping.

Stretch film

Stretch film is manufactured from polyethylene at 500 mm or 750 mm width and a thickness ranging from 20 μm to 30 μm. Most films for grass baleage are 'blown' during manufacturing as a controlled bubble in single or multiple layers (Forristal *et al.*, 1995). During the 'blowing' process, it is possible to add plastic colouring, tackifiers for improved adhesion between plastic layers, and ultraviolet inhibitors (Forristal *et al.*, 1995).

The film must meet a number of requirements to be suitable for grass bales. Such requirements include a minimum of oxygen permeability, uniform thickness and good physical properties. In Sweden it is possible to purchase stretch-film that is officially certified by SP (Sveriges Provnings och Forskningsinstitut). Other countries, such as Ireland, are trying to get a quality standard for plastic films intended for grass baleage.

Among stretch films, white plastic reflects the most heat, whereas black plastic reflects the least. Consequently, bales wrapped in black film may have 20°C higher surface temperature (Lingvall, 1997). Furthermore, a 20°C rise at the bale surface may increase oxygen permeability through the film by as much as 300%, resulting in poor silage. However, in an experiment with different colours of plastic wrap on bales, variation in temperature seemingly had no major effect on silage quality (Itokawa *et al.*, 1995).

The film is pre-stretched simply to make it cling tightly to the bale and to make film layers adhere to one another. Black stretch film often adheres better than white film (Lingvall, 1997). A 70% pre-stretch is generally optimal to minimize oxygen permeability and for the plastic to maintain its elastic properties (Lingvall, 1995), but the film should not be overstretched. The correct stretching of the film also may be determined by measuring the reduction in length or width of the film. With a 70% pre-stretch, the film becomes 1.7 times longer and approximately 20% narrower (Forristal *et al.*, 1995).

Wrappers for single bales

Most wrappers were originally adapted for individual wrapping of round bales. Square bales also were wrapped with these wrappers, but because of problems of rotating square bales smoothly and consistently, purpose built wrappers for square bales have been designed.

Fig. 10.6.   Silage wrapper with rotating table. Graphic kindly provided by Trioplast AB, Smålandsstenar, Sweden.

Fig. 10.7. Silage wrapper with rotating arm. Graphic kindly provided by Trioplast AB, Smålandsstenar, Sweden.

Wrappers generally have a rotating table or a rotating arm (Figs 10.6 and 10.7). Wrappers with a rotating table rotate the bale about its longitudinal axis as the bale is turned on the turntable horizontally. The amount of film overlap is fixed, which means that one turn of the rotating table moves the curved surface of a 120 cm diameter round bale by 20 cm when using a 500 mm film or by 30 cm when using a 750 mm wide film.

The other type of wrapper has a rotating arm, which rotates around the longitudinal axis of the bale. Overlap of film is determined by the relative speed of rotation of the arm and the bale. The speed of the arm and the bale may be controlled separately.

Wrappers for bales in tubes

There are a number of wrappers that wrap a group of bales with stretch film and place them in a row, i.e. tube (Fig. 10.8). These wrap-liners generally work as well for round bales as for square bales. The advantage with the wrap-liners is that the amount of plastic is reduced by approximately 50% compared with individually wrapped bales, as bales in tubes are not wrapped throughout the surface. Disadvantages are that a puncture in the plastic may damage more than one bale and that all bales in the tube may be exposed to air at opening of the tube. In practice, there is no indication that bales have worse quality when wrapped in tubes compared with individual wrapping (Ohlsson, 1996, 1997).

Wrapping

A 500 mm wide film is still used, but the 750 mm film is becoming more common. Film layers are normally placed with a 50–55% overlap. A minimum of four layers should be used if the film is no thinner than 25 µm. But previous and recent experience sometimes found that six layers of a high quality film should be used on grass to achieve a high quality silage and low DM losses (Table 10.4). On baled whole-crop cereals, another two to four layers should be used because the sharp straw otherwise may puncture the film (Ohlsson, 1996). To further stress the importance of using stretch film of high quality and sufficient number layers, Thylén and Nilsson

Table 10.4. Influence of plastic quality and number of plastic layers on oxygen permeability, microbial growth, and dry matter losses in round-bale silage. Film width was 500 mm and thickness was 25 µm. Data are summarized from Lingvall *et al.* (1993).

Film type	Layers of film	Amount of film (kg bale^{-1})	Relative O_2- permeability (%)	Yeast (log CFU g^{-1})	Surface mould (%)	DM losses (g kg^{-1})
Film 1	4	0.783	100	3.9	2	120
Film 1	6	1.094	67	3.0	0	70
Film 2	4	0.765	270	4.4	8	170
Film 2	6	1.102	180	3.6	7	160

Fig. 10.8. Wrap-liner that wraps bales in tubes. Graphic kindly provided by Josefina D'Andrea-Ohlsson, Foulum, Denmark.

(1993) reported that a hole in the plastic of merely 0.02 mm² resulted in 330 l of air entering a bale during a 7-month period.

Increasing film width may be just as important as the total thickness of film on the bale in reducing DM losses and effluent in low-DM herbage because there will be fewer joins of plastic layers on the bale. By using six layers of 750-mm wide film instead of six layers of 500-mm wide film, DM losses of baled grass was reduced by 14% (T. Pauly, Swedish University of Agricultural Sciences, Uppsala, Sweden, 1997, personal communication). Jones (1995) reported that effluent in 185 g kg⁻¹ DM ryegrass was reduced from 46 l t⁻¹ to 28 l t⁻¹ when using four layers of 750-mm film instead of four layers of 500-mm film.

Storage of Bales

It is preferable to select a long-term storage site dedicated for bales only. The location should be chosen considering the distance between storage and feed-out, access when handling the bale, slope and drainage. A bed of fine sand is ideal to avoid puncture by rocks or other sharp objects. The site preferably should be shaded to avoid high surface temperatures of the bales.

Trees close to the storage site should be avoided because they may host birds that may damage the stretch film. As a further protection from bird damage, a net secured approximately 50 cm above the bales should be used. If farm animals are at the storage site, a fence should be erected. As expected, it is not always possible to find an ideal storage site, but precautions against damage by animals and weather could still be made.

Farmers who have baled whole-crop cereals have observed that these bales are more prone to damage by rodents compared with ensiled grass. When whole crops are ensiled in bales it is especially important to choose a storage site that is free of vegetation and to place rat poison next to the bales. Ohlsson (1996) stored bales in a field that was sprayed with glyphosate in the area round the bales to reduce the amount of biomass and make conditions less suitable for rodents.

Handling and storage of bales after wrapping

Because the total film thickness of bales often does not exceed 0.1 mm, handling obviously should be minimized after wrapping. Further, it is better to handle bales shortly after wrapping before pressure from gases inside the bale has built up.

Bales should be gently handled with a purpose-built attachment so as not to damage the plastic or reduce the adherence of plastic layers. Wyss *et al.* (1991) reported that silage quality increased with bale density and for bales stored on the flat ends compared with storage on the curved side. Lingvall (1995) also claimed that round bales are best stored upright because they do not lose shape as readily and because there is more film on the flat ends to protect from damage by sharp objects on the ground. Jones (1995) stated, with support from Swiss and German studies, that bales stored on the flat ends have higher ethanol concentrations and lower aerobic stability compared with bales stored on the curved side. However, bales stored on flat ends have lower DM losses, probably as a consequence of lower effluent production (Jones, 1995).

Bales of low DM concentration should be stored in single layers, to avoid heavy pressures exerted on the seal of bales and to reduce effluent from bales (Lenehan *et al.*, 1997). Drier bales may be stored in up to three layers. While storing bales in layers will save space, there is always a risk of damage from heavy bales in the upper layers. Several layers of bales also make them more difficult to handle.

While it apparently is difficult to agree on which side round bales should be stored when wrapped individually, bales wrapped in groups in a tube only can be stored on the curved side (Fig. 10.8).

Protection and control of bales

It is advisable to inspect the bales regularly for damage to the plastic – daily at the start and then preferably once weekly. Small holes in the plastic may result in large or complete silage losses. Wrapped bales are more easily checked if they are stored close to farm buildings rather than being placed in a field. If the film is damaged, the holes or tears should be repaired rapidly. There is a special tape available to repair the stretch film.

Duration of storage and losses

Early recommendations for storage of big bale silage often were based on short-term storage, 3–4 months, because of variable silage quality and high dry matter losses as storage was prolonged (Mo and Saue, 1988). The change from bagged to wrapped silage combined with improved plastic quality have enabled prolonged storage and nearly reduced storage losses by half (Weddell and Mackie, 1987). Dry matter losses of bales may vary considerably depending on the time and degree of damage to the plastic. On average, however, DM losses are similar in bales as in other types of silos or lower (Table 10.5; Jones, 1995).

Baled grass has been stored for 6–18 months without considerable deterioration of the silage (Gaillard and Zwaenepoel, 1987; Forristal *et al.*, 1995; Ohlsson, 1996). But Savoie (1988) suggests that a total film thickness should be increased with increased duration of storage. A thickness of 100 µm is

Table 10.5. Dry matter losses of silage ensiled in different types of silos. Data from Halvarsson (1991).

Type of silo	DM concentration	Field loss	Storage loss	Total loss
Tower	350	60	80	140
Bunker	250–300	30	150	180
Round bale	400–500	80	70	150

Values are g kg^{-1}.

optimal when storing bales for 3 months, whereas it should be increased to 150 µm when stored for 7 months and 200 µm when stored for 12 months (Savoie, 1988).

Feeding of Baled Silage

Feeding baled silage has several advantages relative to silage produced in bunkers or towers. It is possible to feed bales of differing quality based on the requirements of the animals. Some sheep farmers in the UK, for example, have adopted the feeding of baled silage because it is not exposed to air for as long periods as silage in bunkers or towers.

Round bales require a relatively large area when preparing them for feeding. After the plastic and twine or net have been removed, the bale either has to be 'un-rolled' or divided by purpose-built equipment, such as saws or bale separators before silage can be handled for feeding. The new round balers with mounted knives inside the bale chamber, have made handling of round bales easier, but some type of silage separation is still required because herbage at the outer periphery should not be cut because the bale may lose shape. Square bales are more easily handled than round bales because the bale is already divided into thin sections, making specialized equipment for separating the bale unnecessary. Beyond the steps of opening the bale, silage is handled similarly to conventionally made silage.

The success of feeding round bale silage to dairy cows has varied as a consequence of variable quality and aerobic stability (Mo and Saue, 1988). This type of variation is not unique to baled silage, however. Generally, there are few studies with documented differences in silage intake of cattle or sheep that are related to ensilage technique (Moate *et al.*, 1985). Instead Moseley *et al.* (1988) suggested that intake differences of various silage are related to the physical resistance of breaking down fibres by chewing and that differences in physical resistance are related to leaf : stem ratio and the concentration of acid detergent fibre.

Steinwender (1993) reported that total roughage intake was similar for grass ensiled in bales and in bunkers, whereas feed intake by dairy cows was higher in silage wilted to 600 g kg^{-1} DM compared with 400 g kg^{-1} DM. Milk yield was neither affected by ensilage technique nor wilting, however. In contrast, Hadero-Ertiro *et al.* (1988) reported that ensilage technique does affect animal performance. Dairy cows fed wrapped-bale silage produced more milk than silage from bagged bales, 9.5 l cow^{-1} day^{-1} vs. 8.0 l cow^{-1} day^{-1}.

Randby and Kjus (1989) reported that voluntary intake by steers was good when fed silage from a bunker silo and from round bales, and that live weight gain (LWG) was slightly higher for round-bale silage than silage from the bunker silo (1.13 kg day^{-1} vs. 1.06 kg day^{-1}). Rouzbehan *et al.* (1996b) reported that LWG for 288 kg steers fed baled clover–grass silage was 0.59 kg day^{-1}, or the same as for steers fed pit silage in a study conducted by England and Gill (1985).

While additives have shown positive effects on preservation of baled silage, few additives show large and consistent positive effects on animal performance (Weddell, 1990; Rouzbehan *et al.*, 1996a; Ohlsson and Kristensen, 1996).

Economics of Wrapped Bales vs. Conventionally Stored Silage

The reported cost of using baled grass compared with ensiling in towers or bunkers varies considerably in the literature. A common argument against the use of big-bale silage is its high cost, particularly in high-yielding first-cut crops (Forristal *et al.*, 1995). Some of the differences may be explained by the difficulty of comparing ensilage techniques on a comparable basis and still make comparisons realistic to the farmer. It is not correct, for example, only to consider the contractor's charge when comparing costs of silage-making systems.

In one of the most thorough comparisons of ensilage techniques, Halvarsson (1991) used the replacement value of machines for each silage technique without considering that other machines were already available at the farm and consequently could be used for other tasks. He further assumed that investment of equipment was made to eliminate differences in milk production in a herd of 30 dairy cows in central Sweden. Regardless of the strengths and weaknesses of these assumptions, he found that baled silage was considerably cheaper than ensilage in bunkers and towers (Tables 10.6–10.9). An Irish report (Forristal *et al.*, 1995) stated that baled silage was 43–56% cheaper than using a 200 tonne walled bunker silo including a tank for storage of effluent. By using a simpler 200 tonne bunker silo without walls, costs were the same as baled silage when crop yields were 40 t ha^{-1}, whereas the cost for baled silage was approximately 15% lower when crop yields were 22 t ha^{-1} (Forristal *et al.*, 1995).

Steinwender (1993) reported that ensilage in horizontal silos was the most economical system, while ensilage in towers was less economical than grass baleage. Andrieu *et al.* (1981) reported that presumed economic benefits of ensiling grass in bales might be nullified by inferior nutritive quality. Other studies confirm that economics of baled grass may vary because DM losses of bales may be quite variable.

There are factors other than economics that influence the choice of ensilage technique for the farmer. One such choice is the commitment to a

Table 10.6. Relative input of labour when using different silo types, with ensilage in tower silo as a reference. Investments in equipment have been made to eliminate differences in milk production in a herd of 30 cows with reasonable labour input for each silo type. Data from (Halvarsson, 1991).

| Type of silo | Relative labour input (%) | | | | Total cost[1] US$ kg^{-1} DM |
	Cutting	Harvesting	Feeding	Total	
Tower	100	100	100	100	0.22
Bunker	103	131	138	128	0.23
Round bale	97	126	195	140	0.17

[1]Sum of total costs for machinery, material, and labour. 1 US$ = 7.73 Swedish krona.

Table 10.7. Costs of using ensiled bales at a dairy farm with 30 cows in northern Sweden. The costs of equipment is based on replacement value and consequently does not consider which types of machines are already available on the farm and that may be used for tasks other than ensilage. Data from (Halvarsson, 1991).

Type of machine	Description	Replacement value (US$)
Tractor, 4-wheel drive	70 kW	44,000
Tractor, 2-wheel drive	50 kW	28,000
Front loader (tractor)	1400 kg	6,000
Mower-conditioner	2.4 m cutting width	11,500
Round baler (net-wrap)		16,500
Film wrapper		7,000
Bale loader		1,000
Wagon for transport		7,000
Wagon on rail (silage separation)		5,000
Additional space for silage		14,500
Bale separator		9,000
Conveyor for wagon on rail		2,000
Storage area for bales	450 m^2	3,000
Total investment		154,500

[1]Sum of total costs for machinery, material, and labour. 1 US$ = 7.73 Swedish krona.

Table 10.8. Costs of using bunker silos at a dairy farm with 30 cows in northern Sweden. The costs of equipment is based on replacement value and consequently does not consider which types of machines are already available at the farm and that may be used for tasks other than ensilage. Data from (Halvarsson, 1991).

Type of machine	Description	Replacement value (US$)
2 tractors	50 and 70 kW	72,000
Loader, 4-wheel drive		19,500
Mower-conditioner	2.4 m cutting width	11,500
Chopper		16,000
Wagons and equipment		18,500
Additive pump		500
Silage cutter		2,500
Wagon on rail (silage separation)		5,000
Feeder table		4,000
Additional space for silage		14,500
2 bunker silos	$2.4 \times 5 \times 27$ m (h $\times$ w $\times$ l)	71,000
Total investment		235,000

[1]Sum of total costs for machinery, material, and labour. 1 US$ = 7.73 Swedish krona.

Table 10.9. Costs of using a tower silo at a dairy farm with 30 cows in northern Sweden. The costs of equipment is based on replacement value and consequently does not consider which types of machines are already available at the farm and that may be used for tasks other than ensilage. Data from (Halvarsson, 1991).

Type of machine	Description	Replacement value (US$)
2 tractors	50 and 70 kW	72,000
Mower-conditioner	2.4 m cutting width	11,500
Chopper		16,000
Wagons and equipment		18,500
'Loading table'	8 m length	18,500
Loading fan, electric		9,000
Silage unloader		24,000
Wagon on rail (silage separation)		5,000
Tower silo	6×18 m (w $\times$ h)	57,000
Total investment		231,500

[1]Sum of total costs for machinery, material, and labour. 1 US$ = 7.73 Swedish krona.

long-term investment, which is necessary if the farmer chooses a tower silo or bunker silo. Another choice may be the willingness to depend on higher labour input when choosing grass baleage.

References

Anderson, R. (1985) The conservation and feeding value of high dry matter big bale silage. *Journal of Agricultural Research* 33, 25–30.

Andrieu, J.P., Gaillard, F. and Dulphy, J.P. (1981) L'ensilage d'herbe mi-fanée en balles rondes, utilization par des génisses laitieres d'un an. [The ensilage of semi-dried grass in round bales and its utilization by dairy heifers 1 year old]. Theix, INRA, Bulletin Technique CRZV 46, 31–47 (in French).

Bosma, A.H. and Verkaik, A.P. (1987) Fieldwork for making wilted silage. In: *Eighth Silage Conference, 7–9 September, 1987*. The AFRC Institute for Grassland and Animal Production, Hurley, Berkshire, Poster 18.

Braithwaite, G.D. and Jones, L. (1987) Factors affecting the composition and quality of big bale silage. In: *Eighth Silage Conference, 7–9 September, 1987*. The AFRC Institute for Grassland and Animal Production, Hurley, Berkshire, Poster 12.

Donald, A.S., Fenlon, D.R. and Seddon, B. (1995) The relationships between eco-physilogy indigenous microflora and growth of *Listeria monocytogenes* in grass silage. *Journal of Applied Bacteriology* 79(2), 141–148.

Driehuis, F., Spoelstra, S.F., Cole, S.C.J. and Morgan, M. (1996) Improving aerobic stability by inoculation with *Lactobacillus buchneri*. In: *Proceedings of the 11th International Silage Conference*. University of Wales, Aberystwyth, pp. 106–107.

England, P. and Gill, M. (1985) The effect of fish meal and sucrose supplementation on the voluntary intake of grass silage and live-weight gain of young cattle. *Animal Production* 40, 259–265.

Fenlon, D.R. (1986) Growth of naturally occurring *Listeria* spp. in silage: a comparative study of labouratory and farm ensiled silage. *Grass and Forage Science* 41(4), 375–378.

Fenlon, D.R., and Wilson, J. (1990) Factors affecting the growth and survival of *Listeria monocytogenes* in silage. In: *Proceedings of the 9th Silage Conference, 3–9 September 1990*. Faculty of Agriculture, University of Newcastle upon Tyne, Newcastle upon Tyne, Paper 27.

Fenlon, D.R., Wilson, J. and Weddell, J.R. (1989) The relationship between spoilage and *Listeria monocytogenes* contamination in bagged and wrapped big bales silage. *Grass and Forage Science* 44, 97–100.

Forristal, D., Lenehan, J.J. and O'Kiely, P. (1995) Big bale silage. In: *Cattle Production Seminar for Teagasc Beef Advisors and College Teachers*, October, 1995, Teagasc, Grange Research Centre, Ireland, pp. 1–15.

Forristal, P.D., Lenehan, J.J. and O'Kiely, P. (1996) A study of factors influencing bale density in baled silage systems. In: *Cattle Production Seminar for Teagasc Beef Advisors and College Teachers*, March, 1996. Teagasc, Grange Research Centre, Ireland, pp. 81–82.

Gaillard, F. (1990) The use of stretch film with round bales of dry and wet hay: potential capacity, plastic consumption, preservation results, control of wrapping impermeability. In: *Proceedings of the 11th International Congress on the Use of Plastics in Agriculture, 26 February–2 March, 1990*. New Delhi, India, pp. D.11–D.18.

Gaillard, F. and Zwaenepoel, P. (1987) Round bale silage wrapped in plastic film. *Bulletin Technique du Machinisme et de l'Equipement Agricoles* No. 18, 37–46.

Ganeau, M. (1981) Round bale silage – French experience. *Bulletin d'Information du CNEEMA* 280, 19–21.

Gregory, P.H., Lacey, M.E., Festenstein, G.H. and Skinner, F.A. (1963) Microbial and biochemical changes during the moulding of hay. *Journal of General Microbiology* 33, 147–174.

Hadero-Ertiro, A., Moate, P. Clarke, T. and Rogers, G.L. (1988) A comparison of the feeding value for milk production of pasture silage conserved as round bales either wrapped or bagged in polythene. In: *Proceedings of the Australian Society of Animal Production, 17 May, 1988*, Sydney, p. 410.

Haigh, P.M. (1990) The effect of dry matter content on the preservation of big bale grass silages made during the autumn on commercial farms in South Wales 1983–87. *Grass and Forage Science* 45, 29–34.

Haigh, P.M. (1995) Chemical composition and energy value of big bale silages made in England, 1984–1991. *Journal of Agricultural Engineering Research* 60, 211–216.

Halvarsson, H. (1991) Economic analyses of ensiling in tower silo, bunker silo, or round bales. In: *Ensilage in Round Bales, NJF seminar No. 201, 24–25 October 1991*. Hveragerdhi, p. 10 (in Swedish).

Hogg, R.A., White, V.J. and Smith, G.R. (1990) Suspected botulism in cattle associated with poultry litter. *Veterinary Record* 126(19), 476–479.

Howe, S. (1987) New developments in big bale grass silage. In: Wilkinson, J.M. and Stark, B.A. (eds) *Developments in Silage 1987*. Chalcombe Publications, Marlow, pp. 7–22.

Itokawa, N., Honda, Y. and Kobayashi, R. (1995) Characteristics and quality of wrapped big bale silage under various conservation conditions. *Journal of the Japanese Society of Grassland Science* 40(4), 478–487.

Jones, R. (1995) Big bale silage: Can we afford them? In: Internal Work Document. Institute of Grassland and Environmental Research (IGER), Aberystwyth, pp. 1–7.

Jonsson, A. (1989) *The Role of Yeasts and Clostridia in Silage Deterioration*. Report No. 42, Swedish University of Agricultural Science, Department of Microbiology, 51pp.

Jonsson, A., Lindberg, H., Sundås, S., Lingvall, P. and Lindgren, S. (1990) Effect of additives on the quality of big-bale silage. *Animal Feed Science and Technology* 31, 139–155.

Kim, J.G., Kang, W.S., Han, J.D., Shin, C.N., Han, M.S., Kim, G.Y., Kim, J.G., Kang, W.S., Han, J.D., Shin, C.N., Han, M.S. and Kim, G.Y. (1995) Study on baled silage making of selected forage crop and pasture grasses. I. Discussion on baled silage making as affected by physiological characteristics of the plants. *Journal of the Korean Society of Grassland Science* 15(1), 73–79.

Kjus, O. and Randby, Å. (1987) Ensilering av gras i rundballer. [Ensilage of grass in round bales]. Norwegian Institute of Agricultural Engineering. LTI-trykk No. 68. Ås, 40pp. (in Norwegian).

Koegel, R.G. and Bruhn, H.D. (1971) Inherent causes of spontaneous ignition in silos. *Transactions of the American Society of Agricultural Engineers* 14, 273–281.

Kristensen, V.F. and Skovborg, E.B. (1991) *Winter Wheat for Whole Crop Silage*. I. Report No. 783. Danish Ministry of Agriculture, Danish Institute for Animal Science, 3pp. (in Danish with an English summary).

Landbrugets Maskinoversigt (1997) *Review of Danish Farm Machinery 1997*. Danish Advisory Service – Office for Buildings and Machinery, Udkærsvej 15 Skejby, DK-8200 Århus N., p. 293.

Lenehan, J.J., O'Kiely, P. and Forristal, D. (1997) Effluent production from round bale silage. In: *Summary of Papers Presented at the Agricultural Research Forum, 3–4 April, 1997*, Dublin, pp. 99–100.

Lindberg, H. and Larsson, L. (1993) Storbalar möjliggör bättre foderplanering. [The use of big-bale silage improves feed planning]. *Husdjur* 4, 34–35 (in Swedish).

Lingvall, P. (1993) Ensilage in large bales. In: *Congress Report: Biotechnology for Plants and Animals, Research, Ethics and Policy; Competitive Animal Production with Respect to Environment and Animal Welfare*. Swedish University of Agricultural Sciences, Uppsala, pp. 146–151 (in Swedish).

Lingvall, P. (1995) *The Balewrapping Handbook*. Trioplast AB, Smålandsstenar, 52pp.

Lingvall, P. (1997) Vit plast i täten. [White plastic has a front position]. *Land Lantbruk* 13, March 1997 (in Swedish).

Lingvall, P., Pettersson, C.M. and Wilhelmsson, P. (1993) Influence of oxygen leakage through stretch film on quality of round-bale silage. In: Baker M.J. *et al.* (eds) *Proceedings of the XVIII International Grassland Congress, 8–21 February, 1993, Palmerston North, NZ*. New Zealand Grassland Association, Palmerston North, pp. 600–601.

Lättemäe, P. (1997) Ensiling and evaluation of forage crops – effects of harvesting strategy and use of additives to fresh-cut and wilted crops. PhD thesis, Swedish University of Agricultural Sciences, Uppsala.

Lättemäe, P. and Lingvall, P. (1996) Effect of hexamine and sodium nitrite in combination with sodium benzoate and sodium propionate on fermentation and storage stability of wilted and long cut grass silage. *Swedish Journal of Agricultural Research* 26, 135–146.

Mo, M. and Saue, O. (1988) Recent developments in feed conservation. In: *Proceedings of the 12th General Meeting of the European Grassland Federation, July 4–7, 1988*, Dublin, Ireland, pp. 126–142.

Moate, P.J., Rogers, G.L., Clarke, T. and Cumming, R.B. (1985) A comparison of fine chop silage with bale silage as supplements to pasture fed dairy cows. In: *Recent Advances in Animal Nutrition in Australia*. Conference paper No. 17, November, 24–27, University of New England.

Moseley, G., Jones, E.L. and Ramanathan, V. (1988) The nutritional evaluation of Italian ryegrass cultivars fed as silage to sheep and cattle. *Grass and Forage Science* 43, 291–295.

Nicholson, J.W.G., McQueen, R.E., Charmley, E. and Bush, R.S. (1991) Forage conservation in round bales or silage bags: effect on ensiling characteristics and animal performance. *Canadian Journal of Animal Science* 71(4), 1167–1180.

Ohlsson, C. (1996) Effects of silage additives, harvest date, and particle reduction on quality of baled whole-crop barley – Experiment in 1995. Internal report,

Department of Forage Crops and Potatoes, Research Center Foulum, 8830 Tjele, 15pp.

Ohlsson, C. (1997) Effects of silage additives, harvest date, and particle reduction on quality of baled whole-crop barley – Experiment in 1996. Internal report, Department of Forage Crops and Potatoes, Research Center Foulum, 8830 Tjele, 16pp.

Ohlsson, C. and Kristensen, V.F. (1996) Ensilering i rundballer – Er dette et alternativ til ensilering i almindelig silo. Kan teknikken for rundballeensilering forbedres? [Ensilage in round bales – Is this an alternative to ensilage in conventional silos. Is it possible to improve the technique for round bale silage?] Danish Ministry of Agriculture and Fishery, Research Center Foulum, *SP Report No.* 19, 31–36 (in Danish).

Pedersen, T.A. and Guttormsen, D.M. (1975) Microflora of newly cut grass after addition of liquid manure. Effects of different liquid manure application methods and grass cutting heights. *Acta Agriculturae Scandinavica* 25(4), 337–345.

Petit, H.V., Tremblay, G.F., Savoie, P., Tremblay, D. and Wauthy, J.M. (1993) Milk yield, intake, and blood traits of lactation cows fed grass silage conserved under different harvesting methods. *Journal of Dairy Science* 76(5), 1365–1374.

Pettersson, K. (1988) Ensiling of forages – Factors affecting silage fermentation and quality. PhD Dissertation Report no. 179. Swedish University Agricultural Sciences, Department of Animal Nutrition and Management.

Pfimlin, A. (1993) Market for the use of stretch wrap film for big bale silage in France and the EC. *Plasticulture* 98, 19–25.

Randby, Å. and Kjus, O. (1989) Ensilering i rundballer. [Ensilage in round bales]. *Buskap og avdrått* 41(1), 24–27 (in Norwegian).

Ricketts, S.W., Greet, T.R.C., Glyn, P.J., Ginnett, C.D.R., McAllister, E.P., McCaig, J., Skinner, P.H., Webbon, P.M., Frape, D.L., Smith, G.R. and Murray, L.G. (1984) Thirteen cases of botulism in horses fed big bale silage. *Equine Veterinary Journal* 16(6), 515–518.

Rouzbehan, Y., Galbraith, H., Topps, J.H. and Rooke, J.A. (1996a) The response of sheep to big bale grass silage ensiled with, or supplemented separately with, molassed sugar beet feed. *Animal Feed Science Technology* 59, 279–284.

Rouzbehan, Y., Galbraith, H., Topps, J.H. and Rooke, J.A. (1996b) Response of growing steers to diets containing big bale silage and supplements of molassed sugar beet pulp with and without white fish meal. *Animal Feed Science Technology* 62, 151–162.

Ruxton, G.D. and Gibson, G.J. (1995) A mathematical model of the aerobic deterioration of big-bale silage and its implications for the growth of *Listeria monocytogenes*. *Grass and Forage Science* 50(4), 331–344.

Savoie, P. (1988) Optimization of plastic covers for stack silos. *Journal of Agricultural Engineering Research* 41(2), 65–73.

Shin, C.N. (1990) Studies on developments in bale silage. 1. Effect of dry matter content and methods of sealing on the feed value in small bale Italian ryegrass silage. *Korean Journal Animal Science* 32(7), 385–392.

Skovborg, E.B., Kristensen, V.F. and Andersen, P.E. (1979) Whole crop silage of barley harvested at different stages of maturity. Yield and feeding value to dairy cows. Danish Ministry of Agriculture, *Danish Institute of Plant and Soil Science, Report No.* 1520 Volume 81, 4pp. (in Danish).

Steinwender, R. (1993) Feeding value of round bale silage for dairy cows. *Österreichweite Silagetagung* pp. 111–118.

Sung, K.I., Kim, D.A. and Kim, C.J. (1985) Studies on making and utilization of grass silage – fermentation and feeding value of roll bale silage in accordance with delayed sealing. *Journal of the Korean Society of Grassland Science* 5(2), 116–120.

Thylén, A. and Nilsson, E. (1993) Air penetration into stretch film wrapped silage bales. *Teknik för Lantbruket Report* No. 40. Swedish Institute of Agricultural Engineering, Uppsala, Sweden, pp. 1–10 (in Swedish).

Thøgersen, R., Kjeldsen, A.M., Nielsen, E.S., Strudsholm, F., Jepsen, L., Mikkelsen, M., Ohlsson, C. and Hermansen, J. (1996) Undersøgelse af anaeroba sporer i ensilage. [Investigation on content of anaerobic spores in silage]. *LK-Meddelelse* No. 126, National Danish Advisory Service for Cattle, Udkærsvej 15, Skejby, Århus N, Denmark, 10 pp. (in Danish).

Weddell, J.R. (1990) The effects of a bacterial inoculant additive on big bale silage. In: *Proceedings of the 9th Silage Conference, 3–9 September, 1990.* Faculty of Agriculture, University of Newcastle upon Tyne, Paper 34.

Weddell, J.R. and Mackie, C.K. (1987). A comparison of bagged and wrapped big bale silage – storage losses and animal performance. In: *8th Silage Conference, 7–9 September, 1987.* The AFRC Institute for Grassland and Animal Production, Hurley, Berkshire, Poster 14.

Weissbach, F. and Haacker, K. (1988) Über die Ursachen der Buttersäuregärung in Silagen aus Getreideganzpflanzen. *Das Wirtschaftseigene Futter* 34, 88–99 (in German with English summary).

Wilkinson, J.M., Wadephul, F. and Hill, J. (1996) *Silage in Europe – A Survey of 33 Countries.* Chalcombe Publications, Lincoln, 154 pp.

Wyss, U., Schild, G.J. and Honig, H. (1991) The influence of damage of stretch film on gas content and silage quality in round bale silage. *Landwirtschaft Schweiz* 4(5), 235–239.

Zimmer, E. (1980) Efficient silage systems. Forage conservation in the 80's. In: *Proceedings of the European Grassland Federation.* Occasional Symposium No. 11, Brighton, pp. 186–197.

Zimmer, E. and Wilkins, R.J. (1984) Eurowilt. Efficiency of silage systems: a comparison between unwilted and wilted silages. *Landbauforschung Völkenrode, Sonderheft,* 88pp.

Principles of Grass Growth and Pasture Utilization

11

A.J. Parsons[1] and D.F. Chapman[2]

[1]AgResearch, Grasslands Research Centre, Tennent Drive, Private Bag 11008, Palmerston North, New Zealand
[2]Department of Agriculture and Resource Management, University of Melbourne, Victoria, Australia

Introduction

Imagine a field of grass stretching uniformly before you. In the first part of this chapter, we consider the basic principles of growth and utilization of grass from this whole paddock perspective, whether using cutting, continuous or rotational grazing. Although humans may regard utilization paddock by paddock, animals are constrained to harvesting each paddock bite by bite, using their mouths. Each bite removes only a portion of the herbage present at that location, and all the bites added together may cover only a small portion of the total area available each day. In the second part of this chapter, we consider the constraints to utilization that the mechanism of grazing imposes on the animal and the manager. Finally we look for principles of growth and utilization from a large body of literature on the morphogenesis (birth, numbers, size, fate) of individual plant organs. This third approach – from the viewpoint of an individual plant – is the most difficult from which to perceive pasture growth and utilization, particularly given its heterogeneity. We address this, however, as many concepts for improving pasture growth and persistence are based on exploiting variations in plant organ turnover and fate.

Principles of Growth and Utilization – from a Whole Paddock Perspective

The optimal trade-off between growth and utilization

One notable feature of utilization in the grass crop is that the major harvestable components are the leaves – the very tissues that are essential for the photosynthesis that sustains the growth and dry matter production of the pasture. Moreover, unlike cereals, the grass crop may be harvested repeatedly and the way the crop is harvested has a profound effect on the amount grown. Second, the leaves of many grass species are produced continually but turnover rapidly at a rate dependent primarily on temperature and so are short-lived, e.g. some 33 days for *Lolium perenne* L.; 57 days for *Festuca arundinacea* Schreb. (Lemaire, 1988; Robson *et al.*, 1988). Leaves, or portions of leaves, that are not harvested within this time frame wither, lose nutritive value and in many cases fall from the canopy and become unavailable for harvest. This puts the emphasis of attempts to optimize the utilization of grass very much upon striking a balance between leaving sufficient leaf area behind in the pasture, during a sequence of defoliations and regrowths, to provide for growth but removing sufficient material to maximize yield. This clearly requires that we identify the optimal balance between growth, harvest and senescence. Detailed reviews of both the physiological and morphogenetic basis of grass growth and its response to season and management may be found in Robson *et al.* (1988), Parsons (1988, 1994), Lemaire and Chapman (1996) and Thornley and Johnson (1990).

The effects of the amount of leaf material that is sustained in the pasture, on the balance of the major physiological processes involved in growth and utilization, is described in Fig. 11.1 (Parsons *et al.*, 1983b). In a sward maintained at a high leaf area index (LAI), gross photosynthesis is close to its maximum. The 'losses' of material due to respiration are also close to a maximum, though the proportion of gross uptake lost in respiration (c. 45%) is little altered by the LAI sustained. The rate of gross production of new tissues (variously called net carbon exchange, or net primary production, NPP) is also maximized. However, in order to sustain the high LAI, only a small proportion of the leaf tissue produced can be harvested. A large proportion of the tissue therefore dies unharvested, and the amount harvested (yield ha^{-1}) is necessarily small.

Utilizing a greater proportion of the leaf material that grows causes the crop to approach equilibrium at a lower LAI. The rate of photosynthesis and net carbon exchange is reduced, as is the gross rate of production of tissues. However, the benefits of increasing the *proportion* of leaf tissue that is removed outweigh the effects of reducing the *amount* of leaf grown, and the *amount* of leaf that may be harvested per unit ground area is actually increased. Maximum yield is therefore achieved in a sward at a relatively low LAI, one that is below the optimum for light interception and photosynthesis, but which provides the optimum balance between photosynthesis, yield and leaf death. It is noteworthy that, if we define the efficiency of utilization as the proportion of all grass grown (gross production, NPP) that is harvested, then maximum yield per hectare is achieved at a point when the efficiency of harvest is close to just 50% – i.e. the optimum efficiency of harvest for maximum yield per hectare is around 50%. As much plant material, therefore, dies and contributes to soil organic matter as is harvested. If a greater proportion of leaf material than this is removed, the sward will be sustained at a very low LAI and the *amount* harvested is substantially reduced (Fig.11.1).

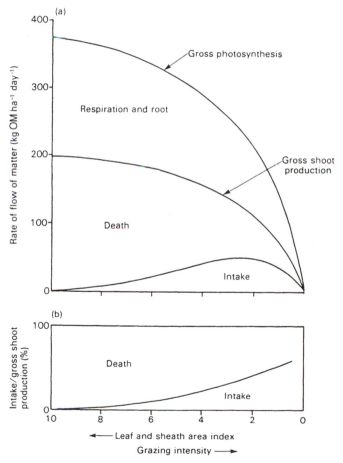

Fig. 11.1. The effect of the intensity of grazing (the average leaf area index at which the sward is sustained) on the balance of the major physiological components of growth and utilization (after Parsons *et al.,* 1983b).

The principles of grass growth and utilization described here can be derived using simple dynamic mechanistic models (e.g. Johnson and Parsons, 1985; Woodward, 1998) which have been used to rationalize field based measurements both of mass flux per unit ground area (e.g. Parsons *et al.*, 1983a, b; King *et al.*, 1984) and of the appearance, turnover and fate of individual leaves and tillers (e.g. Grant *et al.*, 1983; Parsons *et al.*, 1991; Mazzanti and Lemaire, 1994).

Although the principles of grass growth and utilization are most readily conveyed in terms of continuous processes operating at a whole paddock scale, it is soon seen that even under continuous grazing, growth and utilization actually involves a series of discrete defoliations (e.g. at a bite by bite scale), followed by periods of uninterrupted regrowth. We now expand our explanation to demonstrate how the same basic principles emerge under intermittent defoliation or rotational grazing.

When grass swards are allowed to regrow uninterrupted, they do not endlessly accumulate dry matter. Soon after the leaves have come to intercept all incident light, the pasture will reach a ceiling yield. The changes in the major physiological components involved in regrowth from a low to a high LAI, and the origin of the changes in growth rate and the ceiling yield, were described in detail in seminal accounts by McCree and Troughton (1966a, b) and by Robson (1973a, b). The principles are illustrated in Fig. 11.2.

When defoliation removes virtually all leaf tissue, rates of photosynthesis are substantially reduced and respiration rates may initially exceed the uptake of carbon in photosynthesis on a 24-h basis. The sward will consequently have a net loss in weight. New leaf tissue can be produced from 'reserves'. These may be sugars, stored in many cases in sheath bases (Pollock and Jones, 1979; Volenec, 1986), or remobilized structural material and proteins (Davidson and Milthorpe, 1966a, b; Richards, 1993). The period over which new tissue growth can be supported from 'reserves' is very limited, and in any case, the consumption of these resources represents a net *loss* of weight and so is not 'growth' in the common sense. There may be an overriding limitation to regrowth if a large proportion of meristems have been removed (see later section on morphogenetic constraints). In all cases any subsequent accumulation of dry matter must depend on the capacity of the crop to re-establish a leaf area, and to restore inputs from photosynthesis.

Canopy 'gross' photosynthesis (P_{gross}) increases during regrowth as the carbon invested in new leaves enables the crop to intercept an increasing proportion of the available light. However, 'gross' photosynthesis reaches a maximum rate soon after all the available light is intercepted. Thereafter, gross canopy photosynthesis may even decline, as the photosynthetic capacity of the leaves of some grass species declines when successive new leaves, expanding from the base of a dense vegetative sward, develop in shade (Woledge, 1977, 1978).

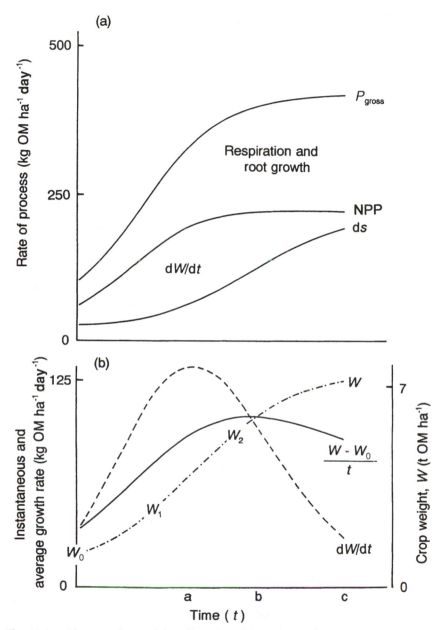

Fig. 11.2. The use of a model to illustrate (a) the effect of the duration of regrowth on the major processes involved in the net accumulation of herbage: gross photosynthesis P_{gross} ; gross shoot production (net primary production (NPP)) and senescence (ds); and (b) the corresponding change in the instantaneous growth rate (dW/dt), the weight of the crop (W) and the average growth rate, $(W - W_0) / t$ (after Parsons and Penning, 1988). See text for explanations of points 'a', 'b' and 'c', and W_1 and W_2.

During regrowth, the rate of respiration of the pasture also increases. Some 25% of gross photosynthesis is consumed in the synthesis of new tissues (this is a relatively fixed biochemical stoichiometry, see review by Thornley and Johnson, 1990), and there is an increase in respiration associated with the 'maintenance' (e.g. active processes and protein turnover) of the increasing mass of live tissue. In the cool temperate grassland species, typified by perennial ryegrass that show a rapid turnover of leaves, a ceiling yield is _not_ achieved because respiration rates come to equal the rates of uptake of carbon in photosynthesis as was once proposed (Donald, 1961). In these grassland species, respiration remains some 40–45% of 'gross' photosynthesis – hence divided roughly equally between 'synthesis' and 'maintenance' (Robson, 1973b). The decline in the rate of accumulation of live tissue and the ceiling yield of live tissue are achieved because there is an increase in the rate of senescence of leaf (per unit ground area) during the regrowth of the crop (ds in Fig. 11.2a). This increase in the rate of senescence largely reflects the increase in the size of leaves currently involved in the turnover of tissue in the sward. During regrowth, changes in the rate of loss by senescence inevitably lag behind changes in the rate of gross photosynthesis and gross tissue production as, at any point in time, the size of leaves currently dying reflects the size of leaves produced previously. Once the rate of gross tissue production has reached a maximum, the rate of senescence 'catches up' and comes to equal the rate of gross production (Fig. 11.2a), and the rate of net accumulation of live tissue declines to zero.

The balance of these processes accounts for the classic sigmoid growth curve in dry weight, W, and the changes in the instantaneous growth rate, dW/dt. In Fig. 11.2b, we draw a further curve which provides us with the basis to identify a theoretical optimum time to revisit and re-harvest the sward (whether at the scale of a paddock or a single bite). This curve depends on the severity of the previous defoliation.

Optimum timing of harvest during regrowth

The principles for understanding the optimal timing of harvest owe much to the Marginal Value Theorem, which is a central tenet of foraging theory in ecology and describes the optimal use of patches of food resources by animals (Charnov, 1976; Stephens and Krebs, 1986). This link between agricultural science and ecology has only recently been recognized.

Starting first with a sward regrowing from a low LAI, harvesting at time 'c' (Fig. 11.2b) would clearly achieve close to the maximum yield from that single period of regrowth ($W - W_0$ is greatest). Harvesting the sward at time 'a' would mean interrupting the growth of the sward at the time when the instantaneous growth rate was at a maximum, but the amount harvested would be reduced. Maximum utilization is achieved if the sward is harvested when the *average growth rate*, $(W - W_0) / t$ is at a maximum (Morley, 1966; Parsons and Penning, 1988) as this takes account not only of the material that has accumulated, but also of the total time that has elapsed. Following defoliation to a low LAI, the average growth rate increases initially, but thereafter changes relatively gradually as the period of regrowth is extended. A *maximum average growth rate* can be identified (at time 'b', Fig. 11.2b) after the time of the maximum instantaneous growth rate, but before the time of the ceiling yield. Note that, graphically, the optimum timing of harvest is seen to be the time at which the plots of dW/dt and $(W - W_0) / t$ intersect – this is a mathematically precise definition.

Optimizing the frequency and severity of defoliation

The effect of the LAI at the start of regrowth (c.f. 'residual' LAI, or the 'severity' of the previous defoliation) on the average growth rate, and the optimal timing of harvest, can be seen in Fig. 11.3. When a sward starts regrowth from a high LAI, photosynthesis and the rate of gross tissue production (NPP) also start higher, but so too does the rate of senescence. As a result, the optimum duration of regrowth, that is the time taken to achieve the maximum average growth rate, is less, and the *value* of the maximum average growth rate may also be reduced. In extreme cases, the average growth rate may decline throughout the period of regrowth. During 50 years of grassland research, there has been much controversy over the fundamental principles of the optimal method to harvest grass: continuously or rotationally; and at what 'frequency' and 'severity' (Brougham, 1956; McMeekan, 1960; Hunt, 1965; Anslow, 1967; Smetham, 1975; Harris, 1978). It has long been recognized that lenient defoliation must be associated with frequent defoliation (Hunt, 1965; Smetham, 1975). Despite major differences in presentation (e.g. re-analyses by Parsons *et al.*, 1988), it is now widely accepted that under rotational grazing, or cutting, maximum yield is achieved most reliably by defoliating the sward to a low residual LAI, and harvesting at the time of the maximum average growth rate (so using relatively long regrowth intervals). This optimizes not only the mass fluxes of material, but also does best to retain the sward structure that ensures this level of yield can be achieved sustainably. Note that earlier widespread analyses argued that the sward should be defoliated, e.g. at the mass W_2 (Fig. 11.2), and reduced to that at W_1. As can be seen in Fig. 11.2, this would not in theory reliably achieve the maximum average growth rate. We can recognize pragmatically that, provided defoliation is severe, the average growth rate changes little over a wide range of regrowth intervals (see again Fig. 11.2). This laxity about the optimal solution allows some tolerance to imprecise grassland management.

    The principles of growth and utilization described in terms of discrete defoliations (at paddock or bite scale) sound different from those described

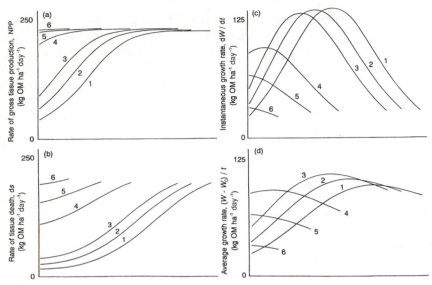

Fig. 11.3. The effect of the 'severity' of defoliation (the initial conditions for regrowth) on (a) the rate of gross tissue production (net primary production), (b) the rate of senescence (ds), (c) the instantaneous growth rate (dW/dt) and (d) the average growth rate, $(W - W_0)/t$, as the duration of regrowth is extended over time (t). The initial leaf area indices for regrowth were 0.5, 0.8, 1.1, 3.4, 5.3 and 6.8 (numbered 1–6, respectively) (after Parsons *et al.*, 1988).

earlier in terms of continuous processes. But the same fundamental relationships between photosynthesis, growth, yield and senescence (as shown in Fig. 11.1) can be shown to apply for all combinations of initial LAI and duration of regrowth when these are re-plotted in terms of the *average* LAI, $(L - L_0)/t$ sustained (Parsons *et al.*, 1988). Apparent small advantages to rotational grazing in terms of mass flux, as in Fig. 11.4a, have proved elusive. This largely reflects that continuous grazing is itself a series of discrete defoliations at a bite scale, and much of the merits of different grazing systems lies more in the capacity they offer to manipulate utilization from the *animals*' viewpoint, and to control heterogeneity.

Principles of Utilization – from an Animal Perspective

Operational difficulties in acquiring food

Although humans regard utilization as a problem to be solved on a paddock by paddock basis, grazing animals are constrained to meeting their dietary requirements bite by bite. The animal is therefore faced with the problem of overcoming the physical constraints of 'searching' and 'handling' discrete

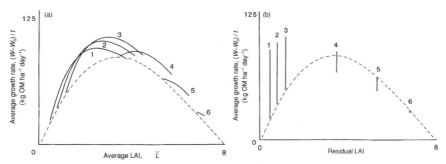

Fig. 11.4. The fundamental relationship (a) between average growth rate and the *average* leaf area index (LAI) of intermittently defoliated pastures regrowing from 6 contrasting residual sward states (same as in Fig. 11.3), and that for continuous grazing (dotted line). Note: there can be *no* simple fundamental relationship between residual sward state *alone* and yield, as a wide range of average rates of yield will be achieved (see b) depending on the duration of regrowth (after Parsons *et al.*, 1988).

items of food. Understanding the principles of utilization and identifying the optimal solution from the animals' viewpoint again borrows much from ecology – notably from the seminal work on functional responses, by Holling (1959).

The development of techniques for monitoring and analysing the jaw movements of animals grazing freely at pasture (e.g. Penning *et al.*, 1984; Rutter *et al.*, 1997) has led to extensive data on the grazing process under a wide range of circumstances. Progress was been made by identifying two kinds of bites during grazing: *prehension* bites during which the jaws open, encompass and remove material from the pasture, and *mastication* bites or chews (Penning, 1986; Champion *et al.*, 1997) during which the material is manipulated before swallowing. In Fig. 11.5a we describe the fundamental way in which the time (and so energy) costs of 'handling' bites varies with the bite mass taken. The concepts were developed initially with sheep. The equivalent for cattle is shown in Fig. 11.6. Of the two costs associated with grazing, first there is a relatively fixed time cost associated with opening and closing the jaw in 'prehension'. The cost is not only remarkable in the way that it varies little as bite mass increases, and is the same when animals graze different plant species, but in that there is very little variance about the mean prehension time. One possible explanation is that the time cost reflects a fixed muscle contraction rate. The fixed cost differs however between cattle and sheep. In sheep, it has been measured at some 0.37 s (Penning *et al.*, 1995). In cattle, with longer jaws (and wider gape) the prehension time cost is greater, typically 0.5–0.7 s, but equally independent of bite mass (Orr *et al.*, 1996; Gibb *et al.*, 1997; see Fig. 11.6). Second, there is an additional time cost associated with 'mastication'. The limited available data for cattle and sheep suggest this increases linearly with bite mass.

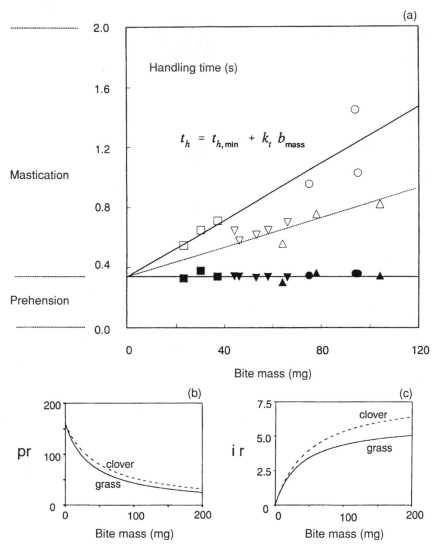

Fig. 11.5. Physical constraints to harvesting grass bite by bite. The fundamental effects of the mass of the bite taken (a) on the time costs associated with prehending and masticating bites, and the consequences of this on the prehension bite rate (b) and instantaneous intake rate (c). Note: small bites are handled less efficiently than large bites and this limits the capacity to compensate for low bite mass by increasing bite rate (after Parsons *et al.*, 1994). The principles are established here using data for sheep grazing grass (solid sloping line, open symbols) or clover (broken sloping line, open symbols). Variations in bite mass from an identical sward were generated by brief periods of fasting (○△●▲) and no fasting (□▽■▼), see Newman *et al.*, 1994. For contrast in cattle see Fig. 11.6.

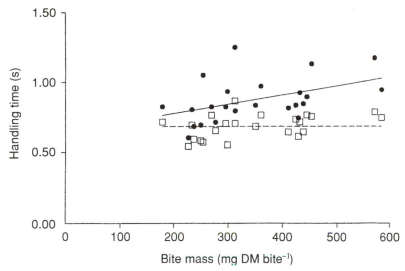

Fig. 11.6. Compared to sheep (Fig. 11.5) the time taken to prehend a bite (broken line, open symbols) is greater for cattle, but cattle appear to spend less time masticating per unit bite mass (solid line, closed symbols). Models propose this is essential to account for the far greater intake rates of cattle compared to sheep (data by courtesy of Gibb *et al.*, 1997).

The relationships between prehension bite rate and bite mass, and between intake rate and bite mass, consistent with Fig. 11.5a, are shown in Fig. 11.5b and c. It can be seen that small bites are handled less efficiently than larger bites, because when grazing small bites there is a larger element of the fixed cost. Although prehension bite rate increases as bite mass declines (see Fig. 11.5b) it is this inefficiency with small bites that explains why the increase in bite rate cannot compensate for the decrease in bite mass, and so why intake rate from short swards (where bite mass is small) is reduced. For further explanation and references, see Parsons *et al.* (1994), Newman *et al.* (1994), Spalinger and Hobbs (1992) and reviews by Ungar (1996), Laca and Demment (1996) and Illius and Hodgson (1996).

Although the principles are unchanged, there are important differences between animals of different body size (e.g. cattle vs. sheep), which interacts with the plant species being consumed. First, the shape of the relationship between intake rate and bite mass (Fig. 11.5c) implies there is only limited scope to increase intake rate by increasing bite mass. The larger bite masses observed in cattle, compared with sheep, cannot therefore be used to explain how cattle achieve such high intrinsic intake rates. To account for the higher intake rates of cattle compared with sheep, the time costs of masticating have to be reduced to less than 2000 s kg⁻¹ for cattle, from the 10,000 s kg⁻¹ (for sheep grazing grass) and 7500 s kg⁻¹ (for sheep grazing clover) seen in sheep. This prediction by a model has subsequently been shown to be the case (Gibb *et al*, 1997; Fig. 11.6). Second, as cattle spend so much less time masticating, there may be less scope for cattle to modify intake rate by reducing mastication time, and so much less scope to benefit from the reduced time required to masticate, e.g. clover compared with grass (Orr *et al.*, 1996; Rutter *et al.*, 1996). The implications of this for identifying the specific constraints faced by different classes of stock, notably high genetic merit cattle, under different grazing managements, has yet to be fully exploited.

Additional costs associated with selective grazing

A further limitation to the capacity of animals to utilize a pasture is that the sward may not contain the desired components (e.g. different plant parts, or species of contrasting nutrient value) in the required proportions. The potential effects of selective grazing on the costs of foraging and on instantaneous intake rate is described in separate models by Spalinger and Hobbs (1992), and has also been addressed by incorporating 'searching' time into the analysis presented above (Parsons *et al.*, 1994; Thornley *et al.*, 1994). Let us consider the worst case, where the alternative food items are distributed on a small scale and spatially at random. Here the potential increase in costs of selective grazing can be substantial (Fig. 11.7). For example, if an animal insists on harvesting a 66% diet of clover (a 2 : 1 ratio of clover to grass) from a sward that contains only 16% clover (a 1 : 5 ratio of clover to grass), then foraging costs can be shown to increase fourfold. This is because to achieve this change in ratio, it can be calculated that only three out of 12 of the potential bites encountered must actually be harvested. Nine out of ten potential bites of grass, or 9 out of 12 potential bites, must be 'passed by'. Few measurements of selective foraging costs at pasture have been made, but it is extremely unlikely that animals would ever entertain such a burden. Rather, these principles can be used to predict the pattern of trade-offs (between selection and its costs) that animals must make to maintain the cost effectiveness of grazing and so why animals appear not to fully utilize all plant components available in a pasture. An account of the cost effectiveness of selective grazing and its impact on daily energy budgets is presented by Thornley *et al.* (1994) and Parsons *et al.* (1994).

When plant species/components are distributed more typically in patches, animals can reduce the costs of selective grazing by using spatial memory (Edwards *et al.,* 1996a) as well as visual cues (Edwards *et al.,* 1997) to locate the preferred food items. This offers theoretical prospects for improving nutrient utilization from pastures by providing more marked spatial arrangements (i.e. heterogeneity) of important alternative forages.

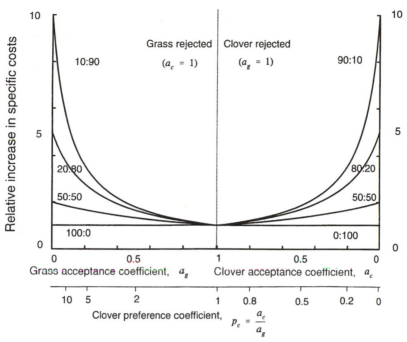

Fig. 11.7. Selective grazing can substantially increase the costs of foraging as calculated here for swards of four contrasting ratios of clover to grass, 100 : 0, 50 : 50, 20 : 80 and 10 : 90. Selective grazing (e.g. for clover, see left side of graph) implies that animals should accept all bites of the preferred species (so *acceptance* coefficients a_c = 1 for clover), but reject a proportion of the potential bites of the less desired forage species (so a_g < 1 for grass). Costs increase substantially with the proportion of less desired bites that is rejected (for full explanation see Thornley *et al.*, 1994).

Pasture utilization at the bite scale

The mass of a bite clearly depends on the area, depth (and 3D shape) of the indentation the bite makes in the sward, and the bulk density (g m^{-3}) of the region of the sward removed. Of these components, one of the most notable features of how domestic ruminants graze, that has only recently become known, is that each bite removes a relatively fixed proportion of the total height of the vegetation. The proportion removed at each defoliation clearly has major implications for utilization of the paddock as a whole, though the relationship is not simple.

It was initially assumed that animals attempted to graze to some fixed maximum *depth*, rather than a fixed *proportion* of total sward depth (see 'ramp' function of Demment *et al.*, 1987; Ungar and Noy-Meir, 1988). The maximum bite depth was assumed to depend on jaw length, and that cattle would therefore graze deeper e.g. than sheep. However recent studies show how under a wide range of managements, the mean proportion of height removed at each bite is relatively fixed (Fig. 11.8) though the proportion varies between studies from some 35% to 70% (see Wade *et al.*, 1989; Ungar *et al.*, 1991; Laca *et al.*, 1992). Bite depths do not differ substantially between cattle and sheep grazing the same sward (e.g. Orr *et al.*, 1997), the major difference in bite mass between cattle and sheep being explained therefore by the larger bite *area* of cattle. Edwards *et al.* (1996b) showed that the proportion removed was also similar when animals grazed from clover or from grass.

As yet there is no definitive explanation for the optimal bite depth being a 'fixed' intermediate proportion, but some insight into why animals only partially exploit sward depth and how they optimize the combination of bite area and depth might be gained from Fig. 11.5. Having paid the 'searching time' costs of getting to the place where the bite is to be taken, and given the fixed time costs of opening and closing the jaw in prehension (note that, at some 30,000–58,000 bites day^{-1}, this cost is appreciable,

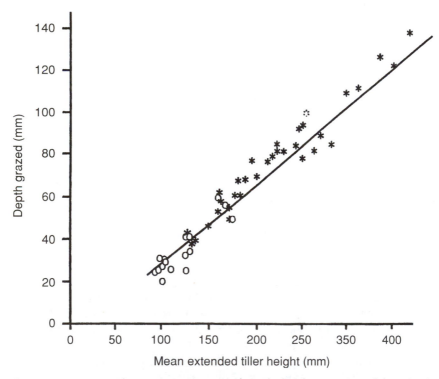

Fig. 11.8. Dairy cattle appear to graze a relatively fixed proportion of the sward height. The relationship between the depth of grazing within the sward and the extended height of grazed tillers is shown for continuous (O) and paddock (∗) grazing systems (from Wade *et al.*, 1989).

approximately 3–6 h!) it follows that the most cost-effective strategy is to take as large a bite as possible, provided that this does not compromise the capacity to handle (masticate and ruminate) it. For a given bite area, taking shallow bites would lead to small bite mass and would not be cost effective energetically. Taking very deep bites could be precluded as the animals would encounter the pseudostem and dead material accumulated at the base of the sward and the increased handling (and ruminating) cost and poorer nutritive value could deter grazing this lower horizon. Illius *et al.* (1995) calculated the energy costs and benefits of prehending and masticating material from swards of five grass species differing in the energy required to sever and handle a bite (a function of both tensile strength and the population density of tillers). They, too, argue for a more long-term cost–benefit basis for the control of biting. They dismiss the notion of simple physical limits to the forces animals can exert being a determinant of bite depth, but did find bite dimensions (including depth) varied depending on energy costs, and between different sized animals.

Utilization at the bite scale to utilization at the paddock scale

We may be no closer to a definitive explanation for the control of bite depth, but in terms of establishing usable principles of utilization, the most important feature is to realize that the proportion removed at each bite station is considerably greater than the proportion that could be removed sustainably across the area of the sward as a whole.

For example, to maintain LAI, leaf loss by defoliation and senescence must match gross leaf production. It can readily be estimated (whether from intake per hectare as a fraction of standing crop mass; or from the appearance, turnover and fate of portions of leaves) that to maintain the optimal LAI, the amount removed averaged homogeneously across the entire grazed area is equivalent to only some 2.5% of the standing mass at any point in the sward. Given that dairy cattle remove closer to 35% at each point they graze, it is readily apparent that animals must confine their grazing to only a small proportion of the total grassland area in any one day – this calculation suggests some 2.5/35 = 7%. This is in keeping with the few measurements made of the spatial distribution of grazing across fields under a wide range of managements (e.g. Wade, 1991; see Fig. 11.9).

The significance of this to utilization is that it is inevitable that heterogeneity will arise in the height of the sward. In practice, encouraging animals optimally to remove ('utilize') all the accumulated herbage, uniformly across the total grazed area sustainably over time, is the overriding intention of the land manager, and is widely seen as a critical determinant in choosing between continuous or rotational grazing. Yet, in principle, all these two systems do is to distribute the heterogeneity in different ways spatially and

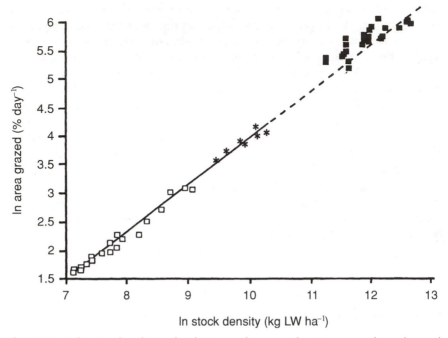

Fig. 11.9. The simple relationship between the area of pasture grazed per day and stock density for continuous (□), paddock (∗) and strip (■) grazing systems (ln, natural logarithm; LW, live weight) (from Wade, 1991).

temporally. This offers contrasting, rather than fundamentally different, opportunities to control utilization.

The role of heterogeneity in utilization

The potential impact of heterogeneity on grass utilization is massive, but at the same time, heterogeneity can be irrelevant, even beneficial to utilization. For instance, the fact that some portions of the total grazed area are currently tall, while others are short, does not in itself imply poor utilization. Indeed rotational grazing is designed specifically to generate such heterogeneity on a paddock-to-paddock scale. In a rotationally grazed system, if the optimum timing of harvest were 14 days, then every paddock in succession could in theory be harvested in sequence at precisely 14 days (there is therefore no spatial or temporal component to the *variance* about the mean defoliation interval). The fact that some areas are tall while others are short simply reflects that the defoliations in the different areas are out of phase. There is no loss in utilization due to heterogeneity. Likewise, the presence of tall and short areas in a continuously grazed pasture need not imply poor utilization. The presence of clearly discernible patches of tall (or otherwise desirable) plant material is actually beneficial in reducing the problems of locating preferred food when grazing selectively (Edwards *et al.*, 1994).

What does matter in either grazing system is if some areas (be they paddocks or patches) are grazed too soon, or too late, relative to the optimum timing of harvest, and so there is a temporal, and spatial, component to *variance* about the mean defoliation interval. Under these circumstances, heterogeneity can be seen to have a direct major effect in reducing utilization. If the timing of harvest on 50% of the total area (be they paddocks or bites) is such that the average growth rate there is reduced to 50% of the maximum, then utilization at the paddock scale is reduced by 25%. Major losses can arise if some areas reach ceiling yield and contribute nothing to average growth rates. One phenomenon that distinguishes detrimental from benign heterogeneity, then, is that tall areas would not 'move around'. Despite concern over its link to utilization, the origin and control of heterogeneity is probably the most poorly understood feature of grassland science. Very few studies (Edwards, 1994) have measured heterogeneity, or make the necessary spatial analyses to determine its relevance.

Principles of Growth and Utilization – from a Plant Morphology Perspective

Much information on the morphology, and morphogenesis, of pastures and pasture plants has been amassed during the quest to *describe* the processes by which pastures accumulate yield. Those processes have been thoroughly reviewed in recent publications (Brock and Hay, 1993; Chapman and Lemaire, 1993; Briske, 1996; Lemaire and Chapman, 1996), and we will not retrace the same ground here in any detail. Rather, we focus on aspects of the *structure* of the sward as determined by plant number/plant size relationships, and draw out some of the implications for management of grass-based pastures that arise from these relationships.

Morphogenesis

Morphogenesis can be defined as the generation and expansion of plant form in space and time (Chapman and Lemaire, 1993). It includes the rate of appearance of new organs such as leaves and tillers; their rate of expansion; and their longevity and fate (intake or senescence). In doing so it considers the consequent population densities (numbers per unit ground area of leaves, tillers) and the size of individual components.

In one sense, a knowledge of the appearance, turnover and fate of plant organs leads to an understanding of growth and utilization that is the same as what we have already described in terms of a mass flux at a whole paddock scale – but simply in a different 'currency' (numbers of leaves and tillers, rather than grams of dry matter). Morphogenesis here is the plant detail of how the flux of carbon is expressed. Thus, there is much in common between the mass flux and morphogenetic approaches to analysing yield, since they both subsume the same basic processes of yield accumulation. However, we note that, in contrast to the mass flux approach, no integrating mathematical models of pasture yield have been constructed using the currencies of leaf and tiller number. We contend that modelling mass flux offers a more powerful way of identifying management systems that provide an optimal solution to both plant and animal requirements (e.g. Parsons and Penning, 1988; Parsons *et al.*, 1988). On the other hand, the morphogenetic approach provides critical understanding of *how* (but not necessarily *why*) different sward structures develop under different grazing regimes and the associated management implications, and a useful conceptual approach to identifying the management needs of different plant genotypes. It also provides the foundation for dealing with the problems of plant persistence under grazing, using either management or breeding strategies or combinations of both.

Size/density relationships

Important relationships between plant size and density operate both within and between genotypes. Within genotypes, sward structure responds dynamically to variation in average LAI. Pastures maintained at low LAI characteristically develop large numbers of small plants/tillers – a state that increases their capacity to survive defoliation (Chapman and Lemaire, 1993; Briske, 1996). Pastures maintained at high LAI even for relatively short periods maintain fewer, larger plants – a state that arguably increases their capacity to compete for light with other plants in a closed canopy.

   Thus the structure of grass-dominant swards revolves around a fundamental, inverse relationship between sward density and plant size (e.g. Grant *et al.*, 1983; Lambert *et al.*, 1986; Davies, 1988) which arises from the morphogenetic processes operating at the individual plant level. The size of new organs produced reflects the canopy environment into which they are emerging as plants adapt to maintain a balance between resource availability and resource allocation patterns that maximizes their long-term fitness. Note, however, that differences in sward structure do not necessarily lead to differences in primary productivity, since size and density can trade off to give more-or-less equal herbage accumulation (e.g. the comparison between continuously grazed swards at different LAI (Bircham and Hodgson, 1983), the comparison between continuously and intermittently grazed pastures (Chapman and Clark, 1984; Brock and Hay, 1993) or between sheep- and cattle-grazed pastures (Lambert *et al.*, 1986; but see Matthew *et al.*, 1995, 1996 for a more-detailed analysis of density–production relationships). It is commonly argued, though, that yield will be more stable in the long term if pastures are managed to encourage density, since recovery of the pasture and attainment of a closed canopy from a low LAI or low mass is more assured and likely to be faster if the number of 'growth units' per unit ground area is greater (e.g. Sheath *et al.*, 1987; Brock and Hay, 1993).

Different genotypes also characteristically develop different sward structures that may lead to different optimal defoliation management requirements. Tall fescue (*Festuca arundinacea*) and perennial ryegrass (*Lolium perenne*) are examples of species with different morphologies and (causally?) different management requirements. Both species maintain a similar, genetically determined constant number of live leaves per tiller. *L. perenne* has a relatively high leaf appearance rate (110°C days) and, hence, a potentially high rate of production of new tillers, and tends to develop dense swards based on a large number of small tillers. *F. arundinacea* has relatively low leaf appearance rates (230°C days) and is characterized by a relatively open sward structure comprised of fewer, larger tillers. At a constant LAI of 3, Mazzanti *et al.* (1994) recorded tiller densities in the range 4000–6000 m^{-2} in tall fescue, whereas perennial ryegrass swards maintained at similar LAI generally support tiller densities in the range 10,000–15,000 m^{-2}.

However, as in the discussion above, none of this implies greater primary productivity of one genotype compared with the other. Indeed, in this example, 4000–6000 tiller m^{-2} of tall fescue and 10,000–15,000 tillers m^{-2} of ryegrass are presumably yielding similarly, since they are each supporting a similar LAI, and therefore, presumably, similar rates of gross photosynthesis. Primary productivity differences would only be expected if factors such as leaf orientation and light interception, respiratory costs, or the rate of regeneration of LAI following leaf removal differed. There is no evidence to suggest that the first two of these factors differ between species but the rate of regeneration of LAI will differ (since there are intrinsic differences in leaf appearance rates), so that each requires different defoliation management to maximize light interception and growth over repeated defoliation sequences in an intermittent grazing system. These relatively subtle genotype × management interactions are not usually accommodated in the design and conduct of comparative agronomic studies (e.g. Johnson and Thomson, 1996). Not surprisingly, such studies have generally shown similar levels of primary productivity, and do not reveal the situations or management regimes options under which yield advantages might be available.

Plant morphogenesis – opportunities

While a lot of descriptive information on plant and pasture morphology is available, the fascinating challenge is to understand why species differ so substantially in the combination of numbers and size of units about which

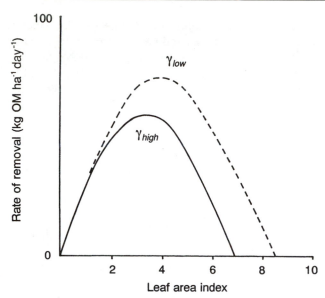

Fig. 11.10. Modifying morphogenesis may increase the maximum average rate of yield and increase the sward state at which this is achieved. See the effect on the rate of removal of material that may be sustained, in swards maintained at each of a range of leaf area index, of a change in the proportion 'γ' of the standing live weight of the grass sward lost per day in senescence. $\gamma_{low} = 0.030$ (longevity =33.3 days leaf^{-1}) : $\gamma_{high} = 0.0375$ (longevity = 26.6 days leaf^{-1}) (from Parsons, 1994).

they operate under a given management, as well as in their capacity to modify their morphology. Least well understood is what are the advantages to production, or to plant fitness, that different strategies for morphogenesis provide. These questions, we suggest, represent important opportunities for future research, and potentially for future grass-based dairy production systems through better management options or improved plant material.

One such avenue is to reduce the rate of senescence (in this case the proportion, 'γ', of the standing crop that is lost in senescence each day) by decreasing the rate of turnover of material. This outcome is shown (Fig. 11.10) to not only increase the maximum possible rate of yield, but also to cause that yield to be achieved at a higher mean leaf area. Increasing the LAI optimum for yield per hectare may lead to a closer coincidence of optimal yield per hectare and per head, which could be particularly important for larger livestock such as dairy cattle.

From Principles to Practice

We have established from first principles how maximizing utilization depends on achieving and sustaining an optimal mean vegetation state averaged over time across the entire grassland area. In addition, for rotational grazing/cutting we can specify a theoretically optimal timing of harvest that combined with severe defoliations helps ensure the maximum average rate of yield. However, grassland managers are accustomed to making decisions based more on stock density, the number of paddocks and the 'speed' of rotation. And yet, in terms of a systems analysis, where the goal is to maximize growth and utilization of grass, these measures are *output* variables – i.e. values that should be determined by, and so emerge from, the main driving factors of weather and plant growth. The more these are fixed, then the more the system is constrained.

Many recent developments in grazing management guidelines recognize that, given the vagaries of weather, seasonal changes in plant growth and in intake demand (e.g. with stage of lactation and calving pattern), it is extremely unlikely that a stock density and paddock management pattern pre-determined at the start of the season will achieve optimal utilization of pasture. Moreover, it is increasingly recognized that no one, single, criteria can sufficiently define the regrowth and performance of the pasture, as for instance, when a residual sward state is used, a huge range of average rates of yield can be achieved depending on how long the sward is allowed to regrow (see Fig. 11.4b). However, the following approaches have been used to integrate the principles of optimal utilization with criteria for practical management decisions.

Under 'continuous' grazing (or where animals have access to one or more paddocks for long periods) practical guidelines have been developed (notably in UK and NZ) based on monitoring and maintaining a given sward state – variously measured in terms of a sward surface height or its equivalent in a trained 'eye-estimate' of herbage mass, or leaf area index. To increase flexibility to meet changing demands for intake, and to allow for some accumulation of feed to hedge against risk, e.g. of dry periods, the guidelines specify a desired seasonal profile of sward state. For management guidelines see Chapter 12.

In practice, sward state is sustained close to the optimum by varying the proportion of the total grass area that is actually grazed: the remaining area is used either as a 'buffer' feed (animals are allowed back in if there are food-deficiencies in the short term), or is cut for conservation. Managements of this nature provide all the necessary signals for making long-term, strategic as well as shorter-term, tactical management decisions e.g. on stock density and fertilizer policy as well as on the overall balance of grazing and conservation/supplementary feeding on the farm.

Under rotational grazing, despite its familiarity, there is no recognized fundamental principle for determining the optimal number of paddocks and the speed of rotation (Jelbart, 1996; Doyle *et al.*, 1996). Moreover, these variables are far from independent of one another, or from measures such as the severity of defoliation, and the optimal timing of harvest in each paddock.

Despite long standing controversy over the differences and relative merits of continuous vs. rotational grazing, some recent practical guidelines valuably recognize their fundamental similarities and the principles of the origin and control of heterogeneity described earlier. For example, in the 'Target 10' system advocated in Australia (Curtis and O'Brien, 1994), grazing decisions under rotational grazing are again made on the basis of what proportion of the total farm area needs to be allocated to grazing each day in order to keep that area (and so each area in turn) at the desired mean state – which in this case requires identifying when animals enter that area (so the optimal timing of harvest at the end of a regrowth), the residual sward state to which it is grazed, and the proportion of the total grass area (could be several paddocks at once) to be opened up in order to achieve this. Note that in this system, the numbers of separate grazing areas (c.f. 'paddocks') and the speed of rotation are indeed emergent variables. Again this general approach provides all necessary signals for decisions regarding the balance of grazing and conservation feeding, with rotational grazing arguably offering greater scope to 'ration' feed allocation. In all year-round grazing systems it can be necessary to restrict feed supply strategically overwinter.

There are many grazing schemes. Many work owing to the personal skills of the manager in responding to continual disruptions to his/her grazing strategy imposed notably by bad weather. What has been attempted over the last 25 years of grassland research is to establish the principles of growth and utilization, as a knowledge of these, rather than a fixed set of axioms and rules, will allow a greater proportion of managers to make the appropriate short-term tactical adjustments to sustain the optimal contribution of fresh forage to dairy production.

References

Anslow, R.C. (1967) Frequency of cutting and sward production. *Journal of Agricultural Science, Cambridge* 68, 377–384.

Bircham, J.S. and Hodgson, J. (1983) The influence of sward conditions on rates of herbage growth and senescence in mixed swards under continuous stocking management. *Grass and Forage Science* 38, 323–331.

Briske, D.D. (1996) Strategies of plant survival in grazed systems: a functional interpretation. In: Hodgson, J. and Illius, A.W. (eds) *The Ecology and Management of Grazing Systems.* CAB International, Wallingford, pp. 37–67.

Brock, J.L. and Hay, R.J.M. (1993) An ecological approach to forage management. In: Baker, M.J. *et al.* (eds) *Proceedings of the XVIIIth International Grassland Congress, 8–21 February, 1993, Palmerston North, NZ.* New Zealand Grassland Association, Palmerston North, pp. 837–842.

Brougham, R.W. (1956) Effect of intensity of defoliation on regrowth of pasture. *Australian Journal of Agricultural Research* 7, 377–387.

Champion, R.A., Rutter, S.M. and Orr, R.J. (1997) Distinguishing bites and chews in recordings of the grazing jaw movements of cattle. In: *Proceedings of the 5th*

Research Meetings of the British Grassland Society. British Grassland Society, Reading, pp. 171–172.

Chapman, D.F. and Clark, D.A. (1984) Pasture responses to grazing management in hill country. *Proceedings of the New Zealand Grassland Association* 45, 168–176.

Chapman, D.F. and Lemaire, G. (1993) Morphogenetic and structural determinants of plant regrowth after defoliation In: Baker, M.J. *et al.* (eds) *Proceedings of the XVIIIth International Grassland Congress, 8–21 February, 1993, Palmerston North, NZ.* New Zealand Grassland Association, Palmerston North, pp. 95–104.

Charnov, E.L. (1976) Optimal foraging, the marginal value theorem. *Theoretical Population Biology* 9, 129–136.

Curtis, A. and O'Brien, G. (1994) *Pasture Management for Dairy Farmers, 'Target 10'.* Victorian Department of Agriculture, Melbourne, 68pp.

Davidson, J.L. and Milthorpe, F.L. (1966a) Leaf growth in *Dactylis glomerata* following defoliation. *Annals of Botany* 30, 173–184.

Davidson, J.L. and Milthorpe, F.L. (1966b) The effect of defoliation on the carbon balance in *Dactylis glomerata. Annals of Botany* 30, 185–198.

Davies, A. (1988) The regrowth of grass swards. In: Jones, M.B. and Lazenby, A. (eds) *The Grass Crop: The Physiological Basis of Production.* Chapman and Hall, London, pp. 85–127.

Demment, M.W., Laca, E.A. and Greenwood, G.B. (1987) Intake in grazing ruminants: a conceptual framework. In: Owens, F.N. (ed.) *Symposium Proceedings: Feed Intake by Cattle.* Agricultural Experimental Station, Oklahoma State University, Stillwater, Oklahoma, pp. 208–225.

Donald, C.M. (1961) Competition for light in crops and pastures. In: Milthorpe, F.L. (ed.) *Mechanisms in Biological Competition.* Cambridge University Press, New York, pp. 283–313.

Doyle, P.T., Stockdale, C.R. and Lawson, A.R. (1996) *Pastures for Dairy Production in Victoria.* Agriculture Victoria, Kyabram, 56pp.

Edwards, G.R. (1994) The creation and maintenance of spatial heterogeneity in plant communities: the role of plant-animal interactions. DPhil thesis, University of Oxford, Oxford.

Edwards, G.R., Newman, J.A., Parsons, A.J. and Krebs, J.R. (1994) Effects of scale and spatial distribution of the food resource and animal state on diet selection: an example with sheep. *Journal of Animal Ecology* 63, 816–826.

Edwards, G.R., Newman, J.A., Parsons, A.J. and Krebs, J.R. (1996a) The use of spatial memory by grazing animals to locate food patches in spatially heterogeneous environments: an example with sheep. *Applied Animal Behaviour Science* 50, 147–160.

Edwards, G.R., Parsons, A.J., Penning, P.D. and Newman, J.A. (1996b) Relationship between vegetation state and bite dimensions of sheep grazing contrasting plant species and its implications for intake rate and diet selection. *Grass and Forage Science* 50, 378–388.

Edwards, G.R., Newman, J.A., Parsons, A.J. and Krebs, J.R. (1997) Use of cues by grazing animals to locate food patches: an example with sheep. *Applied Animal Behaviour Science* 51, 59–68.

Gibb, M.J., Huckle, C.A., Nuthall, R. and Rook, A.J. (1997) Effect of sward surface height on intake and grazing behaviour by lactating Holstein Friesian cows. *Grass and Forage Science* 52, 309–321.

Grant, S.A., Barthram, G.T., Torvill, L., King, J. and Smith, H.K. (1983) Sward management, lamina turnover and tiller population density in continuously stocked *Lolium perenne* dominated swards. *Grass and Forage Science* 38, 333–344.

Harris, W. (1978) Defoliation as a determinant of the growth, persistence and composition of pasture. In: Wilson, J.R. (ed.) *Plant Relations in Pastures*. CSIRO, Melbourne, pp. 67–85.

Holling, C.S. (1959) Some characteristics of simple types of predation and parasitism. *Canadian Entomologist* 91, 385–398.

Hunt, L.A. (1965) Some implications of death and decay in pasture production. *Journal of the British Grassland Society* 20, 27–31.

Illius, A.W., Gordon, I.J., Milne, J.D. and Wright, W. (1995) Costs and benefits of foraging grasses varying in canopy structure and resistance to defoliation. *Functional Ecology* 9, 894–903.

Illius, A.W. and Hodgson, J. (1996) Progress in understanding the ecology and management of grazing systems. In: Hodgson, J. and Illius, A.W. (eds) *The Ecology and Management of Grazing Systems*. CAB International, Wallingford, pp. 429–457.

Jelbart, M. (1996) Putting it all together – integrating efficient management, environmental sustainability and maximum productivity. *Proceedings of the 37th Annual Conference of the Grassland Society of Victoria*, pp. 88–93.

Johnson, I.R. and Parsons, A.J. (1985) A theoretical analysis of grass growth under grazing. *Journal of Theoretical Biology* 112, 345–367.

Johnson, R.J. and Thomson, N.A. (1996) Effects of pasture species on milk yield and composition. *Proceedings of the New Zealand Grassland Association* 57, 151–156.

King, J., Sim, E.M. and Grant, S.A. (1984) Photosynthetic rate and carbon balance of grazed ryegrass pastures. *Grass and Forage Science* 39, 81–92.

Laca, E.A. and Demment, M.W. (1996) Foraging strategies of grazing animals. In: Hodgson, J. and Illius, A.W. (eds) *The Ecology and Management of Grazing Systems*. CAB International, Wallingford, pp. 137–158.

Laca, E.A., Ungar, E.D., Seligman, N. and Demment, M.W. (1992) Effects of sward height and bulk density on bite dimensions of cattle grazing homogeneous swards. *Grass and Forage Science* 47, 91–102.

Lambert, M.G., Clark, D.A., Grant, D.A., Costall, D.A. and Gray, Y.S. (1986) Influence of fertiliser and grazing management on North Island moist hill country. 4. Pasture species abundance. *New Zealand Journal of Agricultural Research* 29, 23–31.

Lemaire, G. (1988) Sward dynamics under different management programmes. *Proceedings of the 12th General Meeting of the European Grassland Federation*, Irish Grassland Association, Dublin, pp. 7–22.

Lemaire, G. and Chapman, D.F. (1996) Tissue flows in grazed plant communities. In: Hodgson, J. and Illius, A.W. (eds) *The Ecology and Management of Grazing Systems*. CAB International, Wallingford, pp. 3–36.

McCree, K.J. and Troughton, J.H. (1966a) Prediction of growth rate at different light levels from measured photosynthesis and respiration rates. *Plant Physiology* 41, 559–566.

McCree, K.J. and Troughton, J.H. (1966b) Non-existence of an optimum leaf area index for the production rate of white clover grown under constant conditions. *Plant Physiology* 41, 1615–1622.

McMeekan, C.P. (1960) Grazing management. In: *Proceedings of the VIIIth International Grassland Congress,* British Grassland Society, Reading, pp. 21–27.

Matthew, C., Lemaire, G., Sackville-Hamilton, N.R. and Hernandez-Garay, A. (1995) A modified self-thinning equation to describe size/density relationships for defoliated swards. *Annals of Botany* 76, 579–587.

Matthew, C., Hernandez-Garay, A. and Hodgson, J. (1996) Making sense of the link between tiller density and pasture production. *Proceedings of the New Zealand Grassland Association* 57, 83–87.

Mazzanti, A. and Lemaire, G. (1994) Effect of nitrogen fertilization on herbage production of tall fescue swards continuously grazed by sheep. 2. Consumption and efficiency of herbage utilization. *Grass and Forage Science* 49, 352–359.

Mazzanti, A., Lemaire, G. and Gastal, F. (1994) The effect of nitrogen fertilisation on the herbage production of tall fescue swards grazed continuously with sheep. 1. Herbage growth dynamics. *Grass and Forage Science* 49, 111–120.

Morley, F.H.W. (1966) Stability and Productivity of Pastures. *Proceedings of the New Zealand Society of Animal Production,* 26, 8–21.

Newman, J.A., Parsons, A.J. and Penning, P.D. (1994) A note on the behavioural strategies used by grazing animals to alter their intake rates. *Grass and Forage Science* 49, 502–505.

Orr, R.J., Rutter, S.M., Penning, P.D., Yarrow, N.H. and Champion, R.A. (1996) Grazing behaviour and herbage intake by Friesian dairy heifers grazing ryegrass or white clover. In: Younie, D. (ed.) *Legumes in Sustainable Farming Systems.* Joint SFS/BGS Conference, Occasional Symposium of the British Grassland Society No. 30, pp. 221–224.

Orr, R.J., Harvey, A., Kelly, C.L. and Penning, P.D. (1997) Bite dimensions and grazing severity for cattle and sheep. In: *Proceedings of the 5th Research Meetings of the British Grassland Society.* British Grassland Society, Reading, pp. 185–186.

Parsons, A.J. (1988) The effect of season and management on the growth of temperate grass swards. In: Jones, M.B. and Lazenby, A. (eds) *The Grass Crop – The Physiological Basis of Production.* Chapman and Hall, London, pp. 129–177.

Parsons, A.J. (1994) Exploiting Resource Capture – Grassland. In: Monteith, J.L., Scott R.K. and Unsworth, M.H. (eds) *Resource Capture by Crops.* Nottingham University Press, Sutton Bonnington, pp. 315–349.

Parsons, A.J. and Penning, P.D. (1988) The effects of the duration of regrowth on photosynthesis, leaf death and the average rate of growth in a rotationally grazed sward. *Grass and Forage Science* 43, 15–27.

Parsons, A.J., Leafe, E.L., Collett, B. and Stiles, W. (1983a) The physiology of grass production under grazing. 1. Characteristics of leaf and canopy photosynthesis of continuously grazed swards. *Journal of Applied Ecology* 20, 11–126

Parsons, A.J., Leafe, E.L., Collett, B., Penning, P.D. and Lewis, J. (1983b) The physiology of grass production under grazing. 2. Photosynthesis, crop growth and animal intake of continuously grazed swards. *Journal of Applied Ecology* 20, 127–139.

Parsons, A.J., Johnson, I.R. and Harvey, A. (1988) Use of a model to optimize the interaction between the frequency and severity of intermittent defoliation and to provide a fundamental comparison of the continuous and intermittent defoliation of grass. *Grass and Forage Science* 43, 49–59.

Parsons, A.J., Harvey, A. and Woledge, J. (1991) Plant/animal interactions in continuously grazed mixtures: 1. Differences in the physiology of leaf expansion and the fate of leaves of grass and clover. *Journal of Applied Ecology* 28, 619–634.

Parsons, A.J., Thornley, J.H.M., Newman, J.A. and Penning, P.D. (1994) A mechanistic model of some physical determinants of intake rate and diet selection in a two-species temperate grassland sward. *Functional Ecology* 8, 187–204.

Penning, P.D. (1986) Some effects of sward conditions on grazing behaviour and intake by sheep. In: Gudmundsson, O. (ed.) *Grazing Research at Northern Latitudes*. Plenum Press, New York, pp. 219–226.

Penning, P.D., Steel, G.L. and Johnson, R.H. (1984) Further development and use of an automatic recording system in sheep grazing studies. *Grass and Forage Science* 39, 345–351.

Penning, P.D., Parsons, A.J., Orr, R.J., Harvey, A. and Champion, R.A. (1995) Intake and behavior responses by sheep, in different physiological states, when grazing monocultures of grass or white clover. *Applied Animal Behaviour Science* 45, 63–78.

Pollock, C.J. and Jones, T. (1979) Seasonal patterns of fructosan metabolism in forage grasses. *New Phytologist* 83, 9–15.

Richards, J.H. (1993) Physiology of plants recovering from defoliation. Session: Plant Growth. In: Baker, M.J. *et al.* (eds) *Proceedings of the XVIIIth International Grassland Congress, 8–21 February, 1993, Palmerston North, NZ*. New Zealand Grassland Association, Palmerston North, Vol. I, pp. 85–94.

Robson, M.J. (1973a) The growth and development of simulated swards of perennial ryegrass. I. Leaf growth and dry weight changes as related to the ceiling yield of a seedling sward. *Annals of Botany* 37, 487–500.

Robson, M.J. (1973b) The growth and development of simulated swards of perennial ryegrass. II. Carbon assimilation and respiration in a seedling sward. *Annals of Botany* 37, 501–518.

Robson, M.J., Ryle, G.J.A. and Woledge, J. (1988) The grass plant – its form and function. In: Jones, M.B. and Lazenby, A. (eds.) *The Grass Crop – The Physiological Basis of Production*. Chapman and Hall, London, pp. 25–83.

Rutter, S.M., Penning, P.D., Champion, R.A. and Orr, R.J. (1996) What can't dairy heifers eats clover quicker than grass? In: Duncan, I.J.H., Widowski, T.M. and Haley, D.B. (eds) *Proceedings of the 30th International Congress of the International Society for Applied Ethology*. Guelph, pp. 134.

Rutter, S.M., Champion, R.A. and Penning, P.D. (1997) An automatic system to record foraging behavior in free-ranging ruminants. *Applied Animal Behaviour Science* 54, 185–195.

Sheath, G.W., Hay, R.J.M. and Giles, K.H. (1987) Managing pastures for grazing animals. In: Nicol, A.M. (ed.) *Livestock Feeding on Pasture*. Occasional Publication No. 10, New Zealand Society of Animal Production, Palmerston North, pp. 65–74.

Smetham, M.L. (1975) The influence of herbage utilization on pasture production and animal performance. *Proceedings of the New Zealand Grassland Association*, pp. 91–103.

Spalinger, D.E. and Hobbs, N.T. (1992) Mechanisms of foraging in mammalian herbivores: new models of functional response. *American Naturalist* 140, 325–348.

Stephens, D.W. and Krebs, J.R. (1986) *Foraging Theory*. Princetown University Press, Princetown, New Jersey, 247pp.

Thornley, J.H.M. and Johnson, I.R. (1990) *Plant and Crop Modelling*. Clarendon Press, Oxford, 669pp.

Thornley, J.H.M., Parsons, A.J., Newman, J. and Penning, P.D. (1994) A cost-benefit model of grazing intake and diet selection in a two species temperate grassland sward. *Functional Ecology* 8, 5–16.

Ungar, E.D. (1996) Ingestive behaviour. In: Hodgson, J. and Illius, A.W. (eds) *The Ecology and Management of Grazing Systems*. CAB International, Wallingford, pp.185–218.

Ungar, E.D. and Noy-Meir, I. (1988) Herbage intake in relation to availability and sward structure: grazing processes and optimal foraging. *Journal of Applied Ecology* 25, 1045–1062.

Ungar, E.D., Genizi, A. and Demment, M.W. (1991) Bite dimensions and herbage intake by cattle grazing short, hand-constucted swards. *Agronomy Journal* 83, 973–978.

Volenec, J.J. (1986) Non-structural carbohydrates in stem base components of Tall Fescue during regrowth. *Crop Science* 26, 122–127.

Wade, M.H. (1991) Factors affecting the availability of vegetative *Lolium perenne* to grazing dairy cows with special reference to sward characteristics, stocking rate and grazing method. PhD thesis, University of Rennes, France.

Wade, M.H., Peyraud, J.L., Lemaire, G. and Comeron, E.A. (1989) The dynamics of daily area and depth of grazing and herbage intake of cows in a five-day paddock system. In: *Proceedings of the XVIth International Grassland Congress*, 4–11 October, 1989, Nice, France. French Grassland Society, Nice, pp. 1111–1112.

Woledge, J. (1977) Effects of shading and cutting treatments on the photosynthetic rate of ryegrass leaves. *Annals of Botany* 41, 1279–1286.

Woledge, J. (1978) The effect of shading during vegetative and reproductive growth on the photosynthetic capacity of leaves in a grass sward. *Annals of Botany* 42, 1085–1089.

Woodward, S.J.R. (1998) Quantifying different causes of leaf and tiller death in grazed perennial ryegrass swards. *New Zealand Journal of Agricultural Research* 41, 149–159.

Grazing Management Systems for Dairy Cattle

12

D.A. Clark[1] and V.R. Kanneganti[2]

[1]Dairying Research Corporation Ltd, Private Bag 3123, Hamilton, New Zealand
[2]USDA-Agricultural Research Service, US Dairy Forage Research Center, Department of Agronomy, University of Wisconsin-Madison, 1925 Linden Drive West, Madison, Wisconsin 53706, USA

Introduction

The scientific study of grazing management received a boost from the writings of Voisin (1959). He strongly advocated the use of rotational grazing with specific intervals of grazing and recovery. These ideas were challenged in the 1950s by McMeekan who developed the concept of small, self-contained farms ('farmlets') to test ideas on farm management. The scale of this approach has often precluded the full replication of treatments, and results from such trials have not received the attention they deserve because of their absence from peer-reviewed scientific journals. However, in New Zealand dairy farming they have played a critical role in the development of an industry based almost solely on grazed pasture. The progression of New Zealand dairying can be traced in the writings of McMeekan (1961), Wallace (1958), Hutton (1968), and Bryant (1990).

The progression of the US dairy industry over the last 50 years follows a much different path. In the 1950s American research showed the advantages to be gained from using land to grow crops such as corn (*Zea mays* L.) and lucerne (*Medicago sativa* L.) for conservation and soybeans for high protein supplement, rather than for grazing dairy cows. In later years, with the availability of heavy machinery for large scale harvesting and processing of crops and with subsequent development of storage technology, dairy cattle grazing declined for the more profitable confinement systems. In the latter system crops were harvested at optimal nutritive value and dairy cows fed indoors with a mixed ration of forage and concentrate that maximized milk production per cow in a predictable way. With this trend, average milk production per cow increased from about 3200 kg in 1960 to 7300 kg in 1994.

The milk price : feed cost ratio has been, and will continue to be, a major determinant of the type of dairy system used. Pasture-based systems are capable of low cost milk production with very high milk output per hectare of land (Penno *et al.*, 1996). In contrast, dairy rations based on crop products are higher cost but are able to support higher per cow milk output than pasture-based dairying. Each system has built up a complex network of technology, technology transfer and industry infrastructure. It is thus proving difficult to capture the benefits from each system and produce a hybrid system that will be more profitable than either of the component systems.

In the USA, grazing systems are gaining in importance on dairy farms as a way of lowering production costs while maintaining high milk production (Fales *et al.*, 1993). Grazing system economics are discussed in detail elsewhere in this book (Chapter 15).

There is a common misconception that seasonal calving cannot be adopted where the market is dominated by fresh milk sales. However, cooperative arrangements within groups of farmers could allow the advantages of seasonal calving, e.g. improved efficiency through batch processes and a better lifestyle, to be achieved, without compromising fresh milk supply. Such a system would require a consideration of the trade-offs between milk production, seasonal milk price variation and feed costs. Such options become increasingly valid as the cost differential between grazed pasture and harvested forage or concentrates increases.

Obviously climatic constraints in many countries make year-round grazing infeasible due to: dormant pastures, snow cover or inadequate load bearing by wet soils. However, if the milk price : feed cost ratio narrows in countries such as the USA it may become necessary to increase grazing either in existing dairy states or by using transhumance to take advantage of different climatic zones and pasture growth patterns.

Irish work (Dillon and Crosse, 1997) has shown that 'hybrid' systems using grazed pasture (3.5 t DM cow^{-1}), pasture silage (1.4 t DM cow^{-1}) and concentrates (0.5 t DM cow^{-1}) on an annual basis, can efficiently use pasture and be highly profitable at a stocking rate of 2.6 cows ha^{-1}. In this system grazed pasture forms the major diet component for 7 months of the year.

There is a worldwide tendency for herd sizes to increase, and concern is sometimes expressed at the distances cows must travel to be milked in dairy grazing systems. However, a square 100-ha farm with 250 cows and with the farm dairy centrally located requires cows to walk a maximum of 700 m at one time. Similarly, a 400 ha farm with 1000 cows would require cows to walk 1.4 km at one time. Thomson and Barnes (1993) have shown that walking up to 8 km day^{-1} does not affect yield, but protein yield was reduced by 4% compared with a control group walking < 0.5 km day^{-1}.

Grazing System Concept

Dairy cattle performance under grazing is positively related to consumption of high quality forage (Cleale and Bull, 1986). High quality forage for grazing purposes often refers to young and leafy forage with 18–24% dry matter, and containing 180–250 g of crude protein, 400–500 g of neutral detergent fibre, and 6.4–7.0 MJ of net energy for lactation in a kilogram of forage dry matter, depending upon stage of plant maturity, species composition and fertilizer application (Cherney and Allen, 1995). Feed use efficiency increases as milk production increases, because proportionately more nutrients are used in milk production than for animal maintenance. As grazed forage is the cheapest source of nutrients, increasing the consumption of high quality forage is critical for profitable dairy production. However, high-producing dairy cows fed forages alone will not reach their genetic potential for milk production, because of lower energy density in forage than concentrates (Donker *et al.*, 1968). Partial substitution of grazed forage with high energy concentrates will increase milk production and maintain adequate body condition (Donker *et al.*, 1968). However, concentrate supplementation should be weighed against its cost in relation to milk prices.

Elements of a Grazing System

A grazing management system is defined as an integrated combination of animal, plant, soil and other environmental components, and the grazing methods by which the system is managed to achieve specific results or goals of a producer (FGTC, 1992). Building a grazing system is a decision-making process based on the knowledge of the components being linked in relation to the lifestyle and profitability goals of a producer within the constraints imposed by the soil and climate conditions on the farm. Consequently, grazing systems are designed as farm- and producer-specific management systems. However, the basic factors and processes involved in system integration are based on principles of forage production in relation to animal production, and therefore, are applicable across the grazing systems. A full understanding of grazing systems requires knowledge of pasture growth and pasture–animal interactions. Further details on these topics are given in Chapter 11, but key concepts relating to intake from pasture are reviewed in the next section.

Intake

Intake determines the amount of nutrients consumed, whereas nutritive value defines nutrient concentration in the forage. Forage intake is primarily a function of forage mass and sward height (Hodgson, 1982). When forage mass or sward height are limiting, intake per bite declines, and biting rate and grazing time increase. However, increases in the latter two factors cannot compensate for the decrease in intake per bite and forage consumption becomes sub-maximal (Hodgson, 1982). Forage mass is defined as the total dry weight of forage above ground per unit area of land (FGTC, 1992). Forage mass available for intake is often expressed in terms of forage allowance which is defined as the forage mass per animal (or unit bodyweight). Forage consumption and animal production generally increase as forage allowance is increased up to two or three times the animal's maximum forage intake (Hodgson *et al.*, 1977). Sward heights lower than 8–10 cm reduce forage consumption by dairy cattle (Hodgson *et al.*, 1977). At higher levels of forage allowance, animals select leafy and digestible forage, and production per animal is maximized. However, production per hectare is always maximized in unsupplemented grazing systems when forage allowances, averaged over the year, are sub-maximal (Wright and Pringle, 1983). In some grazing systems where animal numbers are flexible or supplementary feed is available, forage allowance can be maintained at a relatively constant amount by increasing or decreasing pasture area in response to seasonal variation in forage production. Increasing area may increase forage mass and hence forage allowance by definition, but not animal production, if sward height is limiting intake.

Interactions

It is the interactions between pasture growth, pasture death and animal intake that underpin all grazing management decisions in grazing systems. The aim is to optimize the intake of high quality pasture in any one year without compromising long-term pasture persistence. This will, of course, vary depending on economic constraints. In some circumstances the optimal decision may be to forego grazing altogether.

Forage production

Within overall system constraints a manager's task is to provide adequate amounts of high quality feed on an annual basis. In most cases, a single forage species will not be able to provide a high yield of good quality forage throughout the year, because of species adaptation to a particular season. Cool-season (C_3) species are well adapted to cool climates (35°N or higher latitudes), whereas the warm-season (C_4) species are better adapted to warm climates (35°N or lower latitudes). C_3 species grow best between 15 and 25°C whereas the C_4 species grow best between 30 and 40°C (Nelson, 1996). As a result, forage production is seasonal with duration of peak production varying with species and site specific climatic conditions. Fluctuations in production pattern of a species within a region from year to year are due to variation in soil moisture availability, agronomic practices, or grazing management (Kanneganti and Kaffka, 1995).

Warm-season forages used for grazing purposes include perennial grasses such as switchgrass (*Panicum virgatum* L.), big bluestem (*Andropogon gerardii* Vitman), and Indiangrass (*Sorghastrum nutans* (L.) Nash.), annual grasses such as the sorghums and millets, and legumes such as the lespedezas and annual clovers. Cool-season forages include perennial grasses such as orchardgrass (*Dactylis glomerata* L.), smooth bromegrass (*Bromus inermis* Leyss.), reed canarygrass, timothy (*Phleum pratense* L.), fescues, and perennial ryegrass (*Lolium perenne* L.), and legumes such as red clover (*Trifolium pratense* L.), white clover (*Trifolium repens* L.), lucerne and birdsfoot trefoil (*Lotus corniculatus* L.).

Even though differences in annual forage production among species within a group (C_3 or C_4) are smaller when managed under grazing than when managed under cutting, small variations in species adaptation, production pattern or maturity are of practical use. Early maturing species or varieties of a species may be grown on well drained soils to allow grazing early in the spring, whereas late maturing cultivars may be planted on wet soils to allow grazing later in the spring without affecting forage quality (Casler, 1988). Similarly, other species differences, such as better resistance of birdsfoot trefoil to wet or waterlogged conditions than lucerne (Barta, 1980), or greater tolerance of smooth bromegrass to freezing temperature than tall fescue or perennial ryegrass (Jung *et al.*, 1995), help a producer in matching forages to site-specific conditions of a farm.

Production patterns of cool and warm season forages can be complementary, a feature that is of potential value in maintaining uniform supply of forage. Warm-season forages are more tolerant of high temperatures and drought conditions, and reach peak production about a month or two later than the temperate forages (Matches and Burns, 1985). However, these species groups are best managed in separate pastures because of marked differences in timing of growth, water and nutrient use, and harvest/grazing management (Matches and Burns, 1985). For example, warm-season forages may serve as special purpose crops to provide forage during hot and dry summers in temperate regions where cool-season forages form the main component of a grazing system (Jung *et al.*, 1978). However, factors other than production also need to be considered before planting special purpose crops, such as land availability, species adaptability, ease of establishment, species persistence, labour needs and general compatibility with other farm operations. Further discussion on the management and quality of cool season, and tropical and subtropical grasses is found in Chapters 4 and 5, respectively.

Grass–legume mixtures

Legumes and grasses at similar stages of maturity differ in forage digestibility. With advancing maturity, concentration of structural tissue in leaf and stem of a grass plant increases, whereas in legumes, it is limited mostly to stem (Terry and Tilley, 1964). As a result, cell wall concentration and nutritive quality of leaf are fairly uniform in legumes irrespective of maturity (Terry and Tilley, 1964). Additionally, the proportion of structural tissue is lower in legumes than in grasses at similar stages of maturity (Buxton and Russell, 1988). As a result, legumes are digested at a faster rate and to a greater extent than grasses of similar maturity (Buxton and Russell, 1988). Milk production from legume pasture is often greater than that observed on grass pasture (Thomson *et al.*, 1985).

Production in grass–legume mixed pasture is generally higher and better distributed seasonally than in a single species pasture (Matches, 1979). Legumes provide nitrogen for companion grass production through nitrogen fixation (Matches, 1979; Mallarino *et al.*, 1990). Some combinations of grass and legume species are more compatible in forage production than others. Forage yield is higher and is better distributed across the grazing season when tall fescue is grown in combination with red clover, birdsfoot trefoil or Ladino white clover than when grown alone with fertilizer N (Matches, 1979; Pederson and Brink, 1988). Orchardgrass–white clover, including the Ladino types, and bluegrass–white clover mixtures are more flexible to changes in grazing management while being more persistent than some other mixtures of grasses and legumes (Duell, 1985; Jung and Baker, 1985).

Grazing Management and Pasture Species

In short-growing temperate grasses, much of the leaf area in the canopy is at a lower height than in tall-growing species. Some examples of short-growing temperate grasses are Kentucky bluegrass, perennial ryegrass, tall fescue and orchardgrass, whereas tall-growing species include smooth bromegrass, timothy, and reed canarygrass.

During the vegetative phase of most short grasses, internode elongation is minimal and therefore the apical meristem is close to the soil surface, well protected from grazing. Some internode elongation is observed in the tall-growing species even during the vegetative phase. Assuming that water and nutrients are not limiting, the short grasses can be grazed frequently and to a lower residue, because a few intact leaves that are close to the soil surface escape grazing and continue photosynthesis. Consequently, the short grasses can be managed under both continuous or rotational stocking. However, if most leaf area is removed, adequate rest periods between consecutive grazings must be provided to allow full canopy recovery and to replenish storage reserves (Duell, 1985).

Tall-growing grass species can be grazed frequently, if adequate leaf area is left behind. Close grazing removes elevated growing points and new growth from basal buds is slow, and stand depletion may occur due to weed invasion (Knievel *et al.*, 1971). Tall-growing species are better suited to rotational than to continuous stocking. Similarly, warm-season grasses are intolerant of close grazing as it removes the elevated growing points, and new growth from basal buds is slow due to low carbohydrate reserves (Anderson *et al.*, 1989).

Among the forage legumes, white clover can tolerate continuous stocking and close grazing because growing points on the stolons are located close to the soil surface and are well protected from grazing (Frame and Newbould, 1986). In contrast, lucerne may be grazed close but not frequently, because regrowth has to originate from crown buds using stored reserves, and the crop needs 4–6 weeks to replenish reserves (Smith, 1962). Red clover does not tolerate close grazing because terminal meristems are elevated, but can be grazed frequently, if adequate residual leaf is left. Similarly, birdsfoot trefoil can be grazed frequently if not grazed too closely because it depends predominantly on continuous photosynthetic activity (Hoveland *et al.*, 1990).

Grazing Methods

Grazing method is defined as a procedure or technique of grazing management designed to achieve specific objectives of a grazing system (FGTC, 1992). Grazing methods are based on two types of animal stocking management, continuous and rotational stocking. A grazing system can use a combination of these methods as tools to control sward production in relation to forage consumption and animal production (Harris, 1978). It is important to stress that a wide variety of grazing methods have been shown to result in very similar levels of milk yield per hectare. For example, a review by Ernst *et al.* (1980) showed only a 1.5% advantage in milk yield to rotational grazing compared with continuous stocking. Stocking rate is the primary determinant of milk yield per hectare, as shown by Journet and Demarquilly (1979); their review of 13 stocking rate experiments showed that an increase of one cow per hectare resulted in an average reduction in milk yield per cow of 10%, but an increase of over 20% in milk yield per hectare.

Continuous stocking

Continuous stocking is a method of grazing in which animals are continuously present on a pasture for a long period of time, or throughout the grazing season (FGTC, 1992). The number of animals stocked per unit area may be altered in response to seasonal variation in forage production by either changing the animal numbers or by adjusting the pasture area. Advantages of continuous stocking over rotational stocking are attributed to reduced input costs in fencing, water, and labour for animal movement. A major disadvantage is the requirement for greater skill in monitoring sward growth and in managing grazing pressure (animal units : forage mass; FGTC, 1992) to minimize forage under-utilization during spring and early summer due to high forage growth rate and over-grazing during summer droughts and in winter. Even though a plant may be grazed only intermittently at an interval of 5–25 days under continuous stocking (Clark *et al.*, 1984), some forage species such as lucerne, smooth bromegrass, timothy and most warm-season grasses do not tolerate continuous stocking management (Mitchell *et al.*, 1994). Grazing-tolerant lucerne cultivars with low crowns and spreading characteristics are being developed that may withstand continuous stocking (Brummer and Bouton, 1991). Animal production under well-managed continuous stocking can equal that observed under rotational stocking, provided tolerant species are used (Ernst *et al.*, 1980).

Rotational stocking

In rotational stocking, a pasture is subdivided into units called paddocks, and animals are moved from paddock to paddock in a sequence. Forage in a paddock is grazed down in a short period of time (1–20 days) before the animals are moved into a new paddock. This intermittent stocking allows a rest period for plants to recover from grazing. Duration of a grazing cycle is the sum of rest period and duration of animal stay in a paddock while grazing. The number of paddocks is a function of rest period and duration of animal stay in a paddock. For example, if a rest period is 24 days and the duration of stay is 2 days per paddock, the number of paddocks required is 13 (rest period/duration of stay +1 paddock under grazing).

The rest period should be optimized to maintain the sward in a vegetative state with high leaf : stem ratio while minimizing senescence losses (Parsons *et al.*, 1988). The optimum rest period will vary across the season depending upon the stage of plant development, plant growth rate and defoliation intensity. In spring, when daylength and temperature are increasing, temperate perennial grasses have the potential to flower. Excessive accumulation of flowering stems will markedly reduce forage quality and increase animal rejection which leads to progressive deterioration of sward structure and productivity (Buxton and Marten, 1989). The objective of grazing management during this period is to suppress expression of reproductive development in the sward by grazing the sward frequently and closely. Forage in excess of animal needs can be set aside for conservation to harvest at desired forage quality. During summer and autumn, rest period primarily is a function of growth rate and defoliation intensity in relation to growth and senescence processes (Bircham and Hodgson, 1983; Parsons *et al.*, 1988). An optimum balance between growth and senescence occurs when the average rate of forage production is maximized over the season (Parsons and Penning, 1988; Parsons *et al.*, 1988). A sward producing forage at a faster rate reaches its peak average growth rate sooner than a sward growing at a slower rate (Parsons, 1985). Consequently, rest periods are shorter in summer than in autumn, typically 25 vs. 50 days respectively for New Zealand dairy systems, as growth rates decline with decreasing temperature and light (Kanneganti and Kaffka, 1995). However, in summer drought, variations in rotation length are insufficient to protect pastures from overgrazing and pasture demand must be reduced by drying cows off, reducing stocking rate by removing cull cows, or feeding supplements.

Forage is produced at a faster rate when managed under lax defoliation than when grazed severely, because of higher rates of photosynthesis due to greater amounts of leaf in a laxly grazed compared with a closely grazed sward (Parsons *et al.*, 1988). Consequently, paddocks managed under lax grazing can be grazed more frequently than similar paddocks managed under close grazing. Warm-season grasses, such as switchgrass or big bluestem, do not tolerate close grazing, a minimum residue of 10–15 cm is needed, and they require 4–5 weeks' rest for full recovery (Mitchell *et al.*, 1994).

Strip Grazing

Strip grazing is a variation of rotational stocking, in which animals are moved each day to a new strip, usually within a paddock. This method provides uniform amounts of forage of desired quality on a daily basis, and is a preferred method of managing high-producing dairy cows (Brundage and Sweetman, 1958). This method may not be economical for managing low-producing cows, dry cows or heifers because animal production response may not be high enough to justify the cost of labour involved in animal and fence movement. The number of strips may vary from 10 to 45, depending upon the rest period. The size of a strip may vary to accommodate the seasonal pattern of forage production across the grazing season. In New Zealand it is the preferred method for allocating pasture when demand exceeds supply in winter and early spring.

Leader–follower

The leader–follower method (Archibald *et al*., 1975), or first–last grazing, involves two or more groups of animals grazing a pasture sequentially. The two groups usually have different nutritional requirements. The first group of animals, generally with high nutrient requirement, is allowed first access. The second group, with lower nutrient requirements than the first group, is allowed access after the first group is moved. This method may be applicable on dairy farms where high-producing cows form the leader group and low-producing cows, dry cows or heifers may form the follower groups. This method is labour intensive, but may enhance forage utilization (forage intake : forage produced) while maintaining sward production and forage quality (Mayne *et al*., 1986). However, under high stocking rates for 2 months in early summer, Archibald *et al*. (1975) were unable to show any advantage over a conventional rotational grazing system.

Paddock numbers

Dairy farms are often intensively subdivided. This may reduce the time needed to collect cows for milking, but there is little experimental evidence that it leads to increased system productivity. Recent theoretical analysis of a forage dairy system in late winter and early spring by Woodward *et al*. (1995) showed that per cow intake is likely to be sub-optimal unless rotation rate is closely synchronized to pasture growth rate, and that increased subdivision will not lead to increased productivity unless reasons such as pasture trampling on wet soils mitigate against long grazing durations. For tropical pastures, Reyes *et al*. (1995) showed that stargrass (*Cynodon nlemfuensis* Vanderyst) annual yield was the same for either a 6 or 36 paddock grazing system. Further valuable resources should not be wasted on this topic. Studies on a range of pastures have not shown clear cut advantages to high levels of subdivision.

In designing a new grazing system for dairy cows the initial assumption should be that no internal fences, water troughs or laneways are needed. These facilities should only be added if it can be demonstrated that they will increase milk income more than they increase costs. They should be added until the marginal return from the last paddock or trough equals the marginal cost. In practice, this is hard to calculate but by using the above procedure over-investment of capital can be avoided as well as continuing costs for repairs and maintenance of facilities. In reality, topography, siting of existing facilities, and other constraints will determine the eventual layout chosen. Surveys to get a constant paddock size can greatly assist grazing management decisions.

It is better to have a reduced length of laneway (i.e. larger paddocks) with an excellent surface than increased length of poor quality laneway.

The cost of water troughs can be halved by sharing one trough between two paddocks. Electric fencing is a very cheap means of achieving a flexible number of paddocks on the farm. In many temperate environments dry cows and those in late lactation do not need continual access to water every day. Access to water twice per day can ease grazing management and reduce investment in temporary watering systems.

A preoccupation with paddock number distracts from the major consideration of grazing management. Namely, the allocation of pasture such that the total pasture utilized per year for milk production is maximized.

Where resilient pasture species exist back fencing after grazing is unnecessary. Back fencing may be needed if significant regrowth and grazing of the regrowth is likely to occur, or when soil is wet and damage to plants and poaching is likely. In the latter circumstances, cows should probably only be on the pasture for a restricted time each day.

Stockpiling

Stockpiling forage is defined as allowing forage to accumulate for grazing at a later time (FGTC, 1992), stockpiling is also known as deferred grazing (McCallum *et al.*, 1991). Stockpiling is generally practised to extend the grazing season into autumn and winter when forage production is low. A major advantage of stockpiling forage is that it reduces the harvesting and storage needs of indoor feeding. Quality of stockpiled forage may vary markedly depending upon the date of stockpiling initiation and the subsequent growing conditions (Ocumpaugh and Matches, 1977). Forage yield generally increases but forage quality may decline as the accumulation period increases. Forage production and associated quality are a function of growth and senescence processes which are influenced by the environmental factors and species characteristics (Turner and Begg, 1981; Bircham and Hodgson, 1983). Tall fescue tolerates cold temperatures better and maintains forage quality for a longer period than timothy or bluegrass (Matches, 1979; Van Keuren and Stuedemann, 1979). Legume forage generally cannot be stockpiled without extensive degradation in forage quality due to leaf loss as the legume leaf is easily injured upon exposure to freezing temperature (Mays and Washko, 1960). Stockpiling forage is generally not recommended for high-producing dairy cows due to unpredictable variation in forage quality. However, stockpiling may be a practical option for dry cows or heifers, if soils and weather permit grazing late in the season. Deferred grazing (McCallum *et al.*, 1991) has a place in low input systems where conservation is not practised. Deferred grazing in New Zealand dairy systems refers to the practice of removing surplus areas from grazing in late spring–early summer and allowing pasture to accumulate until required in autumn. It allows feed to be carried forward through the summer at much reduced cost, improves pasture quality on the remaining areas and will provide seed input into pasture in autumn.

Pasture Assessment

Pasture is allocated for grazing based on measurements of standing forage mass (Meijs *et al.*, 1982). In lieu of cutting forage samples to estimate forage mass, sward height may be measured or visually estimated and forage mass can be estimated from known relationships between sward height and forage mass.

Pasture assessment is most important at times when animal demand exceeds pasture growth. Pasture assessment can be used for both long-term feed planning and short-term allocation (Sheath *et al.*, 1987). When allocating a pasture area for four or more days of grazing in a rotational stocking system, significant amounts of forage may be produced during the period of animal stay on a paddock. This effect may be corrected based on an estimate of plant growth rate (Lantinga, 1985).

Problems associated with pasture assessment are: the appropriateness of different methods and their relation to ground level; the need to cut and weigh samples to get calibration equations; the relation of local data to research station data; and the effect of season, species, morphology and operator variation on accuracy (Piggot, 1988).

Methods

There are several methods of pasture assessment currently in use. Visual assessment (Piggot and Morgan, 1985) is the easiest and cheapest, but, in the absence of frequent checks against known standards the method can quickly degenerate to a poorly calibrated guess. However, trained observers who understand the seasonal changes of pasture structure can achieve sufficient accuracy for all commercial farm applications and some experimental situations. The rising plate meter (Holmes, 1974) can be very accurate where specific calibration equations are developed. It effectively measures a 'compressed' sward height. Standard seasonal equations for perennial ryegrass–white clover pastures derived by L'Huillier and Thomson (1988) have been successfully used by farmers, consultants and researchers in New Zealand. K.A. Macdonald (New Zealand, 1997, personal communication) found that trained and untrained observers were equally accurate when using the rising plate meter.

Electronic pasture probes (Campbell *et al.*, 1962) have now developed to the stage where internal calibration equations allow instant readout of predicted pasture mass. They are more expensive than rising plate meters, simple to use, but are more sensitive to environmental conditions.

Other devices to measure sward height are available, e.g. the Hill Farming Research Organization sward stick (Hodgson and Maxwell, 1981). Although height does not allow absolute feed budgets to be calculated, it can be very effective both as a predictor of animal performance and as a management tool.

The method of pasture assessment chosen must be appropriate to the final objective. Consistent, intelligent use of any of the above methods associated with continual evaluation of management outcomes will lead to an improved understanding of grazing management and more efficient use of any given pasture resource.

Forage Conservation and Supplementation

The principles and practices of forage conservation and supplementation are considered in more detail in Chapters 9 and 13, respectively. Forage may be preserved as silage or hay, depending upon the weather conditions at the time of harvesting. Conserved forage may include forage in a pasture, or special purpose crops such as corn for silage (Blaser *et al.*, 1986). Forage conservation in a grazing system can be used as a management strategy to harvest forage in excess of animal needs as hay or silage, to maintain adequate grazing pressure (animal units : forage availability) for good forage utilization, to maintain the sward in a leafy, vegetative state, and to minimize the effects of dung and urine patches on forage consumption in subsequent grazing (Blaser *et al.*, 1986). Pasture area set aside for conservation depends upon winter feed requirement and availability of other crops for silage or hay. If land characteristics, such as slope and drainage permit, a rotation between conserved and grazing areas will maintain good sward population density which is critical for sward structure, longevity and productivity (Jones *et al.*, 1982).

Conserved forage is generally fed during autumn and winter when grazing is not a valid option or pasture is limited in all-year grazing systems. Routine use of conserved forage as a supplement under grazing is generally not recommended, because conserved forage substitutes for grazed forage without increasing animal production (Blaser *et al.*, 1986; Rearte *et al.*, 1986). As conserved forage costs more than the comparable grazed forage, this practice is often uneconomical.

Feed supplements provide additional energy, protein or minerals when grazed forage falls short of animal nutrient requirement. Ideally the objective is to provide a total, balanced diet that is nutritionally adequate to meet the desired animal production while maintaining animals in good health. In well-managed pastures, grazed forage is generally adequate in protein but may be deficient in energy for high-producing dairy cows (Donker *et al.*, 1968). However, a supplement may contain additional protein and/or minerals if the grazed forage is deficient in them.

Despite a large rumen capacity, high-producing dairy cows fed forages alone will not reach their genetic potential for milk production, because of low energy density in forage in relation to animal energy requirement and intake regulation (Donker *et al.*, 1968; Mertens, 1994). Feeding supplements, however, results in varying levels of substitution of grazed forage, in addition to 'true' supplementation brought about by increasing total dry matter consumption (Leaver, 1986). Substitution is defined as a change in forage consumption per unit increase in supplement consumption. Since grazed forage is the cheapest source of nutrients, the objective is to reduce the substitution effect while increasing the supplementation effect. Use of supplement is economical when the value of the extra milk exceeds the supplement cost. Animal production response to supplement varies as a function of animal production potential, physiology, quality of the grazed forage, and supplement characteristics (Owen *et al.*, 1963; Cleale and Bull, 1986).

Milk production seems to peak around 25–30 kg cow^{-1} day^{-1} for New Zealand Holstein-Friesians on solely grazed pasture (Penno *et al.*, 1996). Forage consumption generally is in the range of 1.0–3.0% of body weight, depending upon forage quality and level of milk production (Jurgens, 1978). Under good grazing management, using leafy, grass–legume pasture, forage consumption can be as high as 3% of body weight without any negative impact on milk production of high-producing dairy cows. However, as a general recommendation, forage intake is restricted to 1.8–2.2% of body weight, depending upon forage quality (Adams *et al.*, 1988). The remaining nutrient requirement is provided through concentrates such as grains or mill by-product feeds. Expressing concentrate supplementation as a ratio of milk production, high-producing cows managed under grazing may be supplemented at a rate of 1 kg of concentrate (e.g. corn grain) for 3–5 kg of milk production (Dhiman *et al.*, 1997). Feeding energy-dense supplements to high-producing cows may be economical when grain or other concentrates are available at a low price. However, supplemental feeding should be continuously monitored for its economic justification, because substitution effects increase as the quality of grazed forage increases (Holden *et al.*, 1995).

All Forage Dairying

There is ample experimental evidence (Journet and Demarquilly, 1979; Wright and Pringle, 1983) to show that as stocking rate increases on an individual farm, per cow performance will decrease but per hectare production will increase, and with it the potential to achieve higher profitability per hectare. In dairy systems reliant solely on pasture it is profitability per hectare that becomes the predominant concern because land prices are the main capital cost as opposed to building and equipment costs in housed dairy systems. Bryant (1990) has emphasized that grazing management has had little influence on total milk yield per hectare. However, this work was done on carefully controlled experimental farmlets. In the commercial industry each year there is an enormous economic loss from farmers making incorrect grazing management decisions. These lead to immediate loss of production, loss of body condition score in cows and subsequent anoestrus and rebreeding problems. The latter has more serious consequences in a seasonal dairy system because cows must calve on a 365 day cycle if they are to avoid being culled. Bryant's work (1990) has shown the enormous resilience of New Zealand perennial ryegrass–white clover pastures and the cows that graze them. However, it is possible to move outside the buffering zone of pastoral systems (Stockdale and King, 1980), and this can lead to decreased pasture yield and cow performance. Therefore, to capture the benefits of a stocking rate that can make optimum use of annual pasture yield it is necessary to understand and implement sound grazing management principles. The main principles of dairying at high stocking rates using solely grazed pasture have been expounded by Bryant (1990).

Winter management

At calving in late winter, cow condition score should be 4.5–5 (1–10 scale). Calving date should be approximately 8 weeks before spring pasture growth rate exceeds cow requirements on a per hectare basis. Calving condition score is a function of condition score at drying off in the previous season, and winter feed intake per cow, which is in turn a function of winter stocking rate, winter pasture growth rate and total pasture on the farm. Pasture on farm at calving should be 2000 kg DM ha^{-1} and feed budgeting has a key role to play in autumn and winter to ensure that this target is achieved. The dynamic nature of pasture growth means that several feed budgets must be done through this period.

Spring management

The period from calving to when spring pasture growth rate exceeds cow requirements is the key to a successful season. The aim should be to have at least 1400 kg DM ha^{-1} average pasture on farm when pasture surpluses begin to accumulate. Higher amounts imply that a higher stocking rate could have been carried and will make the late spring control of pasture more difficult if conservation is not practised. Lower amounts will mean cow performance in the first 2 months of lactation will have been compromised to such an extent that production will not recover during the current lactation. It also means that cow condition will have fallen and that anoestrus may be a problem. The pasture on farm at this stage will be a function of pasture on farm at calving, intake per cow, stocking rate and spring pasture growth rate. A feed budget can afford to allow a decrease of 10 kg DM ha^{-1} day^{-1} for 60 days to achieve a decline of pasture on farm from 2000 to 1400 kg DM ha^{-1}. An obvious solution seems to be to apply N fertilizer to increase pasture growth rates or to feed bought-in supplements. It should be remembered, however, that the above system is in place where milk prices are low and feed and other inputs relatively expensive, so a resort to these options may lead to increased milk yield but decreased profits.

Once pasture surpluses begin to accumulate until mid-summer the primary aim of grazing management should be to increase per cow intake. This requires high pasture allowances/high pasture surface height to be offered. This can be achieved equally well by set stocking or by fast rotations of < 20 days. It is only during this period that an unsupplemented dairy cow will achieve her potential pasture intake. It is therefore critical to ensure that all other factors are conducive to high production. Grazing management should aim to leave pasture residuals after grazing of 1600 kg DM ha^{-1}, with average pasture on farm of 2500 kg DM ha^{-1}. This will not maximize pasture growth rates but will maintain high pasture quality. In such a system conservation may be zero or minimal. Where conservation is practised it is important that the decisions to remove paddocks from grazing do not compromise the primary aim of maximizing per cow intake. The importance of this is illustrated by the fact that the No. 2 dairy farmlet trial at Dairying Research Corporation, Ruakura achieves 70% of total lactation milk yield in the first 5 months of lactation (J.W. Penno, New Zealand, 1997, personal communication).

Recent research has shown potential for the late control of perennial ryegrass pastures to encourage the better nutrition of young, daughter tillers in the sward (Matthew *et al.*, 1989), but attempts to translate the findings into benefits in a total system context have so far met with little success.

Summer–autumn management

Summer production in low input systems is governed largely by soil moisture. Extended droughts require timely decisions on culling to reduce stocking rate and allow condition score on the remaining cows to be maintained. Decision on culling and drying off must be taken in such a way that cows do not have to make substantial increases in condition score during the winter, because an extra 150 kg DM cow^{-1} is required to increase condition score by one unit. This will effectively double the feed requirement of a cow for one month during winter.

Mixed Forage–Concentrate Dairying

In New Zealand the proportion of grazed grass in the diet is 90%, but in Europe it ranges from 55–60% in UK, France and Ireland to 35% in Denmark, while it averages 40% in the USA (Agra Europa, 1994). Thus most dairy systems integrate grazed pasture with the feeding of concentrates, conserved forages and by-products. The availability of a range of feeds offers many management opportunities. When supplements are available, set-stocking becomes a more viable option because the supplement can be used to ensure that sward heights do not fall below 6–8 cm (Leaver, 1987). When stocking rates are high, substitution of pasture for concentrates will alleviate overgrazing and subsequent loss in pasture yield. However, if low stocking rates are combined with high levels of supplementary feeding, pasture utilization will be low. Low utilization immediately increases the cost of grazed pasture and also reduces the nutritive value at subsequent grazings, with the risk of losing vulnerable species.

Recent Irish research (Dillon and Crosse, 1997), has shown that profitability of mixed forage–concentrate systems can be increased by the following management practices. Calving date should enable an early start to spring grazing on the entire farm, with grass silage and concentrates used to increase milk yields and avoid overgrazing. In mid-spring 40–50% of the total area should be closed for first cut silage, and grazing management from late April to late June should ensure close grazing to 6 cm sward height to maximize summer pasture digestibility. A second cut of silage should be taken in late July, with the total farm area again available for grazing from mid-August to mid-November. Grass silage will form the ration component in late lactation and when cows are dry.

Similar principles can be used when autumn-calved cows are used for winter milk production, or where all year calving is used, although the proportion of conserved feeds and concentrates will be increased, grazed pasture can still make a substantial contribution to the diet. In mixed forage–concentrate dairying, grazed pasture offers the advantage of decreasing ration costs per cow. However, it will not support the same level of milk yield per cow. The factors to consider in an economic assessment of this trade-off are addressed in Chapter 15.

Sub-tropical and tropical dairy systems

Dairying in sub-tropical and tropical environments involves special challenges to both cows and pastures. However, the principles of grazing management apply equally to temperate and tropical dairy systems. The importance of stocking rate as a primary determinant of milk production was shown for pastures of Coastcross-1 bermudagrass (*Cynodon dactylon*

(L.) Pers), improved stargrass and pangola grass (*Digitaria decumbens* Stent.) by Jerez *et al.* (1986). Their work showed that stargrass fertilized with 400 kg N ha^{-1} year^{-1} and irrigated in the dry season produced 18,133 kg milk ha^{-1} when stocked at 5 cows ha^{-1} with 3 kg concentrate cow^{-1} day^{-1} for the first 70 days of lactation. This compares with New Zealand perennial ryegrass–white clover pasture fertilized with 400 kg N ha^{-1} year^{-1} producing 18,600 kg milk ha^{-1} when stocked at 4.4 cows ha^{-1} with no supplementary feeding (Penno and Clark, 1997).

Senra (1992) has reviewed milk production systems in Cuba showing that milk production ranges from 1300–2700 kg ha^{-1} year^{-1} in extensive systems to 16,000–20,000 kg ha^{-1} year^{-1} for intensive, highly stocked systems based on N fertilized, irrigated grass. Intensive grass–legume mixtures produced 5000–8000 kg milk ha^{-1} year^{-1} and intensive grass systems using glycine (*Neonotonia wightii* (Wight & Arn.) Lackey) as a 'protein bank' produced 10,000–12,300 kg milk ha^{-1} year^{-1}. Attempts to incorporate grass–legume mixtures into tropical dairy systems have generally been unsuccessful. The concept offers substantial cost savings in reduced N fertilizer and grain supplements. However, Davison and Brown (1985) measured a decline of legume content from 28% to 3% in 4 years when a Gatton panic (*Panicum maximum* Jacq.), Tinaroo glycine (*Neonotonia wightii*) and greenleaf desmodium (*Desmodium intortum* (Mill.) Fawc. & Rendle) pasture was stocked with 2 cows ha^{-1}. These twining legumes could recover if pastures were destocked over summer or stocking rate reduced.

In the tropics cows are often housed for all or part of the day to reduce heat stress. There is evidence (Igono *et al.*, 1987) that the use of sprinklers and fans will increase milk production from 23.3 to 25.3 kg cow^{-1} day^{-1} compared with shade alone even under conditions of slight heat stress (27°C, 60% relative humidity).

However, the installation of housing and cooling systems is expensive and increases the requirement for manure disposal. Hongyantarachai *et al.* (1989) showed that cows grazed outdoors in Thailand under moderate heat stress (33°C, 70% relative humidity) had similar milk yields (14.1 vs. 14.2 kg cow^{-1} day^{-1}) and conception rates, lower mastitis infection, higher milk fat content (4.3 vs. 4.0%) but higher liveweight losses compared with cows allowed to graze only at night and restricted to a shaded shed during the day. They concluded that the common practice in Thailand of housing cows during the day and feeding cut pasture was unjustified and expensive. These authors suggest outdoor grazing in the tropics allows selective grazing to improve diet quality compared with 'cut and carry' systems; it allows 40% of excreted nutrients to be returned to the pasture at zero cost. They also did not observe significant damage to soils or pasture when strip grazing occurred at intensities of 300 cows ha^{-1}, although this may not be the case for all soil and pasture combinations.

Conclusions

The proportion of forage to use in a dairy system is a complex decision based on the milk price and the cost of all available feeds, climate, soils, technical knowledge and industry infrastructure. There are established scientific principles related to plant growth, feeding value and intake constraints from grazed pastures that need to be understood and then placed in the context of a specific dairy system. Once principles are grasped it becomes obvious why certain grazing methods are inappropriate for certain species. These principles also explain why, for many grazing systems, the exact grazing method employed will have little influence on the production or profitability of the total enterprise.

In seasonal dairy systems relying solely on grazing it is critical for the correct stocking rate, calving date and drying-off date to be chosen. The grazing method adopted cannot be expected to overcome deficiencies in any of the above. In intensively grazed, low-cost dairy systems the amount of conservation should be minimized. Conserved forage will always be more expensive and of lower feeding value than if the forage had been grazed *in situ*. In all grazed systems where supplementary feed is used, the substitution of supplement for pasture almost inevitably leads to decreased pasture utilization.

References

Adams, R.S., Hlubik, J.G. and Bosworth, S. (1988) Ration programming and feeding tips for dairy cattle on pasture. In: Cropper, J.B. (ed.) *Pasture in the Northeast Region of the United States*. Northeast National Technical Center, Chester, Pennsylvania, pp. 141–150.

Agra Europa (1994) *1994 Dairy Review*. Agra Europe Special Report No. 78. Agra Europe, Tunbridge Wells, Kent.

Anderson, B., Matches, A.G. and Nelson, C.J. (1989) Carbohydrate reserves and tillering of switchgrass following clipping. *Agronomy Journal* 81, 13–16.

Archibald, K.A.E., Campling, R.C. and Holmes, W. (1975) Milk production and herbage intake of dairy cows kept on a leader and follower grazing system. *Animal Production* 21, 147–156.

Barta, A.L. (1980) Regrowth and alcohol dehydrogenase activity in waterlogged alfalfa and birdsfoot trefoil. 1. Flood injury anoxia. *Agronomy Journal* 72, 1017–1020.

Bircham, J.S. and Hodgson, J. (1983) The influence of sward condition on rates of herbage growth and senescence in mixed swards under continuous stocking management. *Grass and Forage Science* 38, 323–331.

Blaser, R.E., Hammes, R.C., Jr, Fontenot, J.P., Bryant, H.L., Poland, C.E., Wolf, D.D., McClaugherty, F.S., Kline, R.G. and Moore, J.E. (1986) *Forage-animal Management Systems*. Virginia Agricultural Experimental Station Bulletin 86-7. Blacksburg, Virginia, 90pp.

Brummer, E.C. and Bouton, J.H. (1991) Plant traits associated with grazing tolerant alfalfa. *Agronomy Journal* 83, 996–999.

Brundage, A.L. and Sweetman, W.J. (1958) Comparative utilization of alfalfa–bromegrass pasture under rotational and daily strip grazing. *Journal of Dairy Science* 41, 1777–1780.

Bryant, A.M. (1990) Optimum stocking and feed management practices. *Proceedings of the Ruakura Farmers' Conference* 42, 55–59.

Buxton, D.R. and Marten, G.C. (1989) Forage quality of plant parts of perennial ryegrasses and relationship to phenology. *Crop Science* 29, 429–435.

Buxton, D.R. and Russell, J.R. (1988) Lignin constituents and cell wall digestibility of grass and legume stems. *Crop Science* 28, 553–558.

Campbell, A.G., Phillips, D.S.M. and O'Reilly, E.D. (1962) An electronic instrument for pasture yield estimation. *Journal of the British Grassland Society* 17, 89–100.

Casler, M.D. (1988) Performance of orchardgrass, smooth bromegrass, and ryegrass in binary mixtures with alfalfa. *Crop Science* 29, 429–435.

Cherney, J.H. and Allen, V.G. (1995) Forages in a livestock system. In: Barnes, R.F., Miller, D.A. and Nelson, C.J. (eds) *Forages Volume I: An Introduction to Grassland Agriculture,* 5th edn. Iowa State University Press, Ames, Iowa, pp. 175–188.

Clark, D.A., Chapman, D.F., Land, C.A. and Dymock, N. (1984) Defoliation of perennial ryegrass (*Lolium perenne*) and browntop (*Agrostis tenuis*) tillers, and white clover (*Trifolium repens*) stolons in set stocked and rotationally grazed hill pastures. *New Zealand Journal of Agricultural Research* 26, 289–301.

Cleale, R.M. and Bull, L.S. (1986) Effect of forage maturity on ration digestibility and production by dairy cows. *Journal of Dairy Science* 69, 1587–1594.

Davison, T.M. and Brown, G.W. (1985) Influences of stocking rate on the recovery of legume in tropical grass–legume pastures. *Tropical Grasslands 19,* 4–10.

Dhiman, T.R., Kanneganti, V.R., Walgenbach, R.P., Massingill, L.J., Wiltbank, M.C., Russelle, M.P. and Satter, L.D. (1997) Production response to feed supplementation of dairy cows in a seasonal calving and grazing system. In: *Proceedings XVIIIth International Grassland Congress, 8–19 June, 1997, Winnipeg and Saskatoon, Canada.* Congress Secretariat, Grasslands 2000, Calgary, Alberta, pp. 29/71–29/72.

Dillon, P. and Crosse, S. (1997) Current and potential production in a profitable dairy system. In: *National Dairy Conference.* Teagasc, Dublin, pp. 2–33.

Donker, J.D., Marten, G.C. and Wedin, W.F. (1968) Effect of concentrate level on milk production of cattle grazing high-quality pasture. *Journal of Dairy Science* 51, 67–73.

Duell, R.W. (1985) The bluegrasses. In: Heath, M.E., Metcalfe, D.S. and Barnes, R.F. (eds) *Forages: The Science of Grassland Agriculture,* 4th edn. Iowa State University Press, Ames, Iowa, pp. 188–197.

Ernst, P., Le Du, Y.L.P. and Carlier, L. (1980) Animal and sward production under continuous grazing management – a critical appraisal. In: *Proceedings of Symposium of the European Grassland Federation.* Wageningen, pp. 119–126.

Fales, S.L., McMurry, S.A. and McSweeny, W.T. (1993) The role of pasture in northeastern dairy farming: historical perspective, trends, and research imperatives for the future. In: Sims, J.T. (ed.) *Agricultural Research in the Northeastern United States: Critical Review and Future Perspectives.* American Society of Agronomy, Madison, Wisconsin, pp. 111–132.

FGTC (Forage and Grazing Terminology Committee) (1992) Terminology for grazing lands and grazing animals. *Journal of Production Agriculture* 5, 191–201.

Frame, J. and Newbould, P. (1986) Agronomy of white clover. *Advances in Agronomy* 40, 1–38.

Harris, W. (1978) Defoliation as a determinant of growth, persistence and composition of pasture. In: Wilson, J.R. (ed.) *Plant Relations in Pastures*. CSIRO, Australia, pp. 67–85.

Hodgson, J. (1982) Ingestive behavior. In: Leaver, J.D. (ed.) *Herbage Intake Handbook*. The British Grassland Society, Reading, pp. 113–138.

Hodgson, J., Capriles, J.M.R. and Fenlon, J.S. (1977) The influence of sward characteristics on the herbage intake of grazing calves. *Journal of Agricultural Science, Cambridge* 89, 743–750.

Hodgson, J. and Maxwell, T.J. (1981) Grazing research and grazing management. *Hill Farming Research Organisation Biennial Report 1979-81*, pp. 169–187.

Holden, L.A., Muller, L.D., Lykos, T. and Cassidy, T.W. (1995) Effects of corn silage supplementation on intake and milk production in cows grazing grass pasture. *Journal of Dairy Science* 78, 154–160.

Holmes, C.W. (1974) The Massey grass meter. Massey University, Palmerston North, New Zealand. *Dairy Farming Annual* 26, 26–30.

Hongyantarachai, S., Nithichai, G., Wongsuwan, N., Prasanpanich, S., Siwichai, S., Pratumsuwan, S., Tasapanon, T. and Watkin, B.R. (1989) The effects of grazing versus indoor feeding during the day on milk production in Thailand. *Tropical Grasslands 23*, 8–14.

Hoveland, C.S., Alison, M.W., Jr, Hill, N.S., Lowrey, R.S., Jr, Fales, S.L., Durham, R.G., Dobson, J.W., Jr, Worley, E.E., Calvert, V.H. II and Newsome, J.F. (1990) *Birdsfoot Trefoil Research in Georgia*. Georgia Agricultural Experimental Station Research Bulletin 396, Athens, Georgia.

Hutton, J.B. (1968) Farming at two cows per acre: 2-At Ruakura, three seasons intensive dairy production. *Proceedings of the Ruakura Farmers' Conference* 20, 215–227.

Igono, M.O., Johnson, H.D., Stevens, B.J., Krause, G.F. and Shanklin, M.D. (1987) Physiological, productive and economic benefits of shade, spray and farm systems versus shade for Holstein cows during summer heat. *Journal of Dairy Science 70*, 1069–1079.

Jerez, I., Menchaca, M.A. and Rivero, J.L. (1986) Evaluation of three tropical grasses. 2. Effect of stocking rate on milk production. *Cuban Journal of Agricultural Science* 20, 231–237.

Jones, M.B., Collett, B. and Brown, S. (1982) Sward growth under cutting and continuous stocking managements: sward structure, tiller density and leaf turnover. *Grass and Forage Science* 37, 67–73.

Journet, M. and Demarquilly, C. (1979) Grazing. In: Broster, W.H. and Swan, H. (eds) *Feeding Strategy for the High Yielding Dairy Cow*. Granada, London, pp. 295–321.

Jung, G.A. and Baker, B.S. (1985) Orchardgrass. In: Heath, M.E., Metcalfe, D.S. and Barnes, R.F. (eds) *Forages: The Science of Grassland Agriculture,* 4th edn. Iowa State University Press, Ames, Iowa, pp. 224–232.

Jung, G.A., Griffith, J.L., Kocher, R.E., Sheaffer, J.A. and Gross, C.F. (1978) Warm-season grasses extend beef cattle forage. *Science in Agriculture* 25(2), 6.

Jung, G.A., Van Wijk, A.J.P., Hunt, W.F. and Watson, C.E. (1995) Ryegrasses. In: Moser, L.E., Buxton, D.R. and Casler, M.D. (eds) *Cool-Season Forage Grasses*. American Society of Agronomy, Madison, Wisconsin, pp. 605–642.

Jurgens, M.H. (1978) *Animal Feeding and Nutrition,* 7th edn. Kendall/Hunt Publishing Company, Dubuque, Iowa, 580pp.

Kanneganti, V.R. and Kaffka, S.R. (1995) Forage availability from a temperate pasture managed with intensive rotational grazing. *Grass and Forage Science* 50, 55–62.

Knievel, D.P., Jacques, A.V.A. and Smith, D. (1971) Influence of growth stage and stubble height on herbage yields and persistence of smooth bromegrass and timothy. *Agronomy Journal* 63, 430–434.

Lantinga, E.A. (1985) Simulation of herbage production and herbage intake during a rotational grazing period. An evaluation of Linehan's formula. *Netherlands Journal of Agricultural Science* 33, 385–403.

Leaver, J.D. (1986) Effects of supplements on herbage intake and performance. In: Frame, J. (ed.) *Grazing*. Occasional Symposium No. 19, The British Grassland Society, Hurley, pp. 79–88.

Leaver, J.D. (1987) The potential to increase production efficiency from animal-pasture systems. *Proceedings of the New Zealand Society of Animal Production* 47, 7–12.

L'Huillier, P.J. and Thomson, N.A. (1988) Estimation of herbage mass in dairy pastures. *Proceedings of the New Zealand Grassland Association* 49, 117–122.

McCallum, D.A., Thomson, N.A. and Judd, T.G. (1991) Experiences with deferred grazing at the Taranaki Agricultural Research Station. *Proceedings of the New Zealand Grassland Association* 53, 79–83.

McMeekan, C.P. (1961). The pros and cons of high stocking rate. *Proceedings of the Ruakura Farmers' Conference* 13, 184–190.

Mallarino, A.P., Wedin, W.F., Goyenola, R.S., Perdomo, C.H. and West, C.P. (1990). Legume species and proportion effects on symbiotic dinitrogen fixation in legume-grass mixtures. *Agronomy Journal* 82, 785–789.

Matches, A.G. (1979) Management. In: Buckner, R.C. and L.P. Bush (eds) *Tall Fescue*. Agronomy Monograph 20, American Society of Agronomy, Madison, Wisconsin, pp. 171–199.

Matches, A.G. and Burns, J.C. (1985) Systems of grazing management. In: Heath, M.E., Metcalfe, D.S. and Barnes, R.F. (eds) *Forages: The Science of Grassland Agriuculture*, 4th edn. Iowa State University Press, Ames, Iowa, pp. 537–547.

Matthew, C., Xia, J.X., Hodgson, J. and Chu, A.C.P. (1989) Effect of late spring grazing on tiller age profiles and summer-autumn pasture growth rates in a perennial ryegrass (*Lolium perenne* L.) sward. In: Desroches, R. (ed.) *Proceedings of the XVIth International Grassland Congress, 4–11 October, 1989, Nice, France*. French Grassland Society, Versailles Cedex, pp. 521–522.

Mayne, C.S., Woodcock, S.C.F., Clements, A.J. and Newberry, R.D. (1986) The effects on milk production of gazing management systems involving preferential treatments of high-yielding dairy cows. In: Frame, J. (ed.) *Grazing*. Occasional Symposium No. 19, The British Grassland Society, Hurley, pp. 114–118.

Mays, D.A. and Washko, J.B. (1960) The feasibility of stockpiling legume-grass pasturage. *Agronomy Journal* 52, 190–192.

Meijs, J.A.C., Walters, R.J.K. and Keen, A. (1982) Sward methods. In: Leaver, J.D. (ed.) *Herbage Intake Handbook*. The British Grassland Society, Hurley, pp. 11–36.

Mertens, D.R. (1994) Regulation of forage intake. In: Fahey, G.C., Collins, M., Mertens, D.R. and Moser, L.E. (eds) *Forage Quality, Evaluation, and Utilization.* American Society of Agronomy, Madison, Wisconsin, pp. 450–493.

Mitchell, R., Anderson, B.A., Waller, S.S. and Moser, L.E. (1994) *Managing Switchgrass and Big Bluestem for Pasture and Hays.* NebGuide G94-1198-A, University of Nebraska, Lincoln, Nebraska, 4pp.

Nelson, C.J. (1996) Plant physiology and developmental morphology. In: Moser, L.E., Buxton, D.R. and Casler, M.D. (eds) *Cool-Season Forage Grasses.* American Society of Agronomy, Madison, Wisconsin, pp. 71–86.

Ocumpaugh, W.R. and Matches, A.G. (1977) Autumn winter yield and quality of tall fescue. *Agronomy Journal* 69, 639–643.

Owen, F.G., Hinders, R.G., Schleusener, P.E. and Van Riper, G.E. (1963) Value of irrigated bromegrass-alfalfa and orchardgrass-alfalfa pastures for lactating dairy cows. *Journal of Dairy Science* 46, 830–834.

Parsons, A.J. (1985) New light on the grass sward and the grazing animal. *Span* 28(2), 47–49.

Parsons, A.J. and Penning, P.D. (1988) The effect of duration of regrowth on photosynthesis, leaf death and the average rate of regrowth in a rotationally grazed sward. *Grass and Forage Science* 43, 15–27.

Parsons, A.J., Johnson, I.R., and Harvey, A. (1988) Use of a model to optimize the interaction between frequency and severity of intermittent defoliation and to provide a fundamental comparison of the continuous and intermittent defoliation of grass. *Grass and Forage Science* 43, 49–59.

Penno, J.W. and Clark, D.A. (1997) Silage for milk production and profit. *Dairyfarming Annual 49,* 50–57.

Penno, J.W., Macdonald, K.A. and Bryant, A.M. (1996) The economics of No. 2 Dairy Systems. *Proceedings of the Ruakura Dairy Farmers' Conference* 48, 11–19.

Pederson, G.A. and Brink, G.E. (1988) Compatibility of five white clover and five tall fescue cultivars grown in association. *Agronomy Journal* 80, 755–758.

Piggot, G.J. (1988) Measuring pasture yield for livestock management on farms. *Proceedings of the New Zealand Society of Animal Production* 48, 219–224.

Piggot, G.J. and Morgan, H.M. (1985) Visual assessment of dry matter yield of pastures on dairy farms. *New Zealand Journal of Experimental Agriculture* 13, 219–224.

Rearte, D.H., Kesler, E.M. and Hargrove, G.L. (1986) Forage growth and performance of grazing cows supplemented with concentrate and chopped or long hay. *Journal of Dairy Science* 69, 1048–1054.

Reyes, J., Garcia Trujillo, R., Senra, A., Vidal, I. and Fonte, D. (1995) A study of two grazing methods. 1. Main productive indices in dairy cows. *Cuban Journal of Agricultural Science* 29, 153–158.

Senra, A. (1992) Milk production systems in Cuba. *Cuban Journal of Agricultural Science* 26, 231–247.

Sheath, G.W., Hay, R.J.M. and Giles, K.H. (1987) Managing pastures for grazing animals. In: A.M. Nicol (ed.) *Livestock Feeding on Pasture.* New Zealand Society of Animal Production Occasional Production No. 10, pp. 65–74.

Smith, D. (1962) Carbohydrate root reserves in alfalfa, red clover, and birdsfoot trefoil under several management schedules. *Crop Science* 2, 75–78.

Stockdale, C.R. and King, K.R. (1980) The effects of stocking rate and nitrogen fertilizer on the productivity of irrigated perennial pasture grazed by dairy cows. 1.

Pasture production, utilization and composition. *Australian Journal of Experimental Agriculture and Animal Husbandry* 20, 529–536.

Terry, R.A. and Tilley, J.M.A. (1964) The digestibility of leaves and stems of perennial ryegrass, cocksfoot, timothy, tall fescue, lucerne, and sainfoin, as measured by an in vitro procedure. *Journal of British Grassland Society* 19, 363–372.

Thomson, D.J., Beever, D.E., Haines, M.J., Cammell, S.B., Evans, R.T., Dhanoa, M.S. and Austin, A.R. (1985) Yield and composition of milk from Friesian cows grazing either perennial ryegrass or white clover in early lactation. *Journal of Dairy Science* 52, 17–31.

Thomson, N.A. and Barnes, M.L. (1993) Effect of distance walked on dairy production and milk quality. *Proceedings of the New Zealand Society of Animal Production* 53, 69-72.

Turner, N.C. and Begg, J.E. (1981) Plant–water relations and adaptation to stress. *Plant and Soil* 58, 97–131.

Van Keuren, R.W. and Stuedemann, J.A. (1979) Tall fescue in forage-animal production systems for breeding and lactating animals. In: Buckner, R.C. and Bush, L.P. (eds) *Tall Fescue*. Agronomy Monograph 20, American Society of Agronomy, Madison, Wisconsin, pp. 201–232.

Voisin, A. (1959) *Grass Productivity*. Crosby Lockwood, London, 353pp.

Wallace, L.R. (1958) Wintering cows on autumn-saved pasture. *Proceedings of the Ruakura Farmers' Conference* 10, 216–227.

Woodward, S.J.R., Wake, G.C. and McCall, D.G. (1995) Optimal grazing of a multi-paddock system using a discrete time model. *Agricultural Systems* 48, 119–139.

Wright, D.F. and Pringle, R.M. (1983) Stocking rate effects in dairying. *Proceedings of the New Zealand Animal Production Society* 43, 97–100.

Supplementation of Cool-season Grass Pastures for Dairy Cattle

13

L.D. Muller[1] and S.L. Fales[2]

[1]Department of Dairy and Animal Sciences, 324 Henning Building, The Pennsylvania State University, University Park, Pennsylvania 16802, USA
[2]Department of Agronomy, 16 Agricultural Science Industries Building, The Pennsylvania State University, University Park, Pennsylvania 16802, USA

Introduction

In the late 1950s the dairy industry in the USA moved from grazing toward confined feeding and management systems. The feeding management system and the control of the nutrients being delivered to dairy cows, as well as other management and economic factors, have contributed significantly to this increased milk production per cow with confinement systems. The increased technical knowledge in dairy nutrition, diet formulation, and accurate delivery of diets to cows, particularly with total mixed rations (TMR), have provided a control of the feeding programme that did not exist with pasture systems.

With relatively stable milk prices during the last 15 years, a well managed grazing system offered an opportunity to reduce the cost of producing forage during the typical 6–7 month pasture season in the midwest and northeastern USA, and has contributed toward increased use of pasture as a forage for dairy cows. Savings in feed costs contribute about half to two-thirds of the decrease in total costs and improved farm profitability with grazing systems. The adoption of management intensive grazing (MIG) systems has been primarily by dairy producers with herds of 50–70 cows. Of the 1200 Pennsylvania dairy producers surveyed, about 16% were using a MIG system (Gripp *et al.*, 1993). A recent Wisconsin study (Jackson-Smith *et al.*, 1996) reported that 14% of the dairy producers in Wisconsin are using intensive grazing, which is a twofold increase in 2 years.

Several studies have reported decreased milk production per cow with MIG system compared with confinement feeding systems. In a survey of 147 dairy farmers in Pennsylvania (Parker *et al.*, 1993), pasture-based dairy farms with 6 months of grazing had approximately 4% lower milk production per cow than year-round confinement systems. However, 8% of the pasture based dairy farms exceeded 9000 kg cow^{-1} year^{-1}, indicating that high levels of milk production per cow could be achieved with pastures (Table 13.1). All of the grazing herds were feeding supplemental grain and many were also feeding supplemental forages.

Grazing systems in other countries, which rely on minimal supplemental feed, have induced dairy producers in the USA to seek the most profitable grazing and dairy production systems. Historically, supplemental feed prices have been lower in the USA than in New Zealand, Ireland, and other countries with grazing-based dairy systems, making it economical to feed more supplements in the USA. In contrast, some large, intensively managed confinement systems in the USA are averaging 12,000–14,000 kg milk cow^{-1}. This presents nutritional and management challenges for grazing systems to reach the genetic potential of the cows. This chapter will focus on the nutrient supplementation of cool-season grass pastures with emphasis on high producing dairy cattle.

Pasture Quality and Nutritional Considerations

Confinement systems with a TMR feeding system allow for known and constant quantities and qualities of forages and nutritionally balanced rations to be offered daily. In the USA, the current average of Holstein cows on production testing programmes is about 8750 kg milk cow^{-1} (Wiggans, 1997). These cows are typically fed about 3000–3500 kg grain cow^{-1} per year with 40–50% of the total dry matter intake (DMI) from grains and concentrates. In contrast to confinement feeding, the amount and quality of pasture available in a grazing system varies throughout the growing season and is influenced by the type of grass, climate, the fertilization programme, and a host of management factors. Pasture availability and nutrient composition change frequently during the typical 6–7 month growing season in the midwest and northeastern part of the USA. Well managed grass pastures often have a total protein content of 200–250 g kg^{-1} and neutral detergent fibre (NDF) content of 400–500 g kg^{-1} (Hongerholt *et al.*, 1997; Kolver and Muller, 1998; Minson, 1990; Van Vuuren *et al.*, 1991; Hoffman *et al.*, 1993; Holden *et al.*, 1994a; Fales *et al.*, 1995).

Generally, the nutrient quality of the pasture is higher than the same plant material harvested as silage or hay (Glenn, 1994). The chemical content and rumen degradability of pastures differ from non-pastured forage. The ruminal degradability of protein in grass is high when harvested

Table 13.1. Summary of dairy farms in Pennsylvania using minimal or intensive pasture systems compared with confinement non-grazing systems[1].

Parameter	Pasture	Confinement
No. of herds	77	70
No. of cows per herd	52	67
Herds > 75 cows, %	7	25
Milk production kg cow^{-1}		
Year 1	7321	7690
Year 2	7655	8020
Herds > 9000 kg cow^{-1}, %	8	12
Herds < 6800 kg cow^{-1}, %	27	15

[1]Parker *et al.*, 1993.

as pasture, and ruminal digestibility is expected to be high when compared with non-pastured forages. Typical average composition of a high quality cool-season grass over the 6 month grazing season is given in Table 13.2. Early spring pastures may often have 250–300 g kg^{-1} true protein (TP) with NDF concentrations of less than 350 g kg^{-1}. Nitrogen fertilization usually increases total protein and soluble protein in pastures.

Pasture as only feedstuff and dry matter intake

Grazing studies from other countries indicate that pasture alone can support milk production of 25–30 kg day^{-1} in the spring (Holmes, 1989; Mayne, 1996). Dairy cows under grazing usually consume less than 3.0% of body weight (BW) as DM, but in high producing cows this may increase to 3.25% of BW (Leaver, 1985). Research studies in the USA with high producing cows fed only pasture are limited. Early lactation cows fed only high quality ryegrass pasture in the spring consumed 19.1 kg of DM or 3.4% of body weight (Kolver and Muller, 1998). Milk production averaged 29.6 kg day^{-1}. However, when compared with cows fed a nutritionally balanced ration TMR with a confinement system, the cows consumed 4.5 kg less DMI per day, produced 14.5 kg less milk per day, and lost substantially more body condition. Although this was a short-term study, results further support the need for strategic supplementation of nutrients and feeds to cows with high genetic merit grazing high quality pastures.

Most studies that have reported DMI values with dairy cows grazing in the USA have included some type of supplemental feed. When supplemental grain is fed, pasture DMI decreases as grain substitutes for pasture, however total dry matter and energy intake are increased. A summary of several studies (Table 13.3) indicates that when adequate high quality

Table 13.2. Average nutrient composition for cool-season grass pasture over a grazing season[1].

Nutrient	Spring	Summer
Total protein (TP), g kg^{-1} DM	210–250	180–220
RUP, g kg^{-1} of TP	200–250	250–300
Sol. P, g kg^{-1} of TP	300–350	250–300
ADF, g kg^{-1} DM	240–280	280–340
NDF, g kg^{-1} DM	400–450	480–550
Hemicellulose, g kg^{-1} DM	170–210	210–250
Cellulose, g kg^{-1} DM	160–200	210–260
NE, Mcal kg^{-1}	1.60–1.70	1.55–1.65
Non-fibre carbohydrate (NFC), g kg^{-1} DM	150–200	120–150
Fat, g kg^{-1} DM	30–40	30–40
Ash, g kg^{-1} DM	80–90	80–90
Ca, g kg^{-1} DM	5.0–7.5	5.0–7.5
P, g kg^{-1} DM	3.0–3.5	3.0–3.5
Mg, g kg^{-1} DM	1.5–2.0	1.5–2.0
K, g kg^{-1} DM	20–35	20–35
S, g kg^{-1} DM	1.6–2.2	1.6–2.2

[1]Summarized from Kolver *et al.*, 1998b; Rayburn, 1991; Hoffman *et al.*, 1993; Holden *et al.*, 1994a; Fales *et al.*, 1995; Hongerholt *et al.*, 1997.
RUP, rumen undegradable protein; ADF, acid detergent fibre; NDF, neutral detergent fibre; NE, net energy.

pasture was available to achieve high DMI and when grain was fed in amounts 'typical' for these high milk production levels, pasture (forage) DMI and total DMI are comparable to the DMI expected with non-grazing cows (NRC, 1989). If one assumes that the daily milk production achieved is about 23–27 kg cow^{-1} when pasture is the only feed, the results from these studies suggest that the milk response to grain feeding in high producing cows would be about 0.8–1.2 kg per 1 kg grain. Hoden *et al.* (1991) reported a 0.6 kg milk response per 1.0 kg grain with cows producing 20–23 kg of milk with the response tending to be greater with high yielding cows.

The response of milk per unit of supplemental grain is expected to be greatest for the first increment of grain fed than for the second or third increments fed. Clearly, pasture alone may not provide an ideal amount and balance of nutrients to achieve the genetic potential of high producing cows, and will probably result in partitioning of body energy reserves toward milk production. However, if maximum reliance on grazing forms the basis of sustainability and profitable dairy farm systems, full exploitation of the potential of grazed grass require development of grazing systems designed to present herbage to the animal that will maximize intake per bite and daily DMI (Mayne and Peyraud, 1996).

Table 13.3. Summary of dry matter index (DMI) from spring grazing studies with lactating cows.

Study	Type of study	Milk production (kg cow^{-1} day^{-1})	DMI (kg cow^{-1} day^{-1})			
			Pasture	Grain	Forage	Total
Holden *et al.*, 1994a	Grass pasture + grain	30.5	15	7.3	—	22.3 (3.0[1])
Holden *et al.*, 1995	Grass pasture + grain + grain and corn silage	29.1 28.8	13.6 10.9	7.7 7.7	— 2.7	21.3 (4.0) 21.3 (4.0)
Hongerholt and Muller, 1998	Grass pasture + grain	38.2	11.4	8.6	—	20.0 (3.6)
Rippel, 1995	Grass[2] + grain Grass/ legume + grain	30.0 30.5	10.5 14.5	8.2 7.7	— —	18.7 (3.5) 22.2 (4.0)
Kolver *et al.*, 1998b	Grass pasture + grain	29.5	9.9	9.1	—	19.0 (3.6)

[1]Total dry matter intake as % of body weight.
[2]Combined data for ryegrass and orchardgrass.

Nutritional limitations and imbalances with pasture

Many research studies have been conducted with supplementary feeding of pasture, however the specific nutrients that limit milk production have been identified in only a few studies. The intake of metabolizable energy (ME) is generally thought to limit milk production of cows fed pasture diets. The supply of protein to the duodenum has also been proposed to limit milk production of cows to approximately 25 kg day^{-1} when fed high quality pasture (Beever and Siddons, 1986).

Lactating cows fed all pasture compared with a TMR had lower DMI (19.0 vs. 23.4 kg of DM day^{-1}), milk production (29.0 vs. 44.1 kg day^{-1}), and body condition score (2.0 vs. 2.5) (Kolver and Muller, 1998). To understand further why the cows grazing pasture in this study produced less milk than did cows fed the TMR and to determine the proportion of this decrease that was attributable to a lower DMI, the Cornell Net Carbohydrate and Protein System (CNCPS) (Fox *et al.*, 1992) model was used to simulate the grazing treatments. This model uses mechanistic and empirical relationships to predict the ME and metabolizable protein (MP) requirements of cattle and the supply of these components from the diet.

Using input data collected in this study and carbohydrate and protein degradation rates used by Kolver *et al.* (1998a), the model predicted that cows fed the TMR would have a ME supply that allowed the production of

43 kg day^{-1} of milk, which compares well with the 44.1 kg day^{-1} of milk that was actually produced. Cows grazing pasture were predicted to have a ME supply that allowed 27.6 kg day^{-1} of milk compared with the 29.6 kg day^{-1} of milk that was actually produced. The difference between the predicted and actual production indicated the amount of energy supplied by tissue mobilization (a loss of one point of body condition scale (BCS; 1 – thin to 5 – fat) in 209 days). These estimates include an adjustment for energy that is associated with grazing and walking activity. The predictions using NRC (1989) and the CNCPS model indicated that the supply of ME was first limiting for milk production of cows fed a high quality pasture as the sole feedstuff rather than the supply of MP or amino acids. The interrelations between protein and energy in terms of ruminal digestion and performance must be considered.

Energy

When pasture is the major source of forage, energy is generally thought to be the most limiting nutrient for profitable milk production and normal reproductive performance. The major sources of supplemental energy are carbohydrates which are often referred to as non-fibre carbohydrates (NFC). The NFC content of cool-season grass pastures, which is a measure of ruminally available carbohydrate, tends to be 150–220 g kg^{-1} of the DM (Table 13.2) compared with the total ration (forage plus grain) needs of about 350 g kg^{-1} of the DM (Fox et al., 1992). In addition, grazing cows require 10–20% more energy over maintenance because of higher levels of activity than cows in confinement (NRC, 1989). The amount of NFC fed from supplemental feeds, primarily grains, to increase the total energy intake on a pasture based system can have long-term effects on energy balance, milk production, body weight and body condition changes, reproductive performance and profitability. The utilization of pasture N and synthesis of microbial protein can be increased by supplying more ruminally available carbohydrates.

Numerous studies have been conducted to measure the milk production response of grazing cows to supplemental energy feeds. However, few have documented the impact of carbohydrate supplements on microbial protein synthesis and utilization of pasture N by lactating cows. Some reviews have been published on the supplementation with various energy sources (Thomson, 1982; Leaver, 1986; Minson, 1990; Kellaway and Porta, 1993). Many of the studies cited in these reviews were not conducted with high producing cows. In addition to the production level, the body condition score, substitution effect, concentrate level, stage of lactation, level of milk production, genetic potential of the cow, and quality of pasture and concentrate are all factors which influence the cow's response to energy supplementation.

Several experiments have demonstrated varying increases in milk production with supplements of energy dense feeds, and the responses are partially related to an increased DMI (Kellaway and Porta, 1993). The decrease in forage intake that occurs with grain feeding is usually less than the increase in grain intake, resulting in an increase in total DMI. Substitution rates of grain for pasture vary, but in general, pasture DMI decreases about 0.5–0.9 kg for each 1 kg of grain fed. The long-term benefits of supplemental energy are usually greater than the short-term benefits (Kellaway and Porta, 1993).

Conrad *et al.* (1983) compared grazing systems of orchardgrass with three levels of concentrate. Feeding concentrate increased the milk yield with both pastures. The milk production response with additional concentrate with orchardgrass pasture averaged about 1 kg of milk per 1 kg of grain supplementation. The estimated DMI from pasture with no supplemental concentrates was approximately 3% of BW. Forage intake decreased and total DMI increased as more grain was fed. The substitution rate of grain for pasture in this study was about 1 kg of grain for 0.8 kg pasture DM. Adding 5.4 kg of shelled corn resulted in 0.6 kg decrease in pasture DMI for each kilogram of corn fed and about 0.8 kg increased milk for each kilogram of corn (Berzaghi *et al.*, 1996). The results of these studies are consistent with the results of four studies with high producing cows summarized in Table 13.3.

In addition to the amounts of energy, the type of supplemental carbohydrate (grain) fed and the method of providing energy needs to be considered. Energetic uncoupling may occur with pasture diets from an under supply of ruminally available energy (or carbohydrate) relative to the release of pasture N in the rumen. Synchronizing the rate of supply of N and energy-yielding substrates to ruminal microbes has been suggested as a means to improve the capture of rumen degradable protein and improving animal performance (Beever and Siddons, 1986; Poppi and McLennan, 1995). Synchronous feeding of supplemental carbohydrate with pasture N vs. feeding grain 4 h after pasture feeding did not alter N status or milk production (Kolver *et al.*, 1998b). Providing grain feeding four times per day by a mobile computerized grain feeder (Gardner *et al.*, 1995) on pasture did not improve milk yield in high producing cows producing 35 kg day^{-1} compared with feeding grain twice daily (Hongerholt *et al.*, 1997). More limited research has studied the supplementation with different types of carbohydrate sources that may vary in the availability of the carbohydrate in the rumen.

Protein

Although the total protein in well managed pastures is high and often exceeds 250 g kg^{-1} (Table 13.2), the protein is highly degradable in the rumen (RDP) to ammonia and may be lost from the rumen and converted to urea in the liver, and eventually excreted in the urine. The ruminal utilization of pasture nitrogen, and subsequent flow to the duodenum, has been reviewed by Beever and Siddon (1986), Minson (1990), and Poppi and

McLennan (1995). Several studies (Beever *et al.*, 1986; Van Vuuren *et al.*, 1990; Berzaghi *et al.*, 1996) have identified the high protein intake, the high rate of protein degradability in the rumen, and low level of NFC in high quality pasture as contributing factors to a low efficiency of N utilization by grazing cows. This has been attributed to an inefficient capture of rumen N as microbial protein and the metabolic cost of synthesizing and excreting additional urea, which then impacts the energy status of the grazing animal. A surplus of RDP can account for losses equivalent to an energy value of 1.5–3.0 kg of milk yield. The results of Holden *et al.* (1994b) with duodenally cannulated cows indicate a pre-duodenal N loss with the highly degradable N in pasture and a loss of N across the wall of the rumen. The high concentrations of ammonia nitrogen in the rumen of grazing cows were in excess of microbial needs for protein synthesis. Thus, the capture of ruminally degraded N is impaired by a limitation of ruminally available energy and/or an excess of ruminally available N. Beever *et al.* (1986) reported that pre-duodenal loss of N occurred when forage contained greater than 25 g N kg^{-1} of the DM, a number exceeded with most cool-season grasses. It was calculated that the duodenal protein supply may only support a daily milk production of approximately 25 kg day^{-1}.

The excess ruminally degraded protein in the pasture may suggest that cows are deficient in protein and amino acids available for absorption from the small intestine. Providing ruminally available energy (carbohydrates) improves the utilization of the high RDP in pastures, improves microbial protein production, and decreases urinary N excretion and the energy cost associated with this excretion. The feeding of rumen undegradable protein (RUP) for high producing cows is often needed in early lactation. Research studies are limited with supplementation of RUP for high producing cows fed pasture as the sole forage. In four different studies (Stobbs *et al.*, 1977; Flores *et al.*, 1979; Rogers *et al.*, 1980; Minson, 1981) in which 0.5–1.0 kg of ruminally protected casein was fed with pasture, milk production increased from 0.5 to 2.4 kg per day, and milk protein percent was increased in three of the four studies. Milk production averaged between 9.6 and 18 kg day^{-1} for the four studies. Kellaway and Porta (1993) reviewed five studies conducted with lower producing cows and concluded that when pasture quality was high, additional protein had no effect on milk yields.

A study with cows producing 35 kg of milk did not find differences in milk yield when cows were fed a grain ration with increased amount of RUP (Hongerholt and Muller, 1998). Multiparous cows did produce more milk protein (1.06 vs. 0.98 kg day^{-1}) when fed the high RUP diet. High quality orchardgrass pasture containing 250 g crude protein (CP) kg^{-1} was the sole forage fed. An evaluation of the results of the study using the CNCPS suggested that ME and not MP was the most limiting nutrient to high producing cows grazing grass pasture. This agrees with the review of Kellaway and Porta (1993) which concluded that when energy was most limiting production, protein supplementation provided little additional response.

Supplementing nutrients to high producing cows

The majority of the dairy producers in the USA provide supplemental feeds primarily to provide energy, perhaps some protein and required minerals. A 2-year study provides information of expected responses of high producing dairy cows when provided high quality, well managed orchardgrass pastures plus a 'nutritionally balanced' grain ration (Fales *et al.*, 1995). This study compared forage production and quality, milk production and animal response, and profitability for three different stocking rates (2.5, 3.2, and 3.9 cows ha^{-1}) using high genetic merit Holstein cows grazing predominantly orchardgrass pasture (Fales *et al.*, 1995). Some key results from the study were:

- Increasing the stocking rate resulted in higher pasture production because more forage was removed which encouraged new growth.
- Increasing the stocking rate tended to increase the nutritional quality of the pasture.
- More pasture was utilized as stocking rate increased.
- Increasing the stocking rate had no effect on milk production per cow because more stored forage was fed when pasture was limiting. Cows averaged about 4500 kg of 35 g kg^{-1} fat corrected milk per cow over the three stocking rates for the 6-month grazing season.
- Total profitability increased directly with stocking rate because of a greater amount of milk produced per hectare.

The results clearly showed that good grazing management, which optimizes pasture intake, the optimum stocking rate, and the management of surplus spring pastures and summer deficits, had a great impact on pasture quality and quantity, and on profitability. Achieving high milk production per cow as well as maximizing the production and utilization of high quality pasture is a challenge and milk production per cow, as mentioned previously, may be lower with a pasture-based than with a confined system (Parker *et al.*, 1993).

The milk yield and composition, and body weight and body condition changes with 48 Holstein cows during each year of the 2-year stocking rate study are shown in Table 13.4. These cows were fed about 1 kg grain to 4 kg milk of a grain ration to supplement the grass pastures, and were fed supplemental forage when available pasture was limiting during the summer. The cows averaged about 38 kg milk cow^{-1} day^{-1} at the start of the trial during both years. Although a confinement group was not directly compared in the study, the lactation curve declined more rapidly in the spring (Fig. 13.1a) and cows gained less body condition (0.3 gain in BCS) for the 6-month study (Fig. 13.1b) than expected when compared with benchmark values for confinement feeding. Milk composition was relatively normal for Holstein cows averaging 36.1 g fat kg^{-1} and 29.9 g protein kg^{-1} over the 2 years (Fig. 13.1c). During the first year's grazing season, total daily DMI per cow was 22.3 kg with 15 kg of pasture DMI (Holden *et al.*, 1994a).

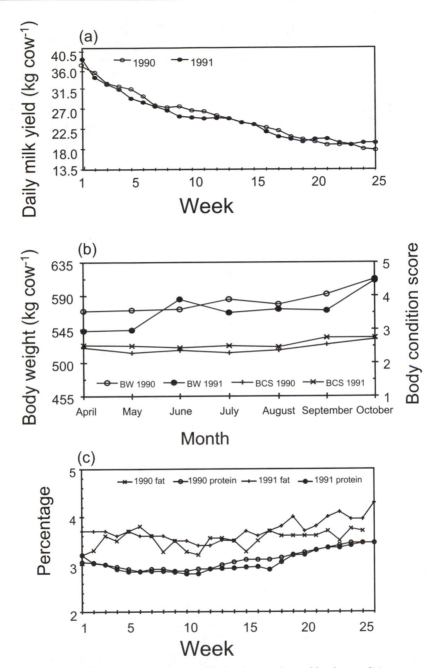

Fig. 13.1. Weekly milk yield (a), monthly body weight and body condition scores (b) and weekly milk fat and protein percentages (c) of Holstein cows intensively grazing mixed grass pasture during two grazing seasons across three stocking rates (Fales *et al.*, 1995).

Table 13.4. Summary of animal performance during 2 years for high genetic merit Holstein cows averaged across stocking rates of 2.5, 3.2 and 3.9 cows ha[-1] (6 month grazing season)[1,2].

Measure	Year		SE
	1990	1991	
Milk production cow[-1]			
Milk, kg season[-1]	4470	4510	60
Milk, kg cow[-1] day[-1]	25.5	24.8	0.35
Milk fat, g kg[-1]	34.8	37.3	0.5
Milk protein, g kg[-1]	30.1	29.7	0.2
3.5% FCM[3], kg season[-1]	4493	4654	73
Body condition changes			
Condition score gain[4]	+0.31	+0.27	
Body weight gain, kg cow[-1]	+44.6	+70.0	

[1]Fales *et al.*, 1995.
[2]Grain was fed daily at the rate of 1 kg 4 kg[-1] milk with a minimum of 4 kg and maximum of 9 kg.
[3]FCM, fat corrected milk.
[4]Increase in body condition according to visual scores on scale of 1 to 5.

There have been few studies for comparison during two grazing seasons with high genetic merit cows on a pasture based system. An earlier study (Hoffman *et al.*, 1993) reported 24 kg milk cow[-1] day[-1] during the 6-month grazing season when cool-season grass pasture was the only forage. Grain fed at rates of 1 kg to 3 kg milk or 1 kg to 4.2 kg milk did not result in differences in milk yield between groups, however body condition scores and weight gains were lower for cows fed the lower amount of supplemental grain.

The two studies (Hoffman *et al.*, 1993; Fales *et al.*, 1995) and those summarized in Table 13.4 show that high levels of milk production can be achieved with high intakes of pasture DM and grain supplemented at about 1 kg to 4 kg milk. No comprehensive data set has been published on the reproductive performance and health of high genetic merit cows on pasture based systems.

Supplemental forages and fibre

Most of the discussion to this point has focused on the protein and energy content of pasture and the supplementation of these nutrients. Conserved forage is often fed as a supplement for grazing dairy cows, particularly when pasture growth and availability may be limiting during certain environmental or climatic conditions. Supplementation can reduce the variation and decrease in forage intake when pasture availability is inadequate and can allow for higher stocking rates and an increased efficiency of land use (Phillips, 1988).

Supplemental forage and fibre may also be beneficial for the ruminal fermentation and for the utilization of nutrients in pasture. High quality spring pastures may be low in NDF with the NDF being highly fermentable in the rumen. A low milk fat is often observed on lush spring pasture and is probably related to a low fibre content and high digestibility (Murphy, 1985). Ruminal pH may decrease to 5.7 after pasture is fed and is often lower than the ruminal pH with cows fed the same forage as harvested as hay or silage (Holden *et al.*, 1994b). When fed grain supplements, mean rumen pH averaged over 24 h was 6.1 and 5.9 (Van Vuuren *et al.*, 1986; Kolver *et al.*, 1998b), however the lowest pH was 5.7 during a 24-hour period. A rumen pH below 6 is associated with a reduced cellulytic activity. The rate of passage of liquids (Holden *et al.*, 1994b; Berzaghi *et al.*, 1996) and solids (Berzaghi *et al.*, 1996) is increased when cows grazed pasture. The high water content of the fresh pasture may be related to the high liquid turnover. The rapid ruminal degradation rates and fermentation of carbohydrates (NFC and structural) may result in high total ruminal volatile fatty acid (VFA) production. All of these changes in ruminal fermentation and digestion would suggest that dietary fibre may be inadequate with high quality pastures.

Many dairy farmers in the USA routinely feed some supplemental forage with pasture (Parker *et al.*, 1993). Limited controlled research studies have examined the effects of feeding supplemental forage to high producing cows when adequate pasture is available. When pasture is available *ad libitum*, offering supplemental forage as a supplement has little effect on total DMI (Phillips, 1988) and the substitution rate of pasture DM for stored forage DM is about 1.0. Milk production was not changed with grass hay inclusion in nine studies reviewed by Phillips (1988). Corn silage supplementation of grass pasture did not improve milk yield or total DMI in Holstein cows averaging 29 kg of milk (Holden *et al.*, 1995).

Considerable interest exists in utilizing alternative feeds including various by-products with pasture based feeding programmes (Holden, 1995). Supplementation of grazing cows with a high fibre grain mixture containing several by-products resulted in a higher milk yield and higher total DMI compared with cows fed a high starch grain mixture (Meijs, 1986).

In contrast, warm environmental temperatures increase the amount of indigestible fibre in the plant (Buxton and Fales, 1994). Thus, the higher fibre content may decrease digestibility and DMI of fresh forage during the summer.

Predicting Limiting Nutrients and Animal Performance with Pasture-based Diets

Simulation models are increasingly being used to understand the many interactions involved in animal production responses. One such model discussed previously is the CNCPS (Fox *et al.*, 1992) which has primarily been used to predict performance and evaluate feeding programmes with diets based on hay and silage. Kolver *et al.* (1998a) tested the predictive ability and potential use of the model for diets based on pasture. Data were obtained from eight pasture studies in the USA and New Zealand in which DMI and animal performance were reported. The model provided reasonably good estimates of changes in body condition score, estimated energy balance, blood urea nitrogen, and milk production. The model under predicted DMI and over predicted ruminal pH. With the difficulty in obtaining accurate measures of intake with grazing cows, the model can be of value if it can provide reasonable predictions of DMI of pasture. Milk production was first limited by the supply of ME when only pasture was fed, but specific amino acids may be limiting milk production when more than 200 g kg^{-1} of the diet consisted of a grain supplement. However, in the study of Kolver and Muller (1998), milk synthesis did not appear to be limited by the supply or profile of amino acids, and the supply of MP did not appear to be limiting until a daily milk production of 36–38 kg was reached (Hongerholt and Muller, 1998). Preliminary evaluations with fibre addition from forages or by-products to pasture based diets indicate an increase in ruminal pH and microbial protein synthesis, and a higher predicted milk yield (unpublished data). Clearly, more use of models, which integrate the results of many research studies, must occur to improve the supplementation of strategies with cool-season grass pastures.

Summary

Well managed pastures that maximize pasture production and quality are of comparable or better quality than most forages harvested with mechanical harvesting systems. One of the biggest challenges of efficiently utilizing pasture for high producing dairy cows is proper supplementation to maximize rumen fermentation and microbial protein synthesis, which in turn contributes to optimum milk production and profit throughout the grazing season. The continual changing of pasture quantity and quality during the grazing season, and the potentially low utilization of protein in pasture by the lactating cow, are major challenges when supplementing pasture. Supplemental feeding of grain (energy), which is the most limiting nutrient in most situations, is needed to maximize cow performance, particularly with high producing cows in early lactation. Protein and amino acids are usually the second limiting nutrient with most grazing situations. The adaptation and use of nutrition models such as the CNCPS is a promising tool to predict animal performance and to understand better the limiting nutrients with a grazing system, and to develop feeding programmes that provide limiting nutrients through supplemental feeding.

References

Beever, D.E., Losoda, H.R., Cammell, S.B.C., Evans, R.T. and Haines, M.J. (1986) Effect of forage species and season on nutrient digestion and supply in grazing cattle. *British Journal of Nutrition* 56, 209–226.

Beever, D.E. and Siddons, R.C. (1986) Digestion and metabolism in the grazing ruminant. In: Milligan, L.P., Grovum, W.L. and Dobson, A.A. (eds) *Control of Digestion and Metabolism in Ruminants*. Prentice Hall, Englewood Cliffs, New Jersey, pp. 479–497.

Berzaghi, P., Herbein, J.H. and Polan, C.E. (1996) Intake, site, and extent of nutrient digestion of lactating cows grazing pasture. *Journal of Dairy Science* 79, 1581–1589.

Buxton, D.R. and Fales, S.L. (1994) Plant environment and quality. In: Fahey, G.C., Collins, M., Mertens, D.R. and Moser, L.E. (eds) *Forage Quality, Evaluation, and Utilization*. American Society of Agronomy, Madison, Wisconsin, pp. 155–199.

Conrad, H.R., Van Keuran, R.W. and Dehority, B.A. (1983) Top grazing high protein forages with lactating cows. In: Smith, J.A. and Hays, V.W. (eds) *Proceedings of the XIVth International Grassland Congress, 15–24 June, Lexington, Kentucky*. Westview Press, Boulder, Colorado, pp. 690–692.

Fales, S.L., Muller, L.D., Ford, S.A., O'Sullivan, M., Hoover, R.J., Holden, L.A., Lanyon, L.E. and Buckmaster, D.R. (1995) Stocking rate affects production and profitability in as pasture systems grazed rotationally by Holstein dairy cows. *Journal of Production Agriculture* 8, 88–96.

Flores, J.F., Stobbs, T.H. and Minson, D.J. (1979) The influence of the legume *Leucaena leucocephala* and formal-casein on the production and composition of milk from grazing cows. *Journal of Agricultural Science, Cambridge* 92, 351–357.

Fox, D.G., Sniffen, C.J., O'Connor, J.D., Russell, J.B. and Van Soest, P.J. (1992). A net carbohydrate and protein system for evaluating cattle diets. III. Cattle requirements and diet adequacy. *Journal of Animal Science* 70, 3578–3596.

Gardner, M.O., Buckmaster, D.R., and Muller, L.D. (1995) Development of a mobile solar powered dairy concentrate feeder. *Applied Engineering in Agriculture* 11, 785–790.

Glenn, B.P. (1994). Grasses and legumes for growth and lactation. In: *Proceedings of the Cornell Nutrition Conference for Feed Manufacturers*. Department of Animal Science and Division of Nutritional Sciences, Cornell University, Ithaca, New York, pp. 1–18.

Gripp, S., Elbehri, A., Yonkers, R., Ford, S. and Luloff, A. (1993) *Summary Report of 1993 Telephone Survey with Pennsylvania Dairy Farm Managers*. A research report for the Pennsylvania dairy profitability project. The Pennsylvannia State University, University Park, Pennsylvania, pp. 19–22.

Hoden, A., Peyraud, J.L., Muller, A., Delaby, L., Faverdin, P., Peccatte, J.R. and Fargetton, M. (1991) Simplified rotational grazing management of dairy cows: effects of rates of stocking and concentrate. *Journal of Agricultural Science* 116, 417–428.

Hoffman, K., Muller, L.D., Fales, S.L. and Holden, L.A. (1993) Quality evaluation land concentrate supplementation of rotational pasture grazed by lactating cows. *Journal of Dairy Science* 76, 2651–2663.

Holden, L.A. (1995) Alternative feeds for dairy cattle on pasture. In: Eastridge, M.L. (ed.) *Proceedings of the 2nd National Alternative Feeds Symposium:Alternative Feeds for Dairy Cattle and Beef.* Ohio State University, Columbus, Ohio, p. 17.

Holden, L.A., Muller, L.D. and Fales, S.L. (1994a) Estimation of intake in high producing Holstein cows grazing grass pasture. *Journal of Dairy Science* 77, 2332–2340.

Holden, L.A., Muller, L.D., Varga, G.A. and Hillard, P.J. (1994b) Ruminal digestion and duodenal nutrient flows in dairy cows consuming grass pasture, hay, or silage. *Journal of Dairy Science* 77, 3034–3042.

Holden, L.A., Muller, L.D., Lykos, T. and Cassidy, T.W. (1995) Effect of corn silage supplementation on intake and milk production in cows grazing grass pasture. *Journal of Dairy Science* 78, 154–160.

Holmes, W. (1989) The utilization of pasture. In: Jarrige, R. (ed.) *Ruminant Nutrition: Recommended Allowances and Feed Tables.* INRA, Paris, pp. 181–188.

Hongerholt, D.D. and Muller, L.D. (1998) Supplementation of rumen undegradable protein to diets of early lactation Holstein cows on grass pasture. *Journal of Dairy Science* 81, 2204–2214.

Hongerholt, D.D., Muller, L.D. and Buckmaster, D.R. (1997) Evaluation of a mobile computerized grain feeder for lactating cows grazing grass pasture. *Journal of Dairy Science* 80, 3271–3282.

Jackson-Smith, D., Barham, B. Nevies, M. and Klemme, R. (1996) *Grazing in Dairyland: The Use and Performance of Management Intensive Grazing among Wisconsin Dairy Farms.* Tech. Report 5, University of Wisconsin, Madison, Wisconsin, p. 10.

Kellaway, R. and Porta, S. (1993). *Feed Concentrates: Supplements for Dairy Cows.* Agmedia, Melbourne, 176pp.

Kolver, E.S. and Muller, L.D. (1998) Comparative performance and nutrient intake of high producing Holstein cows grazing or fed a total mixed ration. *Journal of Dairy Science* 81, 1403–1411.

Kolver, E.S., Muller, L.D., Barry, M.C. and Penno, J.W. (1998a) Evaluation and application of the Cornell net carbohydrate and protein system for dairy cows fed diets based on pasture. *Journal of Dairy Science* 81, 2029–2039.

Kolver, E.S., Muller, L.D., Vargas, G.A. and Cassidy, T.J. (1998b) Synchronization of ruminal degradation of supplemental carbohydrate with pasture nitrogen in lactating dairy cows. *Journal of Dairy Science 81,* 2017–2028.

Leaver, J.D. (1985) Milk production from grazed temperate grasslands. *Journal of Dairy Research* 52, 313–334.

Leaver, J.D. (1986) Effects of supplements on herbage intake and performance. In: Frame, J. (ed.) *Grazing.* Occasional Symposium No. 19, British Grassland Society. University of Reading, Reading, pp. 79–90.

Mayne, S. (1996) Can grazed grass provide for high vs. medium genetic merit cows. In: *Grass and Forage for Cattle of High Genetic Merit, Proceedings of the British Grassland Society.* University of Reading, Reading, pp. 17–22.

Mayne, C.S. and Peyraud, J.L. (1996) Recent advances in grassland utilization under grazing and conservation. In: *Grassland and Land Use Systems, Volume 1.* 16th European Grassland Federation Meeting, pp. 347–360.

Meijs, J.A.C. (1986) Concentrate supplementation of grazing dairy cows. 2. Effect of concentrate composition on herbage intake and milk production. *Grass and Forage Science* 41, 229–236.

Minson, D.J. (1981) The effects of feeding protected and unprotected casein on the milk production of cows grazing ryegrass. *Journal of Agricultural Science, Cambridge* 96, 239–242.

Minson, D.J. (1990). *Forage in Ruminant Nutrition.* Academic Press, New York, 483pp.

Murphy, J.J. (1985). Effect of feeding sodium bicarbonate in the concentrate or beet pulp on milk yield and composition in cows after turnout to pasture in spring. *Irish Journal of Agricultural Research* 24, 143–147.

[NRC] National Research Council (1989) *Nutrient Requirements of Dairy Cattle,* revised, 6th edn. National Academy Press, Washington, DC, 157pp.

Parker, W.J., Muller, L.D., Fales, S.L. and McSweeney, W.T. (1993) A survey of dairy farms in Pennsylvania using minimal or intensive pasture grazing systems. *Professional Animal Scientist* 9, 77–85.

Phillips, C.J.C. (1988) The use of conserved forages as a supplement for grazing dairy cows. *Grass and Forage Science* 43, 215–230.

Poppi, D.P. and McLennan, S.R. (1995) Protein and energy utilization by ruminants at pasture. *Journal of Animal Science* 73, 278–290.

Rayburn, E.D. (1991) *Forage Quality of Intensive Rotationally Grazed Pastures.* Animal Science Mimeo 151. Department of Animal Science, Cornell University, Ithaca, New York, 24pp.

Rippel, C.M. (1995) Dry matter intake and performance of lactating cows grazing grass or grass legume pasture. MS thesis, The Pennsylvania State University, Pennsylvania.

Rogers, G.L., Porter, R.H.D., Clarke, T. and Stewart, J.A. (1980) Effect of protected casein supplements on pasture intake, milk yield and composition of cows in early lactation. *Australian Journal of Agricultural Research* 31, 1147–1152.

Stobbs, T.H., Minson, D.J. and McLeon, M.N. (1977) The response of dairy cows grazing a nitrogen fertilized grass pasture to a supplement of protected casein. *Journal of Agricultural Science, Cambridge* 89, 137–141.

Thomson, D.J. (1982) The nitrogen supplied by and the supplementation of fresh or grazed forage. In: Thomson, D.J., Beaver, D.E. and Gunn, R.G. (eds) *Forage Protein in Ruminant Animal Production.* Occasional Publication No. 6, British Society of Animal Production, London, p. 53.

Van Vuuren, A.M., Van der Koelen, C.J. and Vroons-de Bruin, J. (1986) Influence of level and composition of concentrate supplements on rumen fermentation patterns of grazing dairy cows. *Netherlands Journal of Agricultural Science* 34(4), 457–467.

Van Vuuren, A.M., Tamminga, S. and Ketelaar, R.S. (1990) Ruminal availability of nitrogen and carbohydrates from fresh and preserved herbage in dairy cows. *Netherlands Journal of Agricultural Research* 38, 499–512.

Van Vuuren, A.M., Tamminga, S. and Ketelaar, R.S. (1991) In sacco degradation of organic matter and crude protein of fresh grass in the rumen of grazing dairy cows. *Journal of Agricultural Science, Cambridge* 116, 429–436.

Wiggans, G.R. (1997) *USDA Summary of 1996 Herd Averages.* Fact sheet K-3, National Cooperative Dairy Herd Improvement Program Handbook, National DHI, Washington, DC, 6pp.

Modelling Grass Utilization by Dairy Cattle

14

D.J.R. Cherney[1] and D.R. Mertens[2]

[1]*Department of Animal Science, 327 Morrison Hall, Cornell University, Ithaca, New York 14853-4801, USA*
[2]*USDA-Agricultural Research Service, US Dairy Forage Research Center, 1925 Linden Drive West, Madison, Wisconsin 53706, USA*

Introduction

In the past, grain surpluses and low grain prices often dictated that milk be produced with as little forage as possible, particularly in the USA. As the world situation changes, it is likely that grain surpluses will decrease and maximum utilization of forages will be needed to maintain a competitive dairy industry. In addition, the potential impact of animal wastes on environmental quality may increase the use of forages. Forage production in an integrated farming system has the potential to improve the recycling and balance of nutrients. Grasses may have some advantages over legumes in this area because of their ability to use excess farm generated N (Cherney and Cherney, 1993). There is a need, however, to develop and use models to gain the knowledge needed to maximize the use of grass in dairy farm systems. The infinite variety of combinations of components in a grass farming system preclude the use of traditional research approaches. Instead of evaluating each component individually or with minimum interactions, a systems approach is needed that focuses attention on the interactions.

One of the key subsystems in a dairy farm system is the utilization of grass after it is produced. Animal performance is a function of nutrient intake and availability, nutrient concentration, digestibility and metabolic efficiency (Mertens, 1994). Intake, digestibility, and efficiency of utilization are characteristics of forages that determine animal performance, with variation in intake accounting for 60–90% of the variation in digestible energy or dry matter intake (Mertens, 1994). Models that can predict animal

©CAB INTERNATIONAL 1998. *Grass For Dairy Cattle*
(eds. J.H. Cherney and D.J.R. Cherney)

performance or nutrient requirements to meet production goals can imp-rove the efficiency of utilization of forages. Thus to model grass utilization by dairy cows requires that intake, digestion and efficiency of utilization be included in the model.

All effective scientists and nutritionists are modellers and systems ana-lysts. It is difficult, if not impossible, to investigate a topic or use a system without having some conceptual framework about how and why they operate. These conceptual models serve us well, but may not be as rigor-ous, quantitative or encompassing as we need. Most think of mathematical models when the term modelling is used. Mathematics forces models to be rigorous (not fuzzy) and quantitative, but the basis for modelling is funda-mental understanding of the real system, in our case, the biology of intake, digestion and metabolism of grass forages resulting in meat and milk production.

Intake

Numerous factors influence voluntary intake (Fig. 14.1). How many of these should or can be included in predictive models of voluntary intake is subject to debate. Attempts to predict voluntary intake accurately and rou-tinely have met with frustration, in part because routine laboratory methods do not measure the characteristics of forage that are the true determinants of differences in voluntary intake.

Animal factors

Ingvartsen (1994) compared 23 multiple regression models of voluntary dry matter index (DMI) (kg day^{-1}). Twenty-one of those models included milk yield, either directly or as fat-corrected milk (FCM). The two equations that did not include milk yield (Rook *et al.*, 1991) included milk protein and fat yield. For more complex models listed by Ingvartsen (1994), all those pre-dicting intake when metabolic factors limit intake use either energy retained in milk or milk yield. The Institut National de la Recherche Agronomique (INRA) fill system uses FCM and the Danish fill system uses potential milk yield in the herd, rather than individual milk yields (Ingvartsen, 1994). Clearly milk yield is a critical component driving intake and cannot be excluded from any model predicting voluntary DMI when fill is not limiting in dairy cows.

It has been suggested that the cow be thought of in terms of being an appendage to the udder, rather than the reverse (Brown, 1969). In high pro-ducing dairy cows, energy demands for milk constitute the largest portion of the energy requirements of the cow. DMIs in the range of 3.2–4 times maintenance are not uncommon in cows fed optimal diets (NRC, 1987), and

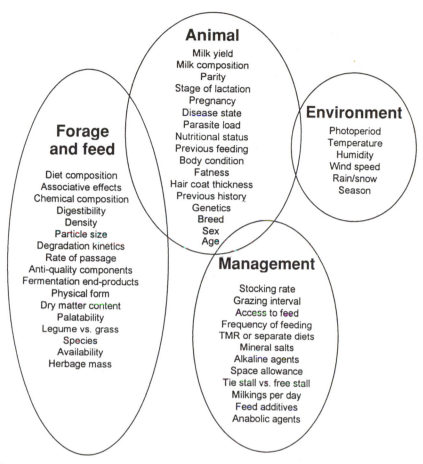

Animal
Milk yield
Milk composition
Parity
Stage of lactation
Pregnancy
Disease state
Parasite load
Nutritional status
Previous feeding
Body condition
Fatness
Hair coat thickness
Previous history
Genetics
Breed
Sex
Age

**Forage
and feed**

Diet composition
Associative effects
Chemical composition
Digestibility
Density
Particle size
Degradation kinetics
Rate of passage
Anti-quality components
Fermentation end-products
Physical form
Dry matter content
Palatability
Legume vs. grass
Species
Availability
Herbage mass

Environment
Photoperiod
Temperature
Humidity
Wind speed
Rain/snow
Season

Management
Stocking rate
Grazing interval
Access to feed
Frequency of feeding
TMR or separate diets
Mineral salts
Alkaline agents
Space allowance
Tie stall vs. free stall
Milkings per day
Feed additives
Anabolic agents

Fig. 14.1.   Factors affecting forage intake, digestion, and utilization.

much of this increased requirement is for milk production. Peak milk production occurs between 4 and 6 weeks post-partum, while peak DMI lags behind and does not peak until 10–14 weeks post-partum.

Liveweight and liveweight change have consistently appeared in models of intake prediction (Forbes, 1995). Lean cows have been shown to consume 20–25% more than fat cows (Bines and Morant, 1983). As with pregnancy, overly fat cows may exhibit reduced intake due to competition between the rumen and body fat for space (Forbes, 1995). Metabolic reasons are also likely to figure into the depressive effect (Scharrer and Langhans, 1990). Body condition score, which is a better estimator of body fat than weight, linear measurements, or combinations of weight and linear measurements (Thomson *et al.*, 1983) has been included in only a few models predicting intake (Ingvartsen, 1994). The Cornell net carbohydrate and protein system (CNCPS; Fox *et al.*, 1992) decreases predicted intake by 27% as empty body fatness increases from 21.3% to 31.5% (body condition score change from 5 to 7) using a simple multiplier effect. The French and Danish fill systems indirectly account for fatness, because intake capacity is determined within sex and breed (Ingvartsen, 1994).

With all forage diets, particularly grass, fill most often limits intake (Moore, 1994). If fill is limiting intake, daily feed intake is limited via volume constraints and stretch receptors (Mertens, 1973). Rate of disappearance of dry matter from the rumen and its reciprocal, retention time in the rumen, are due to digestion and passage. Resistance of forage residues to disappearance from the rumen is a function of physical and morphological characteristics of forage (Cherney *et al.*, 1991).

Retention of particles in the rumen is a function of rate and extent of digestion and rate of passage. Rate of passage is one of the most variable and critical characteristics of the animal's digestive system that affects fibre digestion and intake (Mertens and Ely, 1983). Mertens and Ely (1979), using a modelling approach to fibre digestion and passage in ruminants for evaluating forage quality, concluded that intake of digestible DM is influenced more by indigestible fibre and rate of passage than by rate of fibre digestion. Passage rate is determined by particle size reduction and rate of particle escape, although particles may not escape immediately upon reduction to a size possible for escape because of the particle-retaining capacity of the large particle mat in the rumen (Allen and Mertens, 1988).

For accurate measurement of passage of feeds through the gut, markers that mimic the flows of various dietary components must be utilized. This is possible only with markers that behave in definable and predictable ways (Faichney, 1993). Diets consumed by cattle are heterogeneous in nature, and one marker may not adequately reflect what the animal is eating (Cherney *et al.*, 1991). Once passage data are obtained, suitable mathematical models are required to interpret data. Ellis *et al.* (1994) review some of the methodology for estimating passage kinetics. Finding of suitable markers continues to be a major obstacle in passage work. Passage data for grazing cattle is particularly difficult to obtain. Using probability of escape as the passage rate is another approach used to model passage kinetics (Allen and Mertens, 1988). If the biological functions involved in passage are understood, then they can be mathematically modelled. It also has to be remembered that digestibility is a function of rate of digestion and rate of passage, and that models must integrate these factors when attempting to predict animal performance (Mertens, 1993b).

The NRC (1989) increases energy requirements for cows in the last 2 months of gestation due to the increased needs of the growing fetus at this stage. This increase in the nutrient requirement should increase predicted DMI. On the other hand, Ingvartsen *et al.* (1992) indicated that intake capacity is reduced approximately 0.17 kg per week from the 26th gestation week. Forbes (1995) suggested that even though the rumen becomes compressed during the latter stages of gestation, data do not support a purely physical mechanism of intake depression. It is suggested that oestrogens may be largely responsible for the decreased intakes (Forbes, 1995). Other factors include discomfort, preoccupation with parturition, or other endocrine changes (Forbes, 1995). Whatever the cause, there is ample data to suggest that there is a reduction in intake capacity during late pregnancy, while at the same time the growing fetus is requiring more nutrients. Knowledge of the factors involved would allow for accurate modelling of intake of cows during this period.

Plant factors

There are chemical, physical and morphological characteristics of forages that are related to their retention time in the rumen (Fig. 14.1). Physical and morphological factors such as grinding energy, particle size, structural anatomy, specific density or gravity all are correlated with intake (Minson and Wilson, 1994). Chemical fractions, particularly fibre, lignin and protein have been associated with retention time. Fibre and lignin may be associated with particle reduction in the rumen, with higher fibre resulting in larger particles resulting in longer retention time. Insufficient protein may limit microbial growth, which would result in longer retention time. Many of the physical and morphological characteristics of forages have been excluded from intake predictions because the methods are so laborious as to limit their routine use, there is insufficient data to validate models, or the data is not quantitative in nature, making modelling difficult. The neutral detergent fibre (NDF) content is closely related to many physical characteristics of forages. Until methods of determining physical characteristics improve markedly and can be shown to provide information additional to that obtained with NDF analysis, it is unlikely that physical characteristics will be included in many models.

Because NDF of feedstuffs influences both energy value and fill effect, Mertens (1992) has proposed that it can be used to link the physical and metabolic theories of intake. The relationship between theoretical equations based on physical and metabolic intake controls and experimentally determined regression of intake and ration NDF is shown in Fig. 14.2. The objective of the NDF-Energy Intake System is to identify optimal NDF contents that will maximize forage and fibre intake, while at the same time meeting milk production and body weight change requirements (Mertens, 1992). Mertens (1994) evaluated the NDF-Energy Intake System using summarization data of Jung and Linn (1988). This model demonstrated that effective models need not be complicated or contain many variables. They do need to account for animal, feed and situations under which the cows are fed. The NDF-Energy Intake System assumes that all NDF acts alike, which is not true as density, chewing requirement, digestion rate, and extent of digestion

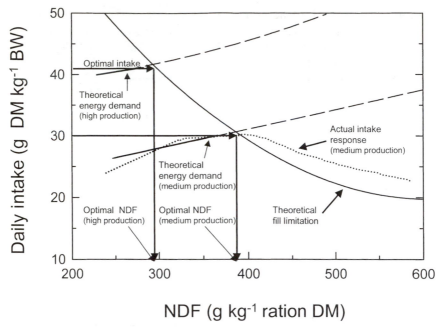

Fig. 14.2.   Illustration of neutral detergent fibre (NDF)-Energy Intake System for predicting intake for optimal and low-fibre rations. Ruminants are theoretically limited by energy demand or by fill. Ration NDF concentration is negatively associated with energy availability and positively correlated with fill. This relationship (shown by arrows) can be used to estimate optimal NDF concentration for a targeted level of animal intake and production. BW = body weight of animal (adapted from Buxton and Mertens, 1995; used with permission.

varies among sources (Mertens, 1992). Despite this, 70–80% of the advantage of balancing dairy rations for carbohydrates can be achieved by balancing for NDF (Mertens, 1992).

Environmental and management factors

Few models include management factors (Fig. 14.1), but these can impact intake significantly under certain situations and should be included in models that will be used under a wide range of conditions. Some effects, such as increased efficiencies of feed utilization by animals implanted with anabolic or oestrogenic compounds can be adequately accounted for by using multiplier adjustment factors (Fox *et al.*, 1992). Other factors, such as milkings per day, may be accounted for through increased energy requirements. Ingvartsen and Anderson (1993) indicated that the Danish-fill system distinguishes between loose-housing and tie-stalls, where intake in loose-housing may be increased by 4%, and that low space allowance decreased intake by up to 8%.

There is no question that environment affects intake. There are numerous data to indicate that animals will reduce intake above thermoneutrality and increase intake below thermoneutrality (NRC, 1981). Most models, which generally are developed under thermoneutral conditions, may not require temperature effects to be included. Another environmental factor, photoperiod, will increase intake by 0.32% per hour (Ingvartsen, 1994). The Danish fill-system, developed in a more northern part of the world where photoperiods change markedly between winter and summer, can be adjusted for photoperiod, but we are unaware of any other models that correct for this. The CNCPS uses environmental effects and researchers have demonstrated effectiveness under certain conditions of environmental stress (Fox and Tylutki, 1997). Animals will adopt behavioural changes in response to environmental stresses, which would tend to mediate responses observed under research conditions (Forbes, 1995). Responses are likely to be more noticeable with cattle grazing grass pastures than those housed inside, in part because these animals are likely to experience more environmental stresses or extremes. It is probable that many responses to environmental factors can be modelled through changes in energy requirements.

Factors affecting animal behaviour, such as access to bunk space, which may be influenced by herd social order, may be difficult to model. Physiological functions may factor in and there is little published information in this area that could be used to quantify these effects. Increasing emphasis on animal comfort and well-being is likely to increase research in this area. It is possible that these effects, like environmental effects, could be modelled through changes in energy requirements.

Pasture factors

Curran and Holmes (1970) concluded that the major obstacle to accurate prediction of intake in grazing cows was the lack of accuracy of measurement of intake. Estimates of intake in the field have improved considerably in the last 20–25 years (Black, 1990), and have allowed for the elucidation of factors involved in predicting intake (Dove, 1996). Despite this, efforts at modelling intake on pasture are still hampered by techniques which allow for simple estimation of diet composition (Dove, 1996). While this is not a modelling problem *per se*, it does illustrate the importance of being able to obtain data. A model will be of little use in practical situations if the parameters needed to run the model are difficult or impossible to obtain.

Dove (1996) indicated that under pasture conditions in Australia, the point at which rumen fill begins to limit intake, rather than the capacity of the animal to use energy, is at around 75% digestibility, so that intake of most grazing animals with abundant pasture available will usually be limited by fill. This suggests that models such as the energy-NDF system described by Mertens (1992) may be of use under grazing situations.

Predicting voluntary intake of animals on pasture is more complicated than in other situations, however, because factors other than fill or energy may limit intake. The amount of energy that must be expended to harvest forage will affect intake prediction, as will previously mentioned environmental effects. Pasture factors, including dry matter content, forage yield, allowance and sward heterogeneity, are factors that will affect intake predictions (Minson and Wilson, 1994; Chapter 11).

Other factors

Intake predictions on pasture become more complicated if animals are supplemented. Depending on the quality of the forage consumed, animals may substitute supplement for forage, there may be an additive effect, or there may be no change at all (Moore, 1994). To predict intake of grazing cows accurately, models must include effects of herbage mass and sward structural characteristics, as well as the substitution between herbage and supplement (Dove, 1996), in addition to forage and animal considerations used for housed animals.

There is a correlation between digestibility and intake that can be explained by factors affecting fill. When fill limits intake, digestibility will be low because cell wall (NDF) will be high. When fill limits intake, then a slowly digesting fibre will result in a greater filling effect. When energy demand controls intake, digestibility will be high because cell wall (NDF) is low. Mertens and Ely (1983), using a dynamic model of fibre digestion and passage, indicated that the relationship between feed intake and digestibility is a result of rate of passage. Their model indicated that an increased rate of passage would result in reduced digestibility, but increased intake. An effective intake model then would not include digestibility alone, but would include the mechanisms of intake control.

Modelling Digestion

Increasing level of intake is associated with reductions in digestibility. When intake of a particular forage is restricted, digestibility increases generally as a result of decreased rate of passage (Grovum and Williams, 1977). Reliance on strictly ruminal kinetics is cautioned here. This is because reductions in total tract digestibilities are often less pronounced than reductions in ruminal digestion due to the increasing contribution of lower tract digestion with increasing intake (Staples et al., 1984). Factors such as increased forage maturity, physical processing, increased level of intake and dietary supplementation of rapidly fermented carbohydrates will shift digestion from the reticulo-rumen to the large intestine (Merchen and Bourquin, 1994). In forage-fed cattle, however, ruminal digestion generally accounts for 90% or more of cellulose digestion, and 60–90% of hemicellulose digestion (Armstrong and Smithard, 1979).

Animal factors

Digestion kinetics are important because they determine the proportion of consumed nutrients that can be used by the animal (Mertens, 1993a). Modelling rate and extent of digestion can be used to (i) simulate aspects of the system that we think are important to aid in our understanding of biological processes, (ii) predict system output, empirically, without considering underlying mechanisms or theories, and (iii) optimize or manipulate the real system (Mertens, 1993a). There are three basic sources of error in measuring digestion kinetics: (i) inaccurate data, (ii) improper mathematical models, and (iii) inappropriate fitting of the data to these models. Three data characteristics that have critical impact on modelling and interpretation of digestion kinetics are (i) the system used to measure kinetic changes, (ii) the specific component of which kinetic information is measured, and (iii) the design of sampling times and replication (Mertens, 1993a).

As we study the process of digestion we see three processes occurring: digestion or hydrolysis, passage and possibly absorption. To model the process of digestion, the mechanisms of digestion, passage and absorption must be described in some way. We can separate the effects of digestion and passage by placing feeds *in situ* or *in vitro* and regulating time of digestion with no passage. This process clearly demonstrates that there is an indigestible fraction that remains in the bag or the filter. When results are plotted, we note that digestion is not instantaneous, so rate becomes important. Finally, we note that there is a lag for some fibre components.

The indigestible fraction is an important kinetic parameter that defines an intrinsic property of the cell wall and affects animal utilization (Mertens, 1993b). Measurement of the indigestible fraction is critical to estimating other kinetic parameters of digestion, the potentially digestible fraction, lag time, and the fractional rate constant. A critical biological concept is that rate of digestion can only apply to a fraction that is potentially digestible, so selection of the appropriate end-point of digestion is essential. Mertens (1977) demonstrated that for NDF residues with lag time < 6 h and digestion rates > 0.06 h^{-1}, indigestible residues can be estimated using a 72-h fermentation time. The 96-h residues are better used to estimate indigestible fractions when digestion rates are < 0.06 h^{-1}.

Plots of NDF residues often exhibit lags before digestion begins (Mertens, 1977). Digestion may be simply described as a sequential two-step process involving attachment or association of microbes with the cell wall followed by digestive hydrolysis of the substrate (Allen and Mertens, 1988). There is evidence of a lag *in vivo*, but other factors such as enzyme limitations due to lack of sufficient numbers of bacteria, or substrate characteristics such as hydration or chemical alteration of the substrate may affect lag *in vitro* (Mertens, 1973). The inclusion of lag in the model is critical to determining rate of digestion, however. Rate of digestion determinations are sensitive to lag, whereas standard *in vitro* dry matter digestibility (IVDMD) determinations are much less sensitive. This suggests that rate determinations are much more sensitive to method and technique used than are IVDMD determinations (Mertens, 1993a). There is a relationship between the disappearance of dry matter and the production of gas *in vitro*, which suggests that accumulated gas pressure readings could be used to measure forage digestion *in vitro* (Schofield and Pell, 1995).

For passage we find two processes occurring, particle size reduction and particle escape from the rumen. There are many factors that regulate the mixing, segregation and passage of particles from the rumen, including particle size reduction, hydration, gas buoyancy, specific gravity or volume, particle shape, and stratification of contents and mixing (Mertens, 1993b). Because so many factors affect passage, simple models of passage are insufficient both mathematically and biologically to represent the functioning rumen in animals. Kinetic analysis can rarely distinguish between two or more distinct rates using excretion data, and only the two most-limiting processes are described. For forage cell-wall particles, this appears to be selective retention of large particles and rate of escape of small particles (Mertens, 1993b).

Which process limits passage, reduction or escape, may be dependent on the particular example being investigated. Passage in high producing cows fed high quality forage with concentrates may be limited by rate of escape (Mertens, 1993b). In cows consuming all forage or high forage diets containing high-fibre forages, particle size reduction may limit intake. Biological processes are associated with each rate in multicompartment models. Approaches other than excretion data, such as measurement of rumen pool sizes, selective marking of components or selective sampling of sites in the digestive system are needed to advance our understanding of passage and our ability to model it.

Plant factors

The objective in measuring digestion kinetics of cell walls is to identify kinetic properties that are intrinsic to the cell wall (Mertens, 1993b). It thus becomes critical to ensure that the system itself is not the limiting factor in digestion. In the case of N supplementation, Cherney and Cherney (1997) indicated that while researchers nearly always cited a procedure used for digestion studies, it was not clear in many situations if N was supplemented in the inoculum buffer. Procedures accounting for 62% of over 100 cited digestion studies in six refereed journals either do not include N in the composition of the buffer, or list it as optional. Factors such as physical processing, intake level and supplement feeding are not accounted for *in vitro* and this may cause discrepancies between *in vitro* and *in vivo* estimates of digestibility (Weiss, 1994). Limitations to using standard *in vitro* or *in vivo* methods to predict *in vivo* digestion become apparent: (i) standard fermentation end points may result in incorrect estimates of maximum digestibilities, (ii) selection of procedures may affect *in vitro* estimates of digestibility, and (iii) the influence of the competition between passage and digestion is not accounted for. It is critical to differentiate between these estimates of digestibility. Estimates of digestibility are usually determined after 48 h of incubation, whereas complete extent of digestion may require longer incubation. The difficulties involved in obtaining data for model development and validation are clear.

Understanding biological mechanisms affecting forage digestion in relation to chemical analyses is critical to accurate prediction of animal response to a diet (Van Soest, 1993). Despite this, most attempts to relate differences in forage cell wall digestion kinetics to cell wall chemical composition have been unsuccessful. A notable exception is lignin, the chemical constituent most often identified as limiting fibre digestibility (Buxton and Mertens, 1995). Lignin is indigestible by ruminants or microbes, and inhibits the digestion of hemicellulose, probably accounting for its close association with digestibility. Notwithstanding this association, every species has a different relationship between lignin content of dry matter and dry matter digestibility (Sullivan, 1959). The interaction between NDF content and lignin content accounts for a large portion of the diversity, and unity of forage populations can be achieved by expressing lignin on a fibre basis (Van Soest, 1993). Some performance models such as the CNCPS include lignin on a fibre basis as one of the measured parameters.

Plant maturity is one of the most important sources of variation in digestion kinetics of forages (Buxton and Mertens, 1995). Cherney *et al.* (1993) characterized changes in digestion kinetics of five temperate perennial grasses with increased maturity and reported that differences in digestion kinetics between grass species were small in relation to changes due to maturity. Data reported by Smith *et al.* (1972) also suggest smaller changes due to species than due to maturity. Cherney *et al.* (1993) noted that potentially digestible fibre increased until shortly after the stem elongated, but that the indigestible fibre increased linearly with harvest date so that most of the increase in NDF with maturity was indigestible (Fig. 14.3).

Differences observed between species may be more attributed to differences between grass and legumes or temperate versus tropical grasses than species *per se* (Buxton and Mertens, 1995). Legumes are generally more digestible than grasses, due in part to leaf : stem ratios. As plants advance in maturity the leaf : stem ratio generally decreases, and this decrease occurs at

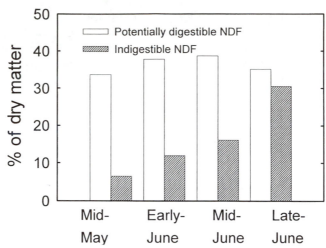

Fig. 14.3. Potentially digestible neutral detergent fibre (NDF) and indigestible NDF of perennial grasses by harvest date averaged over five species and three replicates (*n* = 15; data from Cherney *et al.*, 1993).

a greater rate in grasses than legumes. Dry matter digestibility of temperate forages is usually greater than that estimated for tropical forages. Cool-season temperate grasses often have higher leaf : stem ratios than warm-season tropical grasses and cool-season grasses normally are grown under cooler temperatures. Both of these factors would tend to reduce fibre and improve digestibility. Again, it seems as if the link of fibre and/or lignin on a fibre basis could be key in modelling species effects on animal performance.

Other factors

Dairy cattle fed grass forages are often given supplements of concentrates. Depending on circumstances, effects on digestibility of added concentrate may be positive or negative. Positive effects are most often noted when the supplement provides a nutrient, most often protein, that is deficient in the grass (Merchen and Bourquin, 1994). Negative associative effects on grass digestibility might be expected to occur when high levels of concentrate are fed. Theories to account for negative effects include (i) lag time of digestion increases, (ii) rate of digestion decreases, (iii) potential extent of digestion decreases, or (iv) a combination of the first three hypotheses (Mertens and Loften, 1980). Quantifying a preference for a substrate or proliferation of bacteria may be difficult. The effects of pH on fibre digestion kinetics can be dramatic and may be more quantifiable (Grant and Mertens, 1992;

Fig. 14.4). Few models have attempted to incorporate associative effects; one model, the CNCPS, uses a rumen sub-model which is pH sensitive (Pitt *et al.*, 1996). The model reduces fibre digestion when pH is decreased

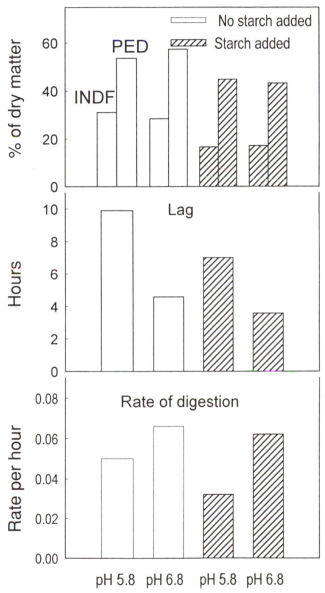

Fig. 14.4. Effect of starch and pH on NDF digestion kinetics of bromegrass (*Bromus inermiss* Leyss) hay. INDF = indigestible NDF; PED = potential extent of NDF digestion; Lag = lag time. Starch is added at 56% of initial DM (data from Grant and Mertens, 1992).

below 6.2 and when effective NDF in the diet is less than 20%, thereby reducing microbial yield (Pitt *et al.*, 1996).

Modelling Metabolism and Utilization

Animal factors

Ruminants cannot convert all of the gross energy that they consume into energy for maintenance and production. There are energy losses in faeces, urine, gas and heat (Fig. 14.5). Of the variation in energy intake among forages, only 5–15% is generally attributed to differences in metabolic efficiency (Buxton and Mertens, 1995). In the high-producing dairy cow, in which intake often limits production, improved understanding and modelling of metabolic effects may be the key to improvements in animal productivity.

Four major factors or events that tend to modify the long-term homeostatic equilibrium of ruminants are age, season, reproduction and lactation (Mertens, 1996). Because these factors occur over the long term, the driving function in a model predicting intake and performance should reflect the coordinated neural, hormonal and physiological changes that influence the desired body energy status for each physiological state. Hormonal and enzymatic regulation is important in metabolic efficiency, but is beyond the scope of this chapter. Readers are referred elsewhere for detailed coverage of this subject (Forbes and France, 1993; Mertens, 1996).

Ruminants absorb much of their energy as volatile fatty acids (VFAs) from the rumen. The principle volatile fatty acids produced by ruminal fermentation of forages are acetate, propionate, butyrate, with some other minor volatile fatty acids. The proportions of these VFAs produced is dependent on the substrates available. Propionate yields more energy than acetate upon ruminal fermentation.

Plant factors

Little propionate is produced when soluble carbohydrate content is low (Minson, 1990), which is more likely to occur with grasses, particularly tropical grasses, than with legumes. Efficiency of utilization is thus lower for grasses with low soluble carbohydrates than when they are higher (Galyean and Goetsch, 1993). There is often also an inverse relationship between rate of carbohydrate fermentation and ammonia accumulation (Cherney *et al.*, 1994), particularly in highly digestible grasses. The conversion of ammonia to urea requires about 12 kcal ME g^{-1} of nitrogen (Van Soest, 1994). Knowledge of how chemical composition of forage affects energy utilization should allow modellers to improve their predictions of animal performance.

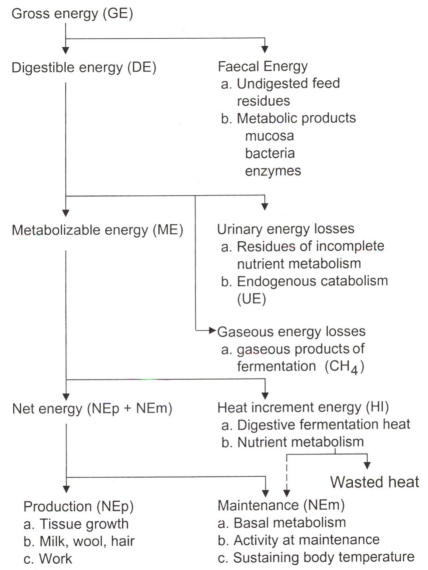

Fig. 14.5. Energy distribution in body processes (adapted from Maynard *et al.*, 1979).

In a series of studies, USDA researchers at Beltsville, Maryland, compared steers fed lucerne (low NDF) with those fed orchardgrass (high NDF) in terms of energy utilization. Steers fed lucerne gained more empty body weight, fat, protein, gross energy and carbon than those fed orchardgrass at equalized intakes (Waldo *et al.*, 1990). Waldo *et al.* (1990) attributed the somewhat lower partial energetic efficiency of orchardgrass to the higher NDF content of orchardgrass compared with lucerne. In another study, Varga *et al.* (1990) also reported that ME utilization was more efficient for steers fed lucerne compared with those fed orchardgrass. Steers fed lucerne had higher daily gross energy intake, and higher DE, urine energy, ME, heat production, and energy retention, but lower faecal energy losses than those steers fed orchardgrass (Varga *et al.*, 1990). Higher digestibilities of dry matter, organic matter, crude protein and neutral detergent solubles, and lower digestibilities of NDF and ADF were associated with steers fed lucerne (Varga *et al.*, 1990). Huntington *et al.* (1988) indicated that interrelationships between N and energy metabolism were responsible for increased efficiency of utilization by Holstein steers fed lucerne compared with orchardgrass in their study.

Portal drained viscera in steers fed orchardgrass had 10% higher oxygen (energy) consumption (Huntington *et al.*, 1988), 10% less incremental energy absorbed as acetate (Glenn *et al.*, 1989) and a greater load of gut contents (Waldo *et al.*, 1990) than steers fed lucerne. This suggests that the viscera are a major energy user and that the more indigestible fibre in a ration, the more energy required for viscera, leaving less energy for production. Greater intake of indigestible residue that must be processed through the gastrointestinal tract, could increase energy and protein requirements for maintenance of the gastrointestinal tract (Galyean and Goetsch, 1993). Greater time spent ruminating with increased indigestible fibre would also result in greater energy expenditure (Coleman *et al.*, 1989). As we gain more understanding of how feeds are utilized by various tissues, data can be incorporated into models to explain differences in efficiency of utilization and animal performance.

The ARC (1980) uses a system based on metabolizable energy (ME), while the NRC (1989) uses a system based on net energy (NE). Actual ME or NE values are difficult and expensive to obtain. The AFRC (1993) has published a number of equations, based on feed type, used to estimate ME of feeds, primarily based on *in vitro* digestibility or fibre measurements. Correlations range from 0.14 for grass silage to 0.90 for barn dried grass hays (AFRC, 1993). Clearly forage factors other than chemical composition or digestibility affect utilization efficiency, which would affect maintenance requirements. A more complete understanding of factors affecting utilization efficiency will require studies with concurrent estimates of energetic efficiency, energy use by specific tissues, and assessment of hormonal patterns (Galyean and Goetsch, 1993). If the objective of the modeller is to explain differences in efficiencies of feed utilization or to account for the 5–15% variation in performance attributed to differences in efficiency, then it will be necessary to have efficiency data in models predicting animal performance.

Conclusion

The efficient utilization of grass after it is produced is key to sustainability for many dairy farming systems. Models that can predict performance are key to improving efficiency. Accurate prediction of intake, which accounts for 60–90% of the variation in animal performance, is critical to this modelling effort. Of the many factors influencing intake, the animal factors of milk production and liveweight change, and the plant factor of NDF appear to most affect predictions of intake, regardless of situation. Other factors certainly affect intake and are important in specific situations. Understanding the kinetics of digestion is crucial, particularly hydrolysis and passage, because they determine the proportion of nutrients in grass that can be used by the animal. Hydrolysis and passage are the primary processes involved in digestion. In modelling digestion, it is important to understand and be able to measure accurately the factors influencing indigestible fraction, rate of digestion, and lag. For passage we must be able to model rate of escape and particle size reduction. Grass chemical composition affects efficiency of metabolic utilization. An improved understanding and modelling of metabolic utilization may be the key to improving performance in the high-producing dairy cow, where intake is often limited by fill.

There are obviously many factors involved in modelling intake, digestion and ultimately animal performance. Some modellers prefer complex models with a low level of aggregation while some researchers (Minson and Wilson, 1994; Forbes, 1995) have suggested that it might be more prudent to use models that are limited to more specific situations. Whatever approach is used, it is important to understand the biology involved before modelling can take place. Understanding the biology can help to determine which data are important to collect, and also to understand the limitations of empirically based models. It is also important to comprehend that biological models must fit actual data. A systems approach involving modelling and focusing on the myriad of interactions involved in the intake, digestion and metabolism of grass forages will allow improved grass utilization efficiency.

Acknowledgements

The helpful review comments of Dr Mary Beth Hall were greatly appreciated.

References

[AFRC] Agricultural and Food Research Council (1993) *Energy and Protein Requirements of Ruminants*. CAB International, Wallingford, 159pp.

Allen, M.S. and Mertens, D.R. (1988) Evaluating constraints on fiber digestion by rumen microbes. *Journal of Nutrition* 118, 261–270.

[ARC] Agricultural Research Council (1980) *The Nutrient Requirements of Livestock*. Technical Review by an Agricultural Research Council Working Party. Commonwealth Agricultural Bureaux, Farnham Royal, 351pp.

Armstrong, D.G. and Smithard, R.R. (1979) The fate of carbohydrates in the small and large intestines of the ruminant. *Proceedings of the Nutrition Society* 38, 283–294.

Bines, J.A. and Morant, S.V. (1983) The effect of body condition on metabolic changes associated with intake of food by the cow. *British Journal of Nutrition* 50, 81–89.

Black, J.L. (1990) Nutrition of the grazing ruminant. *Proceedings of the New Zealand Society of Animal Production* 50, 7–27.

Brown, R.E. (1969) The conversion of nutrients into milk. In: Swan H. and Lewis, D. (eds) *Proceedings Third Nutrition Conference for Food Manufacturers*. Churchill, London, pp. 23–35.

Buxton, D.R. and Mertens, D.R. (1995) Quality-related characteristics of forages. In: Barnes, R.F., Miller, D.A. and Nelson, C.J. (eds) *Forages Vol. II. The Science of Grassland Agriculture*, 5th edn. Iowa State University Press, Ames, Iowa, pp. 83–96.

Cherney, D.J.R. and Cherney, J.H. (1997) In vitro digestion: how reliable and uniform is the technique? In: Williams, M.J. (ed.) *Proceedings of the American Forage and Grassland Council, 13–15 April, Fort Worth, Texas*. American Forage and Grassland Council, Georgetown, Texas, 6, 183–187.

Cherney, D.J.R., Mertens, D.R. and Moore, J.E. (1991) Fluid and particulate retention times in sheep as influenced by intake level and forage morphological composition. *Journal of Animal Science* 69, 413–422.

Cherney, D.J.R., Cherney, J.H. and Lucey, R.F. (1993) In vitro digestion kinetics and quality of perennial grasses as influenced by forage maturity. *Journal of Dairy Science* 76, 790–797.

Cherney, D.J.R., Russell, J.B. and Cherney, J.H. (1994) Factors affecting the deamination of forage proteins by ruminal microorganisms. *Journal of Applied Animal Research* 5, 101–112.

Cherney, J.H. and Cherney, D.J.R. (1993) Annual and perennial grass production for silage. In: *Silage Production from Seed to Animal, Proceedings from the National Silage Production Conference*. NRAES-67. Northeast Regional Agricultural Engineering Service, Ithaca, New York, pp. 9–17.

Coleman, S.W., Forbes, T.D.A. and Stuth, J.W. (1989) Measurement of the plant–animal interface in grazing research. In: *Grazing Research: Design, Methodology, and Analysis*. Crop Science Society of America, Special Publication 16. Crop Science Society of America and American Society of Agronomy, Madison, Wisconsin, pp. 37–52.

Curran, M.K. and Holmes, W. (1970) Prediction of the voluntary intake of food by dairy cows 2. Lactating grazing cows. *Animal Production* 12, 213–224.

Dove, H. (1996) Constraints to the modelling of diet selection and intake in the grazing ruminant. *Australian Journal of Agricultural Research* 47, 257–275.

Ellis, W.C., Matis, J.H., Hill, T.M. and Murphy, M.R. (1994) Methodology for estimating digestion and passage kinetics of forages In: Fahey, G.C., Collins, M.,

Mertens, D.R. and Moser, L.E. (eds) *Forage Quality, Evaluation, and Utilization*. American Society of Agronomy, Madison, Wisconsin, pp. 682–756.

Faichney, G.J. (1993) Digesta flow. In: Forbes, J.M. and France, J. (eds) *Quantitative Aspects of Ruminant Digestion and Metabolism*. CAB International, Wallingford, pp. 53–85.

Forbes, J.M. (1995) Reproduction and lactation. In: *Voluntary Food Intake and Diet Selection in Farm Animals*. CAB International, Wallingford, pp. 186–203.

Forbes, J.M. and France, J. (eds) (1993) *Quantitative Aspects of Ruminant Digestion and Metabolism*. CAB International, Wallingford, 515pp.

Fox, D.G., Sniffen, C.J., O'Connor, J.D., Russell, J.B. and Van Soest, P.J. (1992) A net carbohydrate and protein system for evaluating cattle diets: III. Cattle requirements and diet adequacy. *Journal of Animal Science* 70, 3578–3596.

Fox, D.G. and Tylutki, T. (1997) Accounting for the effects of environment on the nutrient requirements of dairy cattle. *Journal of Dairy Science (Supplement 1)* 80, 150.

Galyean, M.L. and Goetsch, A.L. (1993) Utilization of forage fibre by ruminants: In: Jung, H.G., Buxton, D.R., Hatfield, R.D. and Ralph, J. (eds) *Forage Cell Wall Structure and Digestibility*. ASA-CSSA-SSSA, Madison, Wisconsin, pp. 33–71.

Glenn, B.P., Varga, G.A., Huntington, G.B. and Waldo, D.R. (1989) Duodenal nutrient flow and digestibility in Holstein steers fed fomaldehyde- and formic acid-treated alfalfa and orchardgrass silage at two intakes. *Journal of Animal Science* 67, 513–528.

Grant, R.J. and Mertens, D.R. (1992) Influence of buffer pH and raw corn starch addition on in vitro fibre digestion kinetics. *Journal of Dairy Science* 75, 2762–2768.

Grovum, W.L. and Williams, V.J. (1977) Rate of passage of digesta in sheep. 6. The effect of level of food intake on mathematical predictions of the kinetics of digesta in the reticulo-rumen and intestine. *British Journal of Nutrition* 38, 425–436.

Huntington, G.B., Varga, G.A., Glenn, B.P. and Waldo, D.R. (1988) Net absorption and oxygen consumption by Holstein steers fed alfalfa or orchardgrass silage at two equalized intakes. *Journal of Animal Science* 66, 1292–1302.

Ingvartsen, K.L. (1994) Models of voluntary intake in cattle. *Livestock Production Science* 39, 19–38.

Ingvartsen, K.L. and Anderson, H.R. (1993) Space allowance and type of housing for growing cattle: A review of performance and possible relation to neuro-endocrine function. *Acta Agriculturae Scandinavica, Section A, Animal Science* 43, 65–80.

Ingvartsen, K.L., Anderson, H.R. and Foldager, J. (1992) Effect of sex and pregnancy on feed intake capacity of growing cattle. *Acta Agriculturae Scandinavica, Section A, Animal Science* 42, 40–46.

Jung, H. and Linn, J. (1988) Forage NDF and intake. A critique. In: *Proceedings of the 48th Minnesota Nutrition Conference*. Minnesota Agricultural Extension Service, St Paul, pp. 39–48.

Maynard, L.A., Loosli, J.K., Hintz, H.F., and Warner, R.G. (1979) *Animal Nutrition*. McGraw-Hill Book Company, New York, 602pp.

Merchen, N.R. and Bourquin, L.D. (1994) Processes of digestion and factors influencing digestion of forage-based diets by ruminants. In: Fahey, G.C., Collins,

M., Mertens, D.R. and Moser, L.E. (eds) *Forage Quality, Evaluation, and Utilization*. American Society of Agronomy, Madison, Wisconsin, pp. 564–612.

Mertens, D.R. (1973) Application of theoretical mathematical models to cell wall digestion and forage intake in ruminants. PhD thesis, Cornell University, Ithaca, New York.

Mertens, D.R. (1977) Dietary fibre components: Relationship to the rate and extent of ruminal digestion. *Federation Proceedings* 36, 187–192.

Mertens, D.R. (1992) Nonstructural and structural carbohydrates. In: Van Horn, H.H. and Wilcox, C.J. (eds) *Large Dairy Herd Management*. American Dairy Science Association, Champaign, Illinois, pp. 219–235.

Mertens, D.R. (1993a) Rate and extent of digestion. In: Forbes, J.M. and France, J. (eds) *Quantitative Aspects of Ruminant Digestion and Metabolism*. CAB International, Wallingford, pp. 13–51.

Mertens, D.R. (1993b) Kinetics of cell wall digestion and passage in ruminants: In: Jung, H.G., Buxton, D.R., Hatfield, R.D. and Ralph, J. (eds) *Forage Cell Wall Structure and Digestibility*. ASA-CSSA-SSSA, Madison, Wisconsin, pp. 535–570.

Mertens, D.R. (1994) Regulation of forage intake. In: Fahey, G.C., Collins, M., Mertens, D.R. and Moser, L.E. (eds) *Forage Quality, Evaluation, and Utilization*. American Society of Agronomy, Madison, Wisconsin, pp. 450–493.

Mertens, D.R. (1996) Methods in modelling feeding behaviour and intake in herbivores. *Annales de Zootechnie* 45 (Supplement 1), 153–164.

Mertens, D.R. and Ely, L.O. (1979) A dynamic model of fibre digestion and passage to evaluate forage quality. *Journal of Animal Science* 49, 1085–1095.

Mertens, D.R. and Ely, L.O. (1983) Using a dynamic model of fibre digestion and passage to evaluate forage quality. In: Smith, J.A. and Hays, V.W. (eds) *Proceedings of the XIVth International Grassland Congress, 15–24 June, Lexington, Kentucky*. Westview Press, Boulder, Colorado, pp. 505–508.

Mertens, D.R. and Loften, J.R. (1980) The effect of starch on forage fiber digestion kinetics in vitro. *Journal of Dairy Science* 63: 1437–1446.

Minson, D.J. (1990) *Forage in Ruminant Nutrition*. Academic Press, San Diego, California, 483pp.

Minson, D.J. and Wilson, J.R. (1994) Prediction of intake as an element of forage quality. In: Fahey, G.C., Collins, M., Mertens, D.R. and Moser, L.E. (eds) *Forage Quality, Evaluation, and Utilization*. American Society of Agronomy, Madison, Wisconsin, pp. 533–563.

Moore, J.E. (1994) Forage quality indices: development and application. In: Fahey, G.C., Collins, M., Mertens, D.R. and Moser, L.E. (eds) *Forage Quality, Evaluation, and Utilization*. American Society of Agronomy, Madison, Wisconsin, pp. 967–998.

[NRC] National Research Council (1981) *Effect of Environment on Nutrient Requirements of Domestic Animals*. National Academy Press, Washington, DC, 152pp.

[NRC] National Research Council (1987) *Predicting Feed Intake of Food-producing Animals*. National Academy Press, Washington, DC, 85pp.

[NRC] National Research Council (1989) *Nutrient Requirements of Dairy Cattle*. Revised, 6th edn. National Academy Press, Washington, DC, 157pp.

Pitt, R.E., Van Kessel, J.S., Fox, D.G., Pell, A.N., Barry, M.C. and Van Soest, P.J. (1996) Prediction of ruminal volatile fatty acids and pH within the net carbohydrate and protein system. *Journal of Animal Science* 74, 226–244.

Rook, A.J., Gill, M., Willink, R.D. and Lister, S.J. (1991) Prediction of voluntary intake of grass silage by lactating cows offered concentrates at a flat rate. *Animal Production* 52, 407–420.

Scharrer, E. and Langhans, W. (1990) Mechanisms for the effect of body fat on food intake. In: Forbes, J.M. and Hervey, G.R. (eds) *The Control of Body Fat Content*. Smith-Gordon, London, pp. 63–86.

Schofield, P. and Pell, A.N. (1995) Validity of using accumulated gas pressure readings to measure forage digestion *in vitro*: A comparison of involving three forages. *Journal of Dairy Science* 78, 2230–2238.

Smith, L.W., Goering, H.K. and Gordon, C.H. (1972) Relationships of forage composition with rate of cell wall digestion and indigestibility of cell walls. *Journal of Dairy Science* 55: 1140–1147.

Staples, C.R., Fernando, R.L., Fahey, G.C., Jr, Berger, L.L. and Jaster, E.H. (1984) Effects of intake of a mixed diet by dairy steers on digestion events. *Journal of Dairy Science* 67, 995–1006.

Sullivan, J.T. (1959) A rapid method for the determination of acid-insoluble lignin in forages and its relation to digestibility. *Journal of Animal Science* 18, 1292–1298.

Thompson, W.R., Theuninck, D.H., Meiske, J.C., Goodrich, R.D., Rust, J.R. and Byers, F.M. (1983) Linear measurements and visual appraisal as estimators of percentage empty body fat of beef cows. *Journal of Animal Science* 56, 755–760.

Van Soest, P.J. (1993) Cell wall matrix interactions and degradation-session synopsis: In: Jung, H.G., Buxton, D.R., Hatfield, R.D. and Ralph, J. (eds) *Forage Cell Wall Structure and Digestibility*. ASA-CSSA-SSSA, Madison, Wisconsin, pp. 377–395.

Van Soest, P.J. (1994) *Nutritional Ecology of the Ruminant*. Cornell University Press, Ithaca, New York, 476pp.

Varga, G.A., Tyrrell, H.F., Huntington, G.B., Waldo, D.R. and Glenn, B.P. (1990) Utilization of nitrogen and energy by Holstein steers fed fomaldehyde and formic acid-treated alfalfa or orchardgrass silage at two intakes. *Journal of Animal Science* 68, 3780–3791.

Waldo, D.R., Varga, G.A., Huntington, G.B., Glenn, B.P. and Tyrell, H.F. (1990) Energy components of growth in Holstein steers fed formaldehyde- and formic acid-treated alfalfa or orchardgrass silages at equalized intakes of dry matter. *Journal of Animal Science* 68, 3792–3804.

Weiss, W.P. (1994) Estimation of digestibility of forages by laboratory methods. In: Fahey, G.C., Collins, M., Mertens, D.R. and Moser, L.E. (eds) *Forage Quality, Evaluation, and Utilization*. American Society of Agronomy, Madison, Wisconsin, pp. 644–681.

Economics of Grass for Dairy Cattle

15

K.C. Moore

Department of Agricultural Economics, 226B Mumford Hall, University of Missouri, Columbia, Missouri 65211, USA

Introduction

Providing a ruminant such as a dairy cow a diet of grazed forages just sounds like common sense. In countries such as New Zealand and Australia, milk production has long been associated with a primarily grass-based grazing diet. However, USA dairy farms, blessed with plentiful and low cost grains and by-product feedstuffs, have largely shifted to total mixed ration dairies with little or no grazing by cattle. As milk prices have declined in recent times, coupled with a few years of very high grain prices, many USA dairy producers have expressed interest in pasture-based dairying. This chapter examines the economics of grass-based dairying. The first part of the chapter deals with dairy enterprise budgets. A thorough understanding of the costs and returns involved in milk production is the foundation for many farm management decisions. A description of the costs and returns in dairying, and a comparison of these figures for alternative production scenarios are useful in understanding the trade-offs involved in pasture dairying. The second part of the chapter discusses the reasons behind the renewed interest in grass-based dairying. Here research concerned with the economic consequences of a shift towards a more grazing based diet will be examined. The potential of grass dairying in the USA will be explored, with international competitiveness as one key factor that must be considered. This chapter looks primarily at the economics of pasture dairying in the USA, largely because this is where the current controversy over drylot vs.

grazing dairies exists. But the principles that will be used to analyse the economics behind grass-based dairying are valid in any location.

Dairy Enterprise Budgets

Of the many tools used to evaluate the financial performance of farms, enterprise budgets are some of the most useful. Enterprise budgets describe the costs and returns for a particular production activity on a farm. A grass dairy in New Zealand may require only a few enterprise budgets to fully describe the farm, i.e. milk cows, pasture and hay. Often the forage budgets are collapsed into the dairy production budget, especially if all of the forage is produced and used on the farm. A midwestern USA dairy would likely have more budgets for crops such as corn (*Zea mays* L.) and soyabeans (*Glycine max* [L]. Merr.), and possibly enterprise budgets for specialty hay crops such as lucerne (*Medicago sativa* L.). Enterprise budgets are especially useful in diversified operations, as they help identify which enterprises are profitable and those that need special management attention.

Dairy enterprise returns

Income from a dairy enterprise comes primarily from milk sales, but calf and cull animal sales also contribute to income derived from the milk production enterprise. Table 15.1 presents representative enterprise budgets for milk production from Missouri total mixed ration and grass-based dairies (adapted from Bailey and Hamilton, 1997 and Bailey, 1997). The vast majority of dairying income is from milk marketings, thus the importance of milk production per cow. Moderate sized mixed ration dairies in the midwestern USA often do not reach the milk yields per cow of the large drylot dairies in the south. Grass dairying competes by lowering feed costs via grazing while using supplemental feed to maintain competitive milk production levels. The other primary source of income to a dairy is cattle sales which includes calf and cull animal sales. Some dairies may derive additional income from leasing cattle, co-op patronage dividends, government payments, etc. but these will typically be relatively minor sources of income.

Dairy enterprise costs

Costs involved in a farm production activity fall into two basic categories: variable and fixed expenses. Variable (or operating) expenses are incurred only if production occurs, and as the name implies, vary with the level of production. Example variable costs include feed, fertilizer, supply and fuel expenses. Fixed expenses are incurred regardless of production activity,

Table 15.1. Representative dairy cow production costs and returns per cow for alternative production systems.

	Total mixed[1] ration	Pasture-based[2] dairy
Gross receipts		
Milk sales	$1950	$1950
Cull cow sales	196	137
Young stock sales	85	68
Miscellaneous income	20	0
1. Total gross receipts	$2251	$2155
Operating costs		
Feed (includes waste):		
Grain ration	$ 700	$ 440
Hay and haylage	300	278
Silage and green chop	70	0
Pasture	30	138
Total feed cost	$1100	$ 856
Veterinary and medicine	$ 72	$ 43
Other livestock materials and services	220	142
Machinery costs, feed preparation, etc.	115	72
Utilities	55	22
Other (insurance, property taxes, real estate repairs, custom hire, and miscellaneous)	67	99
Operating interest, 10%[3]	81	62
2. Total cash costs, except labour	$1710	$1296
3. Labour (operator, family, and hired)	$ 305	$ 255
4. Total all variable costs (Line 2 + Line 3)	$2015	$1551
Ownership costs		
Real estate interest, depreciation and taxes	$ 75	$ 86
Breeding herd investment	75	75
Machinery and equipment interest and depreciation	58	56
5. Total fixed costs	$ 208	$ 217
6. Total all costs (Line 4 + Line 5)	$2223	$1768
7. Income above all variable costs (Line 1 minus Line 4)	$ 236	$ 604
8. Total income above total costs (Line 1 minus Line 6)	$ 28	$ 387

[1]Source: Adapted from Bailey (1997), and Siebenborn and Bailey (1996).
[2]Source: Adapted from Bailey and Hamilton (1997).
[3]Operating interest is calculated at 10% of one-half of all operating costs.
All values are US$.

and the costs do not change (in the short-run such as in a given production year) as the amount of output changes. Example fixed costs include depreciation, insurance, interest on loans, and property taxes.

Fixed costs include cash commitments not directly tied to the level of production (e.g. mortgage interest), but also non-cash expenses such as depreciation. Economists will also often include *economic costs*, or charges, which represent returns to all factors of production. There are four basic factors of production: land, labour, capital and management. When an enterprise budget is developed to show economic (or full-ownership) costs, imputed charges are included for land, operator labour (which may include some management return) and capital. Take farmland as an example. If all land used in the operation is owned, and no debt is associated with the land (100% equity financed), then there would be no cost for the land in either variable or fixed cash expenses (i.e. no rent and no interest). Nevertheless, obviously land is used in the dairy enterprise, and the farmer expects some return to this factor of production. The producer would like to see enough profit from the operation so that a fair return is paid for the use of the owned land. *Opportunity costs* are used for this purpose.

Opportunity costs are the value of the foregone opportunity. In the case of cropland or pasture, land could be rented to others instead of being operated by the owner. The foregone rent represents the opportunity cost of farming the land to the owner/operator. Many times enterprise budgets include calculated opportunity costs for all capital inputs whether owned, rented or financed. The reason is that if actual cash costs of all capital assets including land are included in the budget, comparing production efficiency across farms is influenced unfairly by how the business is financed. Land costs would show up for the farms with mortgages or those who rent, but not for debt-free operations. This would cause the debt-free farms to appear much more efficient, even if they really had the exact same feeding efficiency, milk production levels, etc.

Table 15.1 details expenses of representative midwestern USA dairies, both for variable and fixed costs. These costs and returns are based upon Missouri conditions, and would vary as length of growing season, severity of winter, moisture levels, land costs, etc. change with location. Variable costs are easy to understand and generally easy to obtain from producers as they readily see the connection between these expenses and milk production. Feed costs include the cost of purchased feeds and those that are produced on farm. For homegrown feeds, costs are computed using opportunity costs; i.e. forage values for pasture and sales prices for grains and silage. In this sense, the dairy operation is buying the homegrown feeds at prevailing market prices. This is the proper way to handle goods produced and used internally on the farm.

The forage enterprise budget for a dairy (Table 15.2) includes the expenses for pasture production, and the return would be any hay or other harvested forage sales and the rent 'paid' by the dairy cattle. The milk herd

Table 15.2. Forage enterprise budget[1].

Sales to dairy enterprise:	
Grazing sales[2]	$13,807
Hay sales[3]	19,200
Total sales	$33,007
Operating expenses:	
Labour: unpaid family labour (valued at $7/h)	$ 4,200
Rent on land[4]	8,000
Fertilizer[5]	4,000
Seed, spray[6]	1,300
Custom harvesting[7]	8,640
Fuel costs	779
Repairs, parts and supplies	2,037
Truck expense	454
Insurance	352
Operating interest	1,278
Total variable costs	$31,040
Fixed costs:	
Real estate tax, depreciation and interest on buildings	$ 553
Interest and depreciation on machinery and equipment	4,244
Total fixed costs	$ 4,797
Total allocated costs	$35,837
Return to management and risk	−$ 2,830

[1] Source: Bailey and Hamilton (1997).
[2] Assumes grazed forage is valued at US$0.0617 per kg.
[3] Assumes 81 ha of which 40% is harvested for hay. Yields are 6720 kg ha^{-1} valued at US$0.0882 per kg.
[4] Rental rate is US$98.84 ha^{-1}.
[5] Assumes 81 ha at a cost of $49.42 ha^{-1}.
[6] Assumes 81 ha at a cost of US$160.62 ha^{-1} amortized over 10 years.
[7] Assumes 29.2 ha harvested at a cost of US$296.52 ha^{-1}.
All values are US$.

rents the pasture from the forage enterprise at a fair price for the forage consumed by the cattle. The concepts of the costs and value of pasture are further developed in later sections of this chapter. Forage values are used as costs to the milk cows so as not to give the dairy enterprise credit for profits rightfully generated by the forage enterprise. The forage enterprise 'rents' the land from the farm at prevailing rental rates, and 'sells' the hay and grazed forage to the dairy enterprise.

While feed costs account for the majority of variable dairy expenses (Table 15.1), other costs are also incurred. Labour is a major expense, and many farms hire additional help. Unpaid family labour is included in both the forage and the dairy budgets. While this represents an opportunity cost, and not a cash expense nor a tax deductible cost for the farm, it is included when examining all the costs of production. Farm profits are used to pay family living expenses in most cases, so this cost essentially represents the withdrawals from net income for this purpose. Additional variable costs include veterinary expenses, trucking costs, operating interest (which can be an actual cost or an opportunity cost for the capital tied up in variable costs), etc.

Fixed costs (Table 15.1) include both opportunity costs and cash costs. Depreciation is a non-cash expense to a business but is also included in fixed costs. Depreciation represents the annual loss in value of a capital asset used in production. For example, a tractor that is purchased and used on the farm will last for several years. The purchase cost of the tractor may be incurred during one year, but the value of the asset will last for many years. Depreciation expenses are the purchase cost of the asset, spread over its useful life. They capture the annual 'wear and tear' expense of owning a capital asset. Other fixed costs include depreciation plus the opportunity cost of owning land, buildings and other improvements on the farm, and a charge for owning the cow herd. Many farms will have actual cash outlays for fixed asset ownership, typically in the form of interest expenses or rental costs. Opportunity costs are used in enterprise budgets simply to capture the full costs of asset ownership, and to remove any influence of how assets are financed (e.g. equity or debt capital) when comparing budgets across farms.

Looking at the figures in Table 15.1, several items are noteworthy. First is the importance of feed costs in USA dairies, typically over half of total cash expenses. Whether grass-based or not, feed costs are by far the single largest cost category and therefore command much managerial attention. Milk production is held constant across dairy type in these budgets. The amount of supplemental feed a grazing dairy will choose to supply depends on a number of factors including feed prices, length of grazing season, milk price, cow response to supplements, quality of forage diet, type of grass dairy (e.g. seasonal or not), availability of supplements and ability of the farm to handle bulk quantities, facilities, etc. Some farms will have more difficulty maintaining milk production levels than others will. Many are moving away from the standard dairy genetics that are bred for high milk production in confinement. Dairy graziers do not need the highest producing cows, and many are finding that less expensive cows or even beef–dairy crossbreds are more profitable in a grass-based system.

Cull sales are lower in the pasture dairy (Table 15.1) due to an assumed lower cull rate and slightly lower cull cow prices. Gross income for the two alternative production scenarios are very similar, assuming the same level of milk production per animal. Variable costs are much reduced in the grazing dairy, with much of this saving coming from the substitution of pasture for grains that are more expensive and other harvested feedstuffs. Variable cost savings are also assumed due to lower health related costs, variable machinery costs, utilities and interest. Some additional variable costs are incurred for custom hay harvesting. Labour for the grass dairy falls, as less time is required for feed preparation, manure handling, etc. Fixed costs in this example are similar for the two alternative dairies. The mixed ration dairy is using existing facilities that have been on the farm for some time. Thus, depreciation and interest costs are not nearly as high as one would see in a new drylot dairy. The pasture-based dairy is a new start-up, with construction of a new milking centre but with used equipment. Total investment per cow excluding land for the pasture dairy was limited to just US$1914 including US$1000 for the cost of the cow. Fixed costs for converting a drylot dairy to a grazing dairy can be substantially different from the figures shown in these budgets. Investment in land, buildings, machinery and equipment must be managed carefully as fixed costs per cow or per kilogram of milk produced are critical measures of efficiency for a milking operation.

The 'bottom line' in an enterprise budget that includes opportunity costs for land, labour and capital becomes residual returns to management and risk. The use of the term 'residual' means that the remaining figure represents what is left after all costs have been subtracted from income, including opportunity costs. Economists like to construct budgets in this manner so that land, unpaid labour and capital are all explicitly recognized as true costs of production. Accountants are typically more concerned with accrual net farm income, or profit, which is the residual return to unpaid labour, management and equity capital. Farm expenses including hired labour, interest paid and depreciation are subtracted from gross income to arrive at farm profit. Costs for family and other unpaid labour, as well as any wholly owned capital provided by the owner, are not allowed as deductions from income. Profit is essentially labour and management earnings of the family involved, plus a return for the money invested in farm production assets by the owners. Economic cost basis enterprise budgets and a farm income statement prepared by an accountant are similar, except that economic costs include opportunity costs.

The figures in Table 15.1 show the total mixed ration dairy just covering total costs (return to management and risk), and having a small but better margin over variable costs, while the grass-based dairy shows improved net returns in both areas. While these figures are only representative, and numbers for any given farm will differ from those presented, the data do illustrate the reasoning behind the interest in pasture dairying. Milk prices during the last decade have failed to keep up with inflation, so producers are struggling to keep costs down. Grain prices are highly volatile and increase risks for dairies that must purchase such feeds. Machinery and equipment costs continue to rise, making capital replacement decisions difficult in the face of narrow profit margins. Feed costs and production levels are two of the items over which a farm manager has at least some degree of control. In addition, these two figures play a large part in the efficiency and profitability of a dairy. If pasture can be acquired at reasonable cost, this can lower both the amount and risk of feed costs. Management is required to maintain milk production levels, while at the same time keeping fixed costs at a reasonable level.

Evaluating Profitability in the Dairy Business

Enterprise budgets are not the only means of evaluating dairying profitability, and as discussed above they must be used and interpreted with care. Whole-farm planning, simulation modelling, optimization models and a host of other tools are available and important for farm decision making. However, enterprise budgets are used in most of the other decision tools, and serve as a simple and useful information source on their own. A few points are worth making about enterprise budgets. First is that they need a unit of reference, such as per cow, per hectare, per farm, or per kilogram of milk. The choice of unit is up to the decision-maker, and one set of enterprise costs and returns can be presented in several different ways. Budgets typically run for a year, corresponding to the calendar year, or tax year of the business. Too often management focuses on maximizing production per cow or per unit of land. If profit maximization is the goal (and in reality many times it is not the sole or even primary goal of the owner/operator), then *optimizing* rather than maximizing production should be the focus. Enterprise budgets are useful in describing costs and returns from a particular production scenario, but they do not contain information about how changes in the production system will affect net returns. Comparison of budgets representing alternative systems is useful in this case, but optimization on farm must still be performed by the manager. Production per animal, per man or per hectare is less important than optimizing profits for the total farm system.

What Does Grass Really Cost?

If the economics of grass for dairy cattle is to be examined, it is essential that an understanding of the true costs of pastures be provided. All too often pastures are the most under-managed, under-valued and most misunderstood resources on a farm. In countries where grass-based agriculture is the rule, this is not the case. However, in the USA, especially in areas where pastures are interspersed with row crop agriculture, grazing lands often do not command much managerial attention. In this section, we explore the economic costs of pastures. Enterprise budgets provide some of the data necessary, but other factors enter into the real cost of pasture. Understanding these factors and how much pasture really costs is essential when studying the economic consequences behind grazing-based dairies.

Feed costs are the single largest expense in a dairy enterprise and typically represent between 45% and 55% of total cash costs. Managing these costs is essential for producers wanting to improve the profitability of their herds. The first step in controlling these expenses is to know what it costs to supply various feedstuffs. This is straightforward when feeds are purchased, but can be confusing when feed is home raised. Dairies that feed almost exclusively from owned pastures, whether grazed or harvested forages, and do not have much opportunity for other supplemental feeds, do not often look at the cost of grass since this cost is simply rolled into the larger dairy enterprise budget. But in the USA, with our abundance of low cost cereal grains and by-product feedstuffs (from the processing of cereal grains and oilseeds), milk production from a grazed diet must compete with profits from confined feeding of dairy animals. Thus a producer looking into the cost of grass-based dairying, must understand the cost of grass just as they would the cost of any purchased feedstuff.

Pasture budget data from actual farms has been collected and analysed in Missouri for many years (Moore, 1997). While the vast majority of these pastures are used for beef cow–calf production, substantial interest exists in using such land for pasture dairying. Average operating costs (seed, fertilizer and lime; chemicals, machinery operation; repairs; machine and labour hire; utilities; taxes, miscellaneous, and calculated operating interest) for these beef cattle pastures during 1990–94 totalled US$28.61–38.94 ha^{-1}. Pastures used for milk cows would necessitate more costs for fertilizer and lime, and many acres may actually involve renovation into more desirable forage species. There would also probably be new capital charges for additional fence and water system development. When capital expenditures such as pasture renovation or paddock fences are represented in an annual pasture budget, only the annual portion of the expense is included. Such investments provide returns over many years, so it is important to spread the cost of such actions over the life of the investment. Fixed costs (machinery depreciation and interest; real estate depreciation, taxes, and an opportunity cost placed on the value of the land; and imputed owner/operator labour value) bring total annual costs of Missouri pasture to US$113.32–150.81 ha^{-1}.

Many operators view owned pasture as having very low annual cost. This is not the case as these figures show, placing average pasture costs at about US$131 ha^{-1}. Certainly if opportunity costs of land, labour and machinery investment are not included, cash costs of pasture production can be extremely small. However, while opportunity costs (i.e. the value of the foregone opportunity) are not cash costs, they should be included in total costs if producers want to compare purchased vs. raised feed costs equitably. Most people will not place money into an investment without expecting some rate of return. Opportunity costs reflect this return on assets provided to the business by the owner. Land costs are certainly cash expenses for those who rent or have mortgages on their farms. While no longer cash expenses for those who own their farms 'free and clear', there are still the opportunity costs of the capital tied up in those assets.

Pasture costs can and do vary widely. Renting is usually much cheaper than owning pasture; much Missouri pasture can be rented for US$62 ha^{-1} year^{-1}. Costs of owned dairy pasture can exceed US$175 ha^{-1} for intensively managed grass with high fence and water system development costs. Costs per kilogram dry matter produced and utilized is a useful basis for comparing pasture costs. This figure is useful for understanding the true costs of pasture, and can be used to compare the cost of pasture to other alternative feed sources. While quality will differ between pastures, as it does between a given pasture at different times of the year and/or stages of growth, cost per kilogram dry matter provides a uniform standard of comparison. Nutritive value of pastures also differs from that of harvested feeds. Quality differences are best addressed by evaluating the value of alternative feeds in a particular feeding situation. Least cost ration balancing programmes can be used for this purpose. Costs per kilogram dry matter forage consumed by livestock not only depend on the annual pasture costs but on the level of productivity and degree of pasture utilization. For example, US$131 annual cost pasture producing 5605 kg DM year^{-1}, and grazed to only accomplish a 40% utilization rate (i.e. only 2242 kg used), costs US$0.0584 kg^{-1} DM. The same pasture rented for US$62 ha^{-1} costs just US$0.0276 kg^{-1} DM forage consumed at the same 40% utilization rate.

Pasture yields depend on a number of factors including soil type, forage species, fertility level, rainfall, etc. Utilization rate, or how much of the annual production actually gets consumed by livestock, depends on the type of animal, grazing management practices (i.e. continuous vs. rotational grazing), size and shape of paddocks, location of water and shade, etc. As pastures are subdivided and animals are placed in paddocks for shorter periods and rotationally grazed, pasture utilization increases. It is this improvement in the amount (percentage) of forage consumed that can allow fence and water system development costs to be economically feasible. For example, take good quality dairy pasture producing 7848 kg DM year^{-1}, with a US$173 annual cost due to management-intensive grazing development costs, but which in turn allows a 67% utilization rate. Cost per kilogram dry matter consumed is US$0.0329, substantially lower than costs of poorly managed and grazed owned pasture. Annual pasture utilization rates of 85% can be reached with cattle grazed rotationally, which brings the above costs down to just US$0.026 per kg.

Hay or silage enterprise costs are calculated in much the same manner as pasture costs. Machinery operation and investment costs are increased, as are labour charges. Yields are much simpler to measure when forages are mechanically harvested. Once total costs of harvested forages are computed, they can be included in the comparison of feedstuff costs. Storage and feeding costs (including feeding losses or a utilization factor) for harvested forages need to be included in total costs when comparing across alternatives. These expenses, along with the costs to harvest the forage, typically raise the cost of feeding harvested forages to a level substantially above the cost of grazed material. Hay feeding costs can easily reach two to three times as much as the same material efficiently grazed. Costs per kilogram for concentrates vary widely across feeds and across time. Large dairies find it very profitable to pay close attention to by-product feed prices and buy in bulk quantities when opportunities present themselves. Prices generally range anywhere from US$0.09 kg^{-1} DM and higher for by-product feeds and cereal grains.

Ford and Hansen (1994) developed enterprise budgets for alternative dairy feeds based on Pennsylvania figures. Traditional non-intensive use of pasture provided 3318 kg ha^{-1} DM yield (estimated from the nutrient requirements of the cows and milk production levels) for a total cost of US$.026 kg^{-1}. Intensive rotational use of pasture provided 5135 kg ha^{-1} DM at a cost of US$.031 kg^{-1}. Hay production provided 4058 kg ha^{-1} DM but at a cost of US$.095 kg^{-1}, and corn silage yielded 8094 kg ha^{-1} DM with a cost of US$.061 kg^{-1}. Using these figures, the authors modelled two alternative pasture dairies, one with corn for grain and silage and one without, and compared these to an operation using no grazing. Crop mixes for the farms were developed to provide sufficient quality and quantity of feed to maintain average annual milk production per cow at 8209 kg. Annual fence and water costs were included in the two grazed systems. In the pasture system with grain, improved net income resulted from lower crop and general farm expenses as well as less purchased feed. Without grain production, the pasture system had much reduced crop expenses while purchasing more feed. Annual savings from reduced crop machinery provided additional savings. Both pasture systems improved net returns per cow about US$158 over the no pasture system.

Sheppard and Ford (1993) studied the role of pastures in dairy herd expansions. Their analysis used Pennsylvania State University research showing feed cost savings of US$0.50 cow^{-1} day^{-1} (during the 180-day grazing season) due to increased use of pasture. They cite research that examined the net feed value (feeding value in excess of production costs) of various crops. Pasture for grazing lactating cows had a net feed value of US$236.55 ha^{-1} compared with corn silage at US$106.80 ha^{-1} and lucerne silage at US$75.54 ha^{-1}. The authors concluded that pasture could improve the profit in a dairy operation, but equity in the operation remains critically low during an expansion. Ford (1996) computed net feed values for grass pasture, lucerne hay and corn silage. Feeding values were based on as fed values of US$0.011 kg^{-1} corn grain, US$0.31 kg^{-1} soyabean meal, and US$0.12 kg^{-1} legume hay. With an estimated as fed yield of 8969 kg ha^{-1} for grass pasture, net feed value amounted to US$917 ha^{-1}. This compares with net feed values per hectare for lucerne hay and corn silage of US$496 and US$624, respectively. This means that lucerne hay would have to yield

15,471 kg ha^{-1} to equal the feed value per hectare of pasture. A similar break-even yield for corn silage would be 56,278 kg ha^{-1}.

Determining pasture yields makes valuation difficult, but another factor to consider is when that yield occurs. Dry matter production does not come in uniform increments through time. Nor does it match animal demands through a season. Feed budgeting, where management of animals and the grazing system (e.g. fertilization, hay making, rotation speed, etc.) is designed to make full advantage of forage availability by matching production with animal demands, is used widely in countries where grass is the dominant feed source. In the USA, grazing competes with a diet of harvested forages and concentrates. The primary economic advantage of grazing is through providing a lower cost feedstuff. In many areas of the USA, the length of the grazing season limits the amount of time for pasturing cattle. This limits the potential for substituting lower cost grazing for higher cost harvested feeds. Another problem is the seasonal distribution of yields. Many areas would see approximately two-thirds of annual pasture production in the spring, followed by a 'summer slump' period and then roughly one-third of yield occurring in the autumn.

The temporal distribution of pasture yield influences both the management and economics of grazing operations. The spring flush of pasture creates some management challenges which are covered in other chapters of this book. Spring applied nitrogen is common practice for many grass pastures. The benefit is seen as improved pasture production and quality, and in most cases, this does occur. However, specifically when the added yield comes is of major importance. The economic reality is that additional dry matter production during the spring has relatively low value. It provides more yield at a time when animal demand cannot keep up anyway. Most often, this excess production is turned into hay or silage, but this leads to added costs. Intensive early stocking is practised in some areas to utilize and manage this peak in production. However, not all farms have the opportunity or desire to increase animal numbers for short periods. Extending the grazing season through earlier or later grazing opportunities, or providing additional yield during the summer slump period has much more economic value than the same additional yield during spring or autumn. Grazing provides its biggest advantage when substituted for costly harvested feedstuffs. Additional pasture yield during times when hay would otherwise be fed can have two to three times the value of additional pasture yield during surplus periods. Pasture management via cost-effective fertilization, alternative plant species, rotation management, etc. is very critical to the success of grass-based dairies. While annual dry matter production provides one measure important to the valuation of pasture, care must be taken to recognize the importance of when that yield occurs as well.

What is the Economic Benefit of Increased Pasture Grazing?

The next question to turn to is the value of grass in a dairy situation. With the high cost of land, and the expense of developing fence and water systems that allow effective utilization of standing forage, can grazing dairy cattle really improve farm profits? As with many economic questions, the true answer to this question is: It depends! Many different variables enter into the outcome from pasture-based dairying. Having developed an understanding of the cost and income structure of a milk cow enterprise, we can examine the potential impact of increased grazing on net returns.

Feed costs are crucial

The primary benefit of grass-based dairying has been a reduction in the cost of feeding the cow herd. The cows are expected to harvest much of their diet themselves via grazing, with management used in place of expensive equipment. Strategic supplementation of grazing cattle can often be used to increase output and profits simultaneously. With feed expenses being the single most costly item in a dairy, it is logical if not imperative that managers pay close attention to this item. As the cost of grain and the price of machinery used to grow and harvest various feedstuffs have risen, dairy producers have renewed their interest in grazing as an economic alternative to confinement feeding. Recent years have also seen an increase in research devoted to exploring the economics of grass dairying.

The last decade has seen a number of popular press articles describing the successes of dairies turned grass-based systems. The sheer volume of these articles attests to the interest level in pasture dairying and success that some producers have achieved. Testimonials must always be evaluated with a careful eye, but useful information can be obtained from cooperation with actual producers. Brown (1990) reported Pennsylvania farmer data after having worked with several operations switching to grass-based dairying. One farm calculated feed cost savings of US$0.49 cow^{-1} day^{-1}, with another reporting savings of between US$100 and US$150 cow^{-1} year^{-1}. While greater profits due to lowered feed costs were the major advantage cited by the surveyed farmers, reduced machinery expenses, improved animal health, consistently high quality forage, better reproduction, reduced workload, and improved management skills were also some of the benefits mentioned. The main disadvantages cited were the constant need to move cattle, increased dependency on the weather, and higher management requirements. Emmick and Toomer (1991) reported the results of a study of 15 New York State dairy farms that had recently shifted to more reliance on pasture. Total production cost savings averaged US$153 cow^{-1} year^{-1} for the study farms, with a range of between US$40 and US$290 year^{-1}. This amounted to a saving of nearly US$0.035 kg^{-1} of milk produced, averaged across the surveyed operations. Average grazing length for the 15 farms was 178 days, with the lowest cost improvement farm only grazing 70 days.

Comparing costs before and after a given farm has adopted some particular technology is difficult because years differ climatically resulting in different production levels; prices change over time, etc. Similarly, comparing data across samples of farms with different technologies (e.g. grazing versus confinement) is difficult because of heterogeneous management and resources across farms. One can use constant prices, adjust for rainfall and other climactic influences, evaluate data for varying herd sizes, etc. in an effort to account for some of these difficulties. Parker *et al.* (1992) chose to model farm performance of pasture versus drylot dairies in order to avoid some of these difficulties. The authors developed a linked spreadsheet model to compare confinement vs. intensively grazed dairy operations in Pennsylvania. A rolling herd average of 6800 kg of milk was used for both farm types, with the grazing herd on pasture from April to October but even then getting 30–35% of their daily energy requirements from concentrates to maintain milk production and cow body condition. Dry cows and replacements derived up to 95% of their energy needs from pasture during peak grazing. Results showed lower operating costs for the grazing dairy, primarily due to reduced crop production (including hay) expenses, lower purchased feed costs and reduced bedding costs. The overall improvement in income above operating costs per cow was US$121 for the pasture-based animals. Fixed cost changes, reproductive performance effects and labour needs are not included in this figure. Fixed costs would increase for fence and water system development, but these could be offset by reduced machinery and equipment investment and/or use. Supplementation was provided to the grazing-based animals, in part to maintain energy balance and reproductive efficiency. Oestrus detection may improve in a pasture dairy, but labour for breeding may also increase. The authors concluded that the grass dairy could sacrifice up to 467 kg milk cow^{-1} as compared to the confinement dairy and still be equally profitable. Survey data from Parker *et al.* (1993) reported grazing dairies were sacrificing about 365 kg cow^{-1} annual milk production, with an improvement in grazing performance possible over time as most of the surveyed farms had been pasture-based for less than 5 years. This same survey reported estimated operating cost savings of US$101 cow^{-1} for the grazing dairies over confinement dairy costs.

Effects on other costs and returns

While reduced feed expenses per cow or per unit of milk are most important to the ultimate economics of grass dairies, the impacts on other sections of the enterprise budget must be evaluated as well. Milk production will decline in many cases, especially if seasonal milking is adopted. The amount of production that can be sacrificed and leave net income improved is often 10% or less. In countries such as the USA, supplementation of lactating cows on pasture is often economically viable. In locations where grazing seasons are relatively short, pastures may be viewed as an economic supplementation strategy for harvested forages and grains.

Feed costs are not the only variable costs potentially influenced by increased dairy grazing. One estimate places non-feed cost savings due to grazing at US\$55 cow^{-1} year^{-1} (Ford, 1996). Animal health costs are often touted by pasture dairy producers as another advantage of a switch from confinement. Veterinary and medicine costs were reduced for grazing dairies in a survey conducted by Parker *et al.* (1993). Fewer foot problems and reduced mastitis are generally viewed as benefits coming from increased use of grazing. Barn and lot labour, bedding and utilities costs should decline as cows spend more time out on pasture. Breeding costs are influenced both positively and negatively when grazing. Heat detection can be improved by more frequent contact when moving animals, but with animals out on grass it may be harder to monitor. Pasture breeding is challenging, but improved herd health may reduce the number of services required. Machinery repairs and operating costs should decline as more grazing is implemented, and crop production costs would also fall.

Fixed costs are also impacted during a move to grass-based dairying. Fence and water system development can be very expensive. Costs depend on the size of system and number of paddocks, land topography, source and location of water, and a host of other factors. Pastures may need to be developed, resulting in annualized costs for establishing forages. These costs can be offset by a reduction in machinery and equipment inventory, if the plan reduces the amount of crop production. Capital replacement costs may fall as machinery and equipment lasts longer through reduced use. If milk production per cow declines, some dairies have added cows in order to increase farm profits, thus adding the fixed costs associated with more animals. If more land is required for the increased animal numbers, then this adds to fixed costs as well.

Elbehri and Ford (1995) incorporated fixed costs, machinery expense changes and labour demand differences into their monte-carlo simulation model for an intensive grazing Pennsylvania Holstein dairy farm. They modelled different farm scenarios representing ten different forage combinations of corn silage, lucerne hay and haylage, grass hay and haylage, and summer pasture. Economic performances of the farm scenarios were compared for varying criteria, and a risk assessment was also conducted. Milk production was held constant across the ten scenarios and crop costs came from Pennsylvania State University enterprise budgets. Results showed an economic advantage to farms through the adoption of summer intensive grazing. Assuming equal milk production, grazing improved simulated net returns by 14–25% depending on the assumed starting position regarding harvested forage system. This would amount to increased net cash income of between US$8400 and US$12,400 annually. A reduction in total cash expenses was the most significant factor in the income improvement, with cost of milk production falling between US$0.026 kg^{-1} and US$0.031 kg^{-1}. Grazing farms were a better option for risk averse operators, but management level was critical to success. A 5% reduction in milk production per cow could be withstood on pasture dairies before pasture was no longer preferred to harvested forage systems.

Fales *et al.* (1995) researched the effect that stocking rate had on pasture dairy profitability and herbage production. Supplemental feed was provided to the animals during times of inadequate pasture, thus maintaining milk production levels. Milk yield per cow was not influenced by the stocking rate; thus, production per hectare was directly related to stocking rate. Profit per hectare increased with increases in the stocking rate, while profit per cow fell as stocking rate was raised. Optimal stocking rate for any given farm would depend on the resources (e.g. land, cows, etc.) of the particular farm, and which resources are most limiting to production. Rotz and Rodgers (1994) used the DAFOSYM dairy simulation model to compare grazing vs. confinement operations. Pasture reduced the amount of lucerne hay and silage, corn grain, and soyabean meal needed for the milk herd. These savings along with reduced labour and equipment charges resulted in a US$161 cow^{-1} net return advantage for the grazing system.

Murphy *et al.* (1986) collected data from Vermont dairy farms to assess the impact of improved grazing management on farm profits. They compared data from three farms that had switched from continuous grazing to a system of Voisin grazing management during their 5-month grazing season. Net profits per cow were improved an average of US$67 by the improvement in grazing management. A Wisconsin study (Klemme, 1993) of grass-based vs. confinement dairies examined the costs of operations making the switch to intensive pasture grazing. Pasture dairies had about 1500 kg less milk produced per cow, but higher gross farm income due to increased cow numbers. Cash costs were slightly lower for the pasture-based operations. Net returns were about US$0.01 kg^{-1} of milk produced better for the grass operations, due primarily to lower feed costs. Interestingly, fixed costs per kilogram of milk were the same for both types of farms, as land costs kept investment costs high for the pasture operations. Grass dairies labour requirements were similar to those of confinement operations, as more cows had to be added in order to maintain total farm profits in the face of lower output per cow. Concern was expressed as fixed costs on grass dairies averaged 44% of all production expenses, and were continuing to increase. With only slightly better profit per kilogram of milk and lower milk production per cow, grass dairies were turning to more cows and land in order to maintain overall farm income levels. This was pushing fixed costs up instead of the desired decrease in these expenses.

Effects on dairy receipts from intensive pasture use are very dependent on the particular system in operation. Milk production per cow is the single most important economic trait in a dairy. The genetics of many of our animals has been developed under the assumption of diets based on harvested forages and concentrates that meet all the nutrient requirements of the cattle. Such cows will not perform the same on grass, and selection of animals needs to be made on factors other than their genetic potential to produce milk in confinement. Maintenance of adequate milk production from a strictly grazing diet is unlikely in many situations. Harvested feeds must be provided during times of insufficient pasture availability. Strategic supplementation for lactating cows on grass will be economically advantageous in many cases. Milksolids (i.e. fat and protein) can be negatively affected when cows are placed on grass, reducing the total amounts paid for these components in the milk cheque. However, milk quality can improve with reduced cell counts as herd health improves on pasture. If crop production is reduced as pastures are developed, income from hay and grain sales will go down.

The Economic Potential of Grass Dairying

Declining profit margins have led many USA dairies to look for ways to remain viable. With little control over the price of milk, controlling costs or expanding into larger units are the two primary choices for those desiring greater net income. Expansion is feasible for farms with adequate access to capital and labour, and that possess the management skills necessary to adopt technologies and lower their per unit cost of milk. However, many small to medium sized operations are looking for an alternative to more debt and more cows. The cost structure of a dairy operation is markedly different between grazing and drylot farms. Both variable and fixed costs are affected. Expenses will vary across farms, and the costs of a converted dairy can be substantially different from those of a start-up grass dairy. Production per cow will often fall when shifting from a mixed ration to a grazing dairy. However, keeping the goal of net farm income in mind, cost reductions can offset this loss. A return to pasture-based milk production has been an option that many have chosen and more are considering. Grazing can help some farms lower costs and improve profits. The continuing trend of declining profit margins will force all farms continually to face the challenge of getting better or getting bigger.

Management is the key to having a successful pasture dairy due to the multitude of issues that impact success. Lowering feed costs while maintaining milk production levels is crucial, so that the all-important figure of feed cost per kilogram of milk produced is competitive. Length of grazing season will limit how long many farms can graze cattle, and will thus limit the potential for reducing feed costs. Costs and availability of supplemental feedstuffs will further limit the potential of pasture dairies in some locations. Fixed costs become another critical factor in grass dairy success. The motivation behind some existing drylot dairy conversions to a pasture-based system has been due to increasingly high fixed costs. Machinery and equipment costs for harvesting and feeding forages, along with land and cattle costs have become barriers to expansion for some mixed ration dairies. Some elect to trade fence and management for machinery and equipment, but have not reduced fixed costs. Many do not reduce their machinery and equipment inventory substantially, while others have simply substituted fixed costs through additional pasture land and cows. To make a grass-based dairy successful and sustainable, management must carefully monitor and adjust the levels of both fixed and variable costs while keeping income above the total of both.

Grass-based dairying has proved to be very profitable in many areas of the world. New Zealand dairies rely almost entirely on pastures and harvested forages, and they provide milk on the world market for very low cost. Cash cost of New Zealand milk production was compared with that of average USA producers by Brookes (1996). New Zealand farmers were able to produce one kilogram of milk for US$0.1184 as compared with USA dairies at US$0.2988 kg^{-1}. Climate, relative input prices, and economies of scale are factors along with grazing that allow New Zealand to produce milk so efficiently. As the global marketplace continues to open and expand, dairy producers around the world will increasingly be competing against each other in terms of costs of production. USA dairies will continue to search for strategies to lower per unit costs of milk production. Grain prices, land costs, machinery prices and all other costs and returns will be evaluated. In the balance lies the ultimate economics of grass for dairy cattle.

Acknowledgements

The helpful review comments of Dr Steve Ford were greatly appreciated.

References

Bailey, K. (1997) Projected dairy cow production costs and returns in 1998. In: *Missouri Farm Financial Outlook 1998*. University Outreach and Extension, Department of Agricultural Economics, University of Missouri-Columbia, p. 80.

Bailey, K. and Hamilton, S. (1997) Economics of grass-based dairying in Missouri: A preliminary study. In: *Missouri Farm Financial Outlook 1998*. University Outreach and Extension, Department of Agricultural Economics, University of Missouri-Columbia, pp. 64–69.

Brookes, I.M. (1996) New Zealanders make nearly 2-1/2 times their U.S. counterparts. *Hoard's Dairyman* 139(3), 179.

Brown, R.J. (1990) Farmer experiences with intensive grazing. In: *Proceedings of the Dairy Feeding Systems Symposium*. Publication 38, Northeast Regional Agricultural Engineering Service, Ithaca, New York, pp. 228–235.

Elbehri, A. and Ford, S.A. (1995) Economic analysis of major dairy forage systems in Pennsylvania: the role of intensive grazing. *Journal of Production Agriculture* 8, 501–507.

Emmick, D.L. and Toomer, L.F. (1991) The economic impact of intensive grazing management on fifteen dairy farms in New York state. In: *Proceedings of the 1991 Forage and Grassland Conference*. American Forage and Grassland Council, Georgetown, Texas, pp. 19–22.

Fales, S.L., Muller, L.D., Ford, S.A., O'Sullivan, M., Hoover, R.J., Holden, L.A., Lanyon, L.E. and Buckmaster, D.R. (1995) Stocking rate affects production and profitability in a rotationally grazed pasture system. *Journal of Production Agriculture* 8, 88–96.

Ford, S. (1996) Grazing looks better as dairy profits tighten. In: *Farm Economics*. Department of Agricultural Economics and Rural Sociology, Pennsylvania State University, University Park, Pennsylvania, July/August, 4pp.

Ford, S. and Hansen, G. (1994) Intensive rotational grazing for Pennsylvania dairy farms. In: *Farm Economics*. Department of Agricultural Economics and Rural Sociology, Pennsylvania State University , University Park, Pennsylvania, May/June, 4pp.

Klemme, R. (1993) *An Economic Comparison of Grass-Based and Confinement Dairying in Wisconsin*. Wisconsin Rural Development Center and the Center for Integrated Agricultural Systems, University of Wisconsin, Madison, 11pp.

Moore, K.C. (1997) Managing beef cow feed costs: Dry matter production costs of pasture, hay and stockpiled pasture. In: Williams, M.J. (ed.) *Proceedings of the American Forage and Grassland Council*. American Forage and Grassland Council, Georgetown, Texas, 6, 228–232.

Murphy, W.M., Rice, J.R. and Dugdale, D.T. (1986) Dairy farm feeding and income effects of using Voisin grazing management of permanent pastures. *American Journal of Alternative Agriculture* 1, 147–152.

Parker, W.J., Muller, L.D. and Buckmaster, D.R. (1992) Management and economic implications of intensive grazing on dairy farms in the northeastern states. *Journal of Dairy Science* 75, 2587–2597.

Parker, W.J., Muller, L.D., Fales, S.L. and McSweeny, W.T. (1993) A survey of dairy farms in Pennsylvania using minimal or intensive pasture grazing systems. *Professional Animal Scientist* 9, 77–85.

Rotz, C.A. and Rodgers, J.R. (1994) Grazing is profitable option for this Pennsylvania farm. *Hoard's Dairyman* 137(6), 643.

Sheppard, G.L. and Ford, S.A. (1993) The role of pastures in dairy herd expansions. In: *Expansion Strategies for Dairy Farms Regional Conference*, Mercer, Pennsylvania.

Siebenborn, E. and Bailey, K. (1996) 1995 Missouri dairy enterprise business earnings and costs with projections for 1997. *Farm Management Newsletter FM 96-3*, University Extension, Department of Agricultural Economics, University of Missouri-Columbia, 9pp.

Index